D0883250

Palaeoethnobotany

The prehistoric food plants of the Near East and Europe

Palaeoethnobotany

The prehistoric food plants of the Near East and Europe

JANE M. RENFREW

Figures drawn by Alan Eade

COLUMBIA UNIVERSITY PRESS
New York 1973

TO COLIN

First published 1973 by Methuen & Co Ltd, London,
and Columbia University Press, New York
Printed in Great Britain

Library of Congress Cataloging in Publication Data

Renfrew, Jane M
Palaeoethnobotany.
(Studies in prehistory)
1. Man, Prehistoric—Food. 2. Ethnobotany.
3. Plant remains (Archaeology) I. Title.
GN799.F6R46 581.6'32'093 72-12752
ISBN 0-231-03745-7

Contents

List of Plates

List of Figures and Maps

Preface

This book arose out of my PhD dissertation *Palaeoethnobotany and the Neolithic Period in Greece and Bulgaria* submitted to the University of Cambridge in 1969. It has been considerably expanded and brought up to date, and does not include the detailed accounts of the Greek and Bulgarian samples.

The need for a book describing the main features of cereal grains, seeds, fruits and nuts most frequently discovered in archaeological contexts in Europe and the Near East has long been evident. It would be too much to expect there not to be a large number of omissions and defects, but it is hoped that by amassing all this information and gradually adding to it, an agreed palaeoethnobotanical corpus for identifications can be established.

Throughout the research and the preparation of this book I have received help, advice and encouragement from a large number of people, which it is my pleasant duty to acknowledge here.

Professor J. G. D. Clark and Professor A. R. Clapham supervised me during my research and I have greatly benefited from their wisdom and encouragement.

Mr Alan Eade, Cereals Section, N.I.A.B., has spent hundreds of hours making the carefully measured drawings which add so much to the value of this volume, and for which I am most grateful.

Mr Alan Eade, Prof. R. Riley of the Plant Breeding Institute, Trumpington, and the Curator of the Botanical Gardens, Cambridge, have all helped me to build up a comparative collection of botanical specimens.

It is a pleasure to thank the following people for help at various stages of the research: Dr W. van Zeist, Dr M. Hopf, Mme Villaret-von Rochow, Professor Geoffrey Dimbleby, Professor Sir Joseph Hutchinson, Mr E. S. Higgs, Dr F. Merton, Dr I. Rorison, Dr E. Rothwell, Miss C. Quartley, Miss Mary Cra'ster, Dr G. D. H. Bell and the late Dr G. P. Carson.

For help in working on the Greek palaeoethnobotanical material I must thank Mr P. Megaw, as Director of the British School of Archaeology at Athens, Mr D.

French, Professor R. Rodden, Dr D. Theochares, Professor J. L. Caskey, Mme Sakalariou, Professor Bakalakis, Professor John Evans, Professor A. C. Renfrew and Dr Peter Warren. In 1963 and 1964 I received a travel grant from New Hall, Cambridge, towards this research in Greece. In 1968 and 1969 I received a Sheffield University research grant for work at Sitagroi. In 1969 I was able to visit the Rice University excavation at Chaga Sefid, Deh Luran, Iran, through the generosity of the National Science Foundation of America.

The Bulgarian research was made possible by the British Council and a scholarship from the Bulgarian Government for the summer of 1966. In Bulgaria I received much help from Professor Dimitrov (Sofia), and from Dr G. Georgiev and Dr V. Mikov in the National Museum, Sofia; I also received help from Mr P. Detev (Plovdiv), Mr Nikolov (Stara Zagora), the Curator of the Nova Zagora Museum, Mme Toncheva (Varna), Mr Zlatarski (Dâlgopol), the Curator of the Roussé Museum, Dr Angelov (Tirnovo) and Mr Bogdan Nikolov (Vratsa).

From 1964 to 1967 this research was supported by a State Studentship from the Department of Education and Science.

In the production of the book I acknowledge with gratitude the help I have received from the following: Mr P. R. Morley, Mr L. P. Morley, and Mr Wilson (photographic printing); Mrs Drake and Mrs P. R. Morley (typing), and Miss Janice Price who has guided it carefully through the press.

My husband has been a constant source of inspiration and good advice throughout this research.

ACKNOWLEDGEMENTS

The author and publisher are grateful to the following for permission to reproduce copyright material:

Mr Norbert Schimmel for the frontispiece and dustjacket; W. A. Casparie for pl. 46; Dr K. E. Behre for pl. 47; Dr Evan Guest for the illustrations in the Glossary; Dimitrios Harissiades and George Rainbird Ltd for pl. 43b; North Holland Publishing Company for pl. 8; Paul Elek Productions Ltd for pl. 48b; John Wiley & Sons for pl. 48a; National Institute of Agricultural Botany for pl. 17 and 18.

Photographs for the plates were taken by several people to whom I am most grateful: Cambridge Instrument Co. for pl 2, Professor A. C. Renfrew for pl. 4, 5, 7; School of Agriculture, Cambridge University, for pl. 25; Mr Peter Morley for pl. 26, 27, 28, 29, 30, 31, 32, 34, 35, 37, 38, 40, 43, 44, 45; and Mr D. Tloupas in Larissa for pl. 36, 39, 41 and 42. The remaining plates are based on the author's photographs.

Chapter 1

Introduction: the Development of Research

Palaeoethnobotany may be defined as the study of the remains of plants cultivated or utilized by man in ancient times, which have survived in archaeological contexts. In this book it is confined particularly to the study of the remains of food plants – cereal grains, seeds, fruits and nuts which were utilized for food in prehistoric times in Europe and the Near East. Plants which were used for other purposes – as raw materials for textiles, dyes, herbal remedies, and for building or constructional purposes – lie outside the scope of the present work.

The survival of ancient plant remains has continued to capture the interest of botanists, plant geneticists, agriculturalists and archaeologists since C. Kunth (1826) studied the desiccated fruits, grains and seeds found by J. Passalacqua in the tombs of ancient Egypt, and O. Heer (1866) examined the material from the waterlogged prehistoric villages of Switzerland (see pl. 1) discovered in the dry winters of 1853–4. Once it was clear that this kind of material could survive from remote periods of antiquity many studies were undertaken. They took two forms: either (and most frequently) they were reports on the species present at a particular site, or they were studies of the evolution of a particular species in prehistoric times – for example O. Heer's study 'Über den Flachs und die Flachskultur im Alternum' (1872). In the 1870s and 1880s Heer and Messikomer were continuing the work on finds from the alpine lakeside settlements, Deiniger, Staub and Schröter began work further east in Central Europe at such sites as Lengyel, Aggtelek and Butmir, and the ancient Egyptian material occupied Unger, Braun and Schweinfurth. At this time too Professor L. Wittmack of Berlin began working on the plant remains from Germany, Italy, Greece, Anatolia and Peru. This work in the Old World was to some extent consolidated in two major syntheses: G. Buschan's *Vorgeschichtliche Botanik* (1895) and E. Neuweiler's paper 'Die Prähistorischen Pflanzenreste Mitteleuropas' (1905).

In the first half of this century palaeoethnobotany continued to develop in Europe with the work of E. Neuweiler, F. Netolitzky, N. Arnaudov, K. Maly, E. Hoffman, A. Schulz, A. Fietz, H. L. Werneck, G. F. L. Sarauw, G. Hatt, K. Jesson and E. Schiemann. A large amount of this research was excellently summarized by K. and F. Bertsch in their book *Geschichte unserer Kulturpflanzen* (1949). The work carried out during this period in Egypt was brought together comprehensively by Täckholm, Täckholm and Drar in their *Flora of Egypt* (1941 f.).

Since 1950 European palaeoethnobotanists have been concerned not only with finds in their own regions, but also with palaeoethnobotanical finds from the Near East which throw direct light on the evolution and domestication of the major food plants of this part of the Old World, and their introduction into Europe. The possibility of being able to use palaeoethnobotanical evidence in the study of the origins of crop plants was realized in the 1880s by A. de Candolle in his *Origine des plantes cultivées* (1883); and amongst others it was used by E. Schiemann in her article 'New results on the history of cultivated cereals' (1951). Research in the Near East on these problems has been carried out chiefly by Dr Hans Helbaek of Copenhagen, working together with the prehistorians Professor R. J. Braidwood in Iraq, Professors F. Hole and K. Flannery in south-west Iran, Dr D. Kirkbride in Palestine and Mr J. Mellaart in Anatolia. His results and conclusions have been drawn together in a large number of papers of which the following are perhaps the most important: 'The palaeoethnobotany of the Near East and Europe' (1960), 'Ecological effects of irrigation in ancient Mesopotamia' (1960), 'First impressions of the Çatal Hüyük plant husbandry' (1964), 'Early Hassunan vegetable food at Es-Sawwan near Samarra' (1965), 'Pre-pottery neolithic farming at Beidha' (1966), 'Commentary on the phylogenesis of *Triticum* and *Hordeum*' (1966), 'Plant collecting, dry-farming and irrigation agriculture in prehistoric Deh Luran' (1969) and 'The plant husbandry of Hacilar' (1970). Other important analyses of Near Eastern plant remains have been made by van Zeist and Botteima (1966) at Tell Ramad, and van Zeist and Casparie (1968) at Tell Mureybit. Maria Hopf has analysed remains from Jericho (1969). This work has recently been summarized by Renfrew (1969) and by Lissitsina (1970).

Besides his work on the earliest domesticated plants in the Near East Helbaek has, since 1938, been working on the ancient plant remains in Europe, and has examined material from later Near Eastern contexts and from Egypt. His long series of papers is of exceptional interest. He has developed techniques and principles for identification of plants preserved in a wide range of circumstances, some of which are found described in his papers.

At the present time there are a number of other leading workers in this field in Europe; among them mention must be made of Dr Maria Hopf who has worked on material from Greece, Bulgaria, Yugoslavia, Germany and Spain – her report on the fruits and seeds from the neolithic settlement at Ehrenstein (1968) sets a new standard for palaeoethnobotanical publication with its comprehensive discussions and excellent illustrations. Madame Villaret von Rochow continues work on the Swiss material, Dr Z. Tempir, Dr E. Opravil and Dr F. Kühn are working in Czechoslovakia, and Dr B. Hartyani, Dr G. Novaki and Dr A. Patay in Hungary. In Poland there are a number of workers foremost amongst whom are Dr M. Klichowska and Dr K. Wasylikova. In East Germany Dr K-D. Jager and Dr J. Schultze-Motel are the leading palaeoethnobotanists, whilst Dr K. E. Behre, Dr K. H. Knörzer and Professor U. Willerding are working in West Germany. Dr W. van Zeist is based in Holland, Dr H. Hjelmquist in Sweden and Dr M. Follieri works in Italy. In Britain Mrs H. H. Clark has worked on a number of palaeoethnobotanical finds, and is also working on material from North Africa; the present author has worked mainly on finds from Greece, Bulgaria, Yugoslavia and Iran; and members of the British Academy Research Project into the Early History of Agriculture are working on finds from prehistoric sites in southern Europe and the Near East.

In the past few years there have been several important developments. They spring directly from the setting up of the International Work Group for Palaeoethnobotany in 1968, which held its first Symposium at Kacina Castle outside Prague in October of that year, and met together for the second Symposium in Budapest in April 1971. The first development is the annual publication of a bibliography for palaeoethnobotany, not only of the Near East and Europe, but also for the Far East and the Americas. This work was begun by Dr J. Schultze-Motel and has been published so far in *Die Kulturpflanze*; it gives not only a list of recent publications but also a synthesis of the results contained within them for the origin and history of cultivated plants. Another useful project which has been started by Professor U. Willerding is the publication of well-documented distribution maps for each species of cultivated plant at different prehistoric and early historic periods in a given area. Willerding's article 'Vor- und frühgeschichtliche Kulturpflanzenfunde in Mitteleuropa' (1970) is an excellent beginning to a project which should in time embrace the whole of Europe and the Near East. The Symposia of the International Work Group for Palaeoethnobotany fulfil an important need in bringing together a number of palaeoethnobotanists who were previously working in comparative isolation.

If palaeoethnobotany first developed as a study in its own right in Europe it has now spread widely throughout the world. In Russia most palaeoethnobotanical identifications have been made by leading plant geneticists – Dr F. Ch. Bachteev, Dr G. Lissitsina, Dr M. Jakubciner and Dr A. I. Mordvinkina. In India palaeoethnobotany began with the identification of plant remains from prehistoric sites in the 1930s and it has been summarized by K. A. Chowdhury in his article 'Plant remains from pre- and proto-historic sites and their scientific significance' (1965), and by F. R. Allchin in 'Early cultivated plants in India' (1969). Palaeoethnobotanical research has reached China also: the results of work so far are summarized by W. Watson in 'Early cereal cultivation in China' (1969) in which he refers to the careful work of Ting Ying, 'Examination of rice husk found in red baked earth of the neolithic period in the Chiang Han plain' (*K'ao ku hsüeh pao* No. 4, p. 31 f.).

The first work on the palaeoethnobotanical riches of the New World was done by a Frenchman, Saffray, who published the results of his examination of a Peruvian mummy bundle in 1876. This was closely followed by Alphonse Tremeau de Rochebrune's article 'Recherches d'ethnographie botanique sur la flore des sépultures péruviennes d'Ancón'. It was the rich plant remains from the Ancón cemetery which first attracted Professor Wittmack's attention. He wrote an excellent, well-illustrated chapter on 'Plants and fruits' in W. Reiss and A. Stübel's *The Necropolis of Ancón in Peru* (1880–7). This was later followed by his more comprehensive *Die Nutzpflanzen der alten Peruaner* (1888). Since the end of the nineteenth century a number of important works on Peruvian palaeoethnobotany have been published: the syntheses of Costantin and Bois (1910), Safford (1917), Harms (1922) and Yacovleff and Herrera (1934–5) for example. All this work has been splendidly continued by Margaret A. Towle in her book *The Ethnobotany of Pre-Columbian Peru* (1961).

Palaeoethnobotany was slow to develop in North America. The first significant study was that of Melvin R. Gilmore 'Vegetal remains of the Ozark Bluff-Dweller culture' (1931). He identified the remains of sixty-eight species preserved in a dry rock shelter. Gilmore was also responsible for the founding of the Ethnobotanical Laboratory at the University of Michigan Museum of Anthropology in 1930.

In 1936 two important papers were published by Volney H. Jones: a study of the aboriginal cotton of the south-west United States (the first specialist study of a single crop plant in the U.S.A.), and a study of the desiccated plant remains from Newt Kash Hollow rock shelter. Between 1940 and 1960 Volney H. Jones produced a large number of reports on material from the south-west United States, working from the Ethnobotanical Laboratory in Michigan.

Since 1940 there has been a marked specialization of work on the early history of various New World crops, in particular maize. It has been intensively studied by E. Anderson (1942) in Arizona, P. Mangelsdorf and C. E. Smith Jnr (1949) in New Mexico; Mangelsdorf and Galinat have worked on the extensive collections made by MacNeish from the caves of Taumalipas and the Tehuacan valley in Mexico (1956); in addition Nickerson has studied ancient maize from South Dakota, and Blake and Cutler material from Arkansas (see R. A. Yarnell 1969, 221 f.). The cucurbits have been well studied by Thomas A. Whitaker (1948, 1949); and aboriginal beans were worked on by Lawrence Kaplan in the material from Taumalipas and Tehuacan in Mexico, and the Cordova Cave, New Mexico. Cotton has received special attention from Kate Pat Kent (1962) and from C. E. Smith Jnr, who worked on the Tehuacan finds (1965).

Apart from the specialization into studying separate crops which is much more advanced in America than elsewhere, there has developed the special study of coprolites preserved in desiccated conditions. Volney H. Jones reported the contents of faeces from the Newt Kash Hollow (1936), and R. L. Fonner (1957) analysed others from Danger Cave and Juke Box Cave, Utah. Coprolite analysis has been developed by Eric O. Callen working on material from Huaca Prieta, Peru, and Taumalipas and Tehuacan, Mexico (1969).

Regional studies of plant utilization and exploitation have developed since the appearance of George F. Carter's *Plant Geography and Culture History in the American Southwest* (1945). The Tehuacan Archaeological–Botanical Project, directed by R. S.

MacNeish, and studies such as 'Aboriginal relationships between culture and plant life in the Upper Great Lakes region' by Richard A. Yarnell (1964) have done much to foster this line of approach.

The present work, however, is concerned primarily with the palaeoethnobotany of the Near East and Europe in prehistoric times. In this region there is a need for an agreed, systematic, botanical corpus of the criteria which should be used to identify palaeoethnobotanical material. Without this the interpretation of the results of palaeoethnobotanical investigations cannot proceed meaningfully. The aim of this book is to bring these criteria together in one place, where they can be readily available to the students of the palaeoethnobotany, archaeology and agriculture of the Near East and Europe. Since this corpus of information is drawn from so wide a field it will inevitably be inadequate on certain topics, and it is hoped that corrections and additions may be incorporated into subsequent editions.

As will be seen, palaeoethnobotany concentrates chiefly on the identification of features of the gross external morphology of the seeds and fruits, and only to a lesser extent on the study of cell patterns of the epidermis and other structures: although this may develop into a useful line of approach in the future. Examination of cereal grains under the electron scanning microscope has shown that not only the outlines of the cells but their surface relief also may be an important diagnostic feature (pl. 2): this is a field which requires much further research.

We are here concerned chiefly with the identification of ancient plant remains on the basis of comparison of their gross morphological features with those of their modern counterparts. This study begins with a survey of the conditions in which the material survives and the distortions which may be caused to the form of the grains and seeds. There follows a discussion of the basis for identifying the species of the temperate cereals: wheat, barley, oats, rye, broomcorn millet and Italian millet. For each there is an account of the main palaeoethnobotanical finds, the genetic history, the environmental requirements and the possible uses which they may have had for prehistoric man. The pulses – horsebeans, peas, lentils, bitter vetch, grass peas and chickpeas – are similarly treated. Then follows a description of flax – probably first cultivated for its oily seeds. The chapters on fruits and nuts are chiefly concerned with those species which were cultivated in prehistoric Europe: vine, olive, fig, apple, pear. The wild fruits and nuts which were also extensively used are considered in a little less detail. There follows a survey of the utilization of drug plants in prehistoric times in the Near East and Europe. The final part of this section of the book takes the form of a catalogue of the edible weed seeds most frequently encountered on prehistoric sites. The two concluding chapters are concerned with the food values of the plants cultivated and utilized in prehistoric times in the Near East and Europe; and with giving a brief account of the origins and development of agriculture in the Near East and its spread to and development within Europe. In many cases the finds of ancient plant remains have raised new problems, setting in question accepted genetic relationships between species of individual plant families – for example among wheats and barleys. In other directions the evidence highlights the importance of groups of plants which have hitherto received comparatively little attention – for example, pulses were present in the Near East on most of the earliest farming sites, yet their importance agriculturally and as a source of food has scarcely been mentioned.

It seems clear that the chief cultivated crops of temperate Europe (wheat, barley, peas and lentils) were first domesticated in the Near East, and were introduced into Europe before 5000 B.C. (Renfrew 1969). But subsequent agricultural development led to the domestication probably within Europe of several other crops – oats, rye and some of the millets for example.

The earliest agriculturalists concentrated on the cultivation of annual crops – cereals and pulses – and relied on local wild trees to provide them with fruits and nuts to supplement their diet; it was not until this type of agriculture had become well established that orchard husbandry began to be practised. This development was a much longer-term undertaking. The results of planting, grafting, pruning and cultivating the soil around the trees would not become available for five to ten years afterwards, and then might continue to benefit several generations. This must reflect a mature economic system and relative social security. It is perhaps no coincidence that the vine and the olive do not come into cultivation before the end of the fourth millennium B.C. in the Near East and south-east Europe. The fruit trees of temperate Europe – apples, pears, cherries and plums – did not come into cultivation until the early bronze age a millennium later.

The interpretation of the results of palaeoethnobotanical research clearly is of great importance to prehistorians: it is equally certain that they must be based on a complete understanding of the nature of the material, how it has been preserved and how the identifications are made. These considerations must now be studied in greater detail.

Chapter 2

The Survival of the Evidence

> There was, doubtless, a time when the acorn, the fern root, the earthnut and many other native esculents, afforded directly that nutriment to the human race, they now give to our domestic animals; when perhaps the first rude inhabitants of our islands lived chiefly upon such herbs, seeds and roots as they could find in the forest, and whose properties they learned by the slow and dearly-bought experience of ages. How far it may have been so, to what extent such resources were made use of by those early races, we know not, for they passed away before the dawn of tradition, their history unrecorded, and their very existence only attested by a few fragments of chipped flint or half-charred wood and bones.
>
> PIERPOINT JOHNSON 1862, i–ii

At the time Pierpoint Johnson was writing, an exceptional number of seeds and fruits of wild and cultivated plants which had been harvested for food in prehistoric times, were being uncovered and carefully examined by archaeologists and a distinguished palaeoethnobotanist, Oswald Heer, in Switzerland. It was the severe droughts of 1854 and the following summers which exposed around the shores of the Swiss lakes a hitherto unsuspected number of prehistoric settlements, which had been preserved under water, partly buried in peat and lake sediments (Bibby 1957, 221 f.); Pierpoint Johnson's pessimism was to prove unfounded. In these Swiss deposits Heer found 'burnt fruits and seeds (which) unquestionably belong to the age of the lake dwellings; and a portion of them are in very good preservation, for the process of burning has not essentially changed their form' (see pl. 1). The same remark applies both to the apples and the corn, as the outer rind for the most part has neither separated nor shrivelled up. Many of the remains of plants, however, 'have been preserved in an unburnt state'. These appeared almost fresh-looking and could only be distinguished from their modern counterparts (Heer 1866, 336 f.) by

> the inside portions – the germ and albuminous part have disappeared, and only the burnt cellular part, which forms the seed-shell or pericarp has remained. . . . Thus the seeds of raspberries, goosefoot, . . . etc. are hollow inside, and in some cases only a brown powder indicates the remains of the former contents: the same remark applies to the stones of cherries and sloes, to hazelnuts, cornels, the seeds of pond-weeds etc.

Once the possibility of the survival of this evidence had become established in Europe, a century of work began to identify and evaluate the significance of various species of wild and cultivated plants found in association with the cultural remains of prehistoric man especially from the neolithic period onwards.

This work was essentially based on the careful identification of the seeds and fruits found. Heer observed that preservation by carbonization or in a waterlogged state had not essentially changed their form, and established that the species found were mainly those available at the present day. This same phenomenon had already been established in 1826 by Kunth examining the 'mummified' plant remains from ancient Egyptian tombs explored by Joseph Passalacqua at the beginning of the century (Kunth in Passalacqua 1826, 227): he wrote

> Les fruits et les fragments de plantes découverts par M. Passalacqua dans les tombeaux de l'ancienne Egypte, appartiennent presque tous à des végétaux que l'on rencontre encore aujourd'hui dans ces contrées. La comparaison la plus scrupuleuse des parties analogues ne m'a laissé entrevoir aucune différence. Il me parait par conséquent prouvé que la végétation de ces deux époques est parfaitement identique.

Thus early in the development of palaeoethnobotany three different means of survival of this ancient plant material had been recognized: by carbonization, by waterlogging and by desiccation.

As the work on plant remains continued new evidence was discovered: Netolitzky in 1906 wrote a study of the plant remains in coprolites (*Die Vegetabilien in den Faeces. Eine mikroskopische-forensistische Studie*) anticipating by nearly half a century the excellent work now being done by Callen, Cameron and others on pre-hispanic material from the dry caves and camp sites of America (Callen 1963, 186). Coprolites have only occasionally been recognized in prehistoric sites in Europe, for example, at Muldbjerg (Troels-Smith 1959, 593–4).

Shortly afterwards Netolitzky discovered seeds of *Echinochloa colonum* (L.) Link in the intestines of prehistoric mummies found at Nag'el Deir near Girga, Egypt (Netolitzky 1912; Täckholm, Täckholm and Drar 1941, 449). Seeds preserved in human internal organs have since been recovered from some of the iron age bog burials in Denmark (Helbaek 1950; 1958A; 1963, 178), where they have been found to be extraordinarily well preserved – the starch has kept its specific agglomerate structure and the ability to stain with iodine, and in some cases the epidermis, inner integument and even some of the protein cells of cereal grains may survive intact in these conditions (Helbaek 1963, 179).

One important source of evidence, that of impressions of grains and seeds in baked clay, only came to be fully appreciated as the study of ancient plant remains in Scandinavia and Britain developed during this century, largely through the work of G. F. L. Sarauw and of Dr Hans Helbaek. Helbaek found, especially in Britain (Jessen and Helbaek 1944, 9), that hand-made prehistoric pots often contained impressions of grains which had accidentally been incorporated in the damp clay during the manufacture of the vessel, and that during firing the grain had been burnt out leaving a more or less closed depression on the surface of the vessel, which, depending on the fineness of the clay, usually retained enough of the detailed features to allow an accurate identification of the original grain, and to give an indication of its size when moist. Later, impressions were to be recognized in the sun-baked mudbrick from prehistoric sites in the Mediterranean and Near East (idem 1963, 183).

Occasionally associated with impressions of grain in mudbrick or daub, and in coarse pot sherds, there occur silica skeletons of parts of the epidermis and glumes – especially of grasses whose epidermal cells contain a high proportion of silica deposited during growth. This silicon remains deposited even after the organic material has disappeared due to burning or natural decay, and when embedded in mudbrick or baked clay it can survive indefinitely. Helbaek found such skeletons in the ʿAmouq A deposits in north Syria (idem 1960C, 540) and he also found fragments of silica skeletons in deposits of ash and carbon dust at Jarmo (idem 1959, 188).

Occasionally palaeoethnobotanical material survives unexpectedly: among the magnificent organic remains preserved by ice in the frozen Scythian tombs at Pazyryk in the Altai mountains of Siberia a deposit of hemp seeds was found in Barrow 2 associated with a copper censer under a cone-shaped miniature tent covered with a felt rug (Artamonov 1965, 106; Rudenko 1970, 284 f.). A large number of almond nuts were recovered from the 4th-century shipwreck off Kyrenia, Cyprus. They had been preserved in an amphora saturated with salt water.

In order, then, to be able to identify palaeoethnobotanical material it is necessary first to understand the changes which the material undergoes in the different conditions outlined above, so that the palaeoethnobotanist can be in a position to reconstruct in his mind the appearance of the seeds, grains and fruits in their fresh state. There follows a detailed discussion of the most common forms of preservation encountered by palaeoethnobotanists.

Carbonization

A great quantity of palaeoethnobotanical material is preserved in carbonized form: the grains and seeds (and occasionally complete ears and fruits) were reduced to carbon while more or less retaining their characteristic shape, as a result of overheating near a hearth, or in a parching oven (the ears of hulled wheats require parching before they can successfully be threshed or winnowed (Harlan 1967)), or by a house fire in the area in which they were stored.

It has been suggested that deposits of grain may have become carbonized spontaneously by a slow natural process taking place at normal temperatures (Percival 1936; Biffen 1934). Biffen believed that carbonization was due to the action of anaerobic bacteria and that it took place fairly rapidly. The bacteria would have obtained their oxygen from the cellulose in the cell walls – incidentally reducing them to charcoal. It is interesting to note that both charred and uncharred grains may sometimes be found in the same storage pit, for example, at Fayum, Egypt (Täckholm, Täckholm and Drar 1941, 251).

Helbaek (1963, 181) maintains that:

> carbonization requires heat, and deposition of grain under conditions such as are met with in excavations would very soon result in putrefaction and complete destruction if the grain

was not heated beyond roasting point. There is too little cellulose and too much easily decaying starch in a cereal grain for the fermentation of heat to turn it to charcoal.

Spontaneous combustion may occur when, for example, a rick of green hay ferments internally to such an extent that the temperature rises to ignition point: the fire thus generated, it is true, may cause carbonization of plant material – but it has been argued that carbonization cannot occur without fire (Dimbleby 1967, 101).

If the grains were subjected to direct contact with the fire they would surely be reduced to dust: it seems likely that deposits of carbonized grain did not actually catch fire during the conflagration, but were enclosed in spaces protected from direct contact with the burning timbers of the structure, and there were slowly heated, like charcoal in a kiln (Helbaek 1955B, 655). Occasionally one comes across material which has been much distorted by fire – possibly due to sudden contact with intense heat, or to the

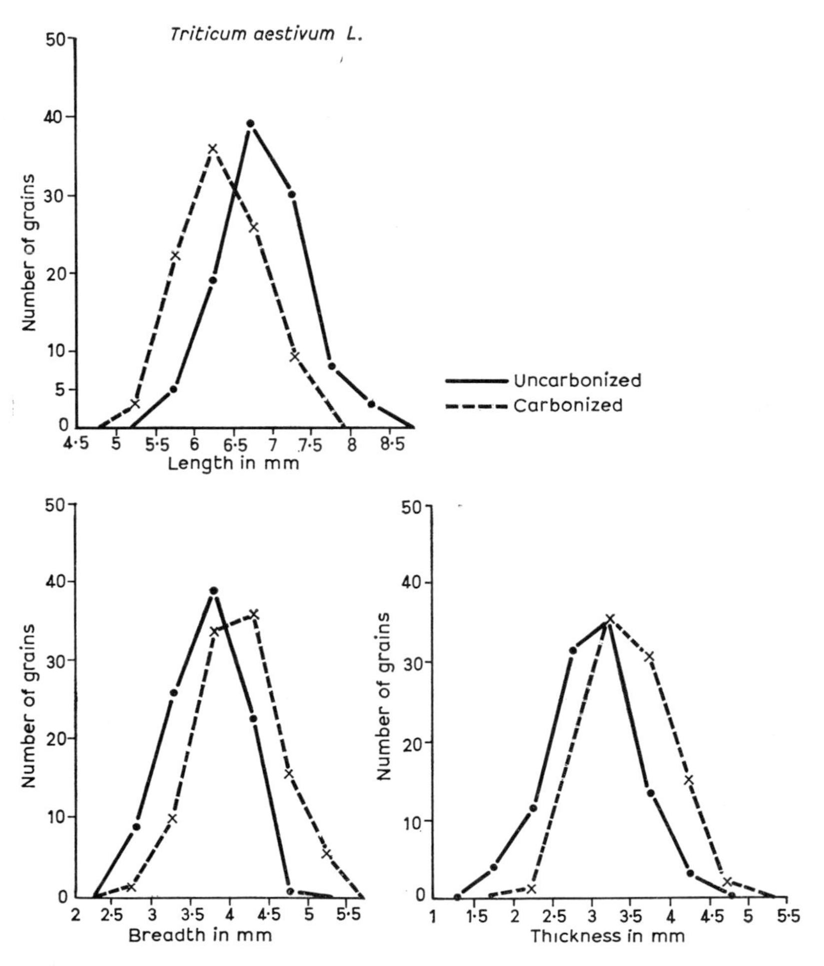

Fig. 1 Triticum aestivum L. Histograms showing changes in dimensions due to carbonization of wheat grains.

grain not being completely ripe, or the heap of grains becoming congealed together with a tar-like substance exuded during carbonization. In cases where the shape of the grains has become obviously distorted it is not wise to place too much emphasis on their measurements or proportions in identifying the material.

The technique of identifying carbonized plant remains is based on the careful comparison of the minute details of morphology preserved in the carbonized grains and

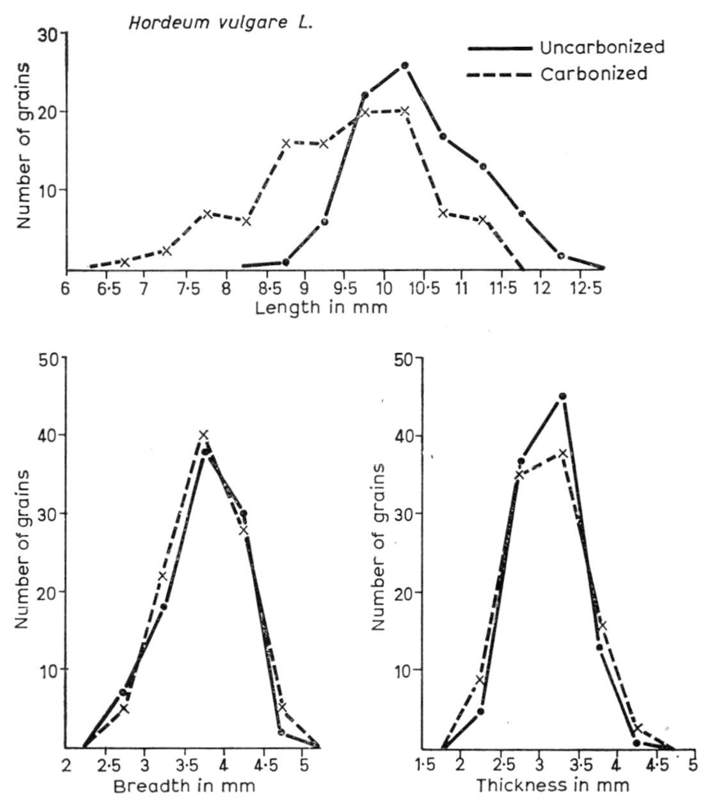

Fig. 2 Hordeum vulgare L. Histograms showing changes in dimensions due to carbonization of hulled barley grains.

seeds with the corresponding parts of fresh plants of the same species. The palaeo-ethnobotanist must, however, be aware of the changes in shape, size and proportions caused by heat in order to visualize the original appearance of the deformed and often mutilated carbonized grain, to be able to identify it (cf. Helbaek 1963, 181).

In order to have some first-hand experience of these changes 100 grains each of *Triticum aestivum*, *Hordeum vulgare*, *Avena sativa* and *Secale cereale* were measured, then placed in four bottles, one for each species, and baked in a gas oven at 200°C for 12 hours. The grains were then measured again. The results for each species are presented graphically in figs. 1–4. In all cases it can be seen that the length of the grains decreases with carbonization (though to only a small extent in the case of hulled barley).

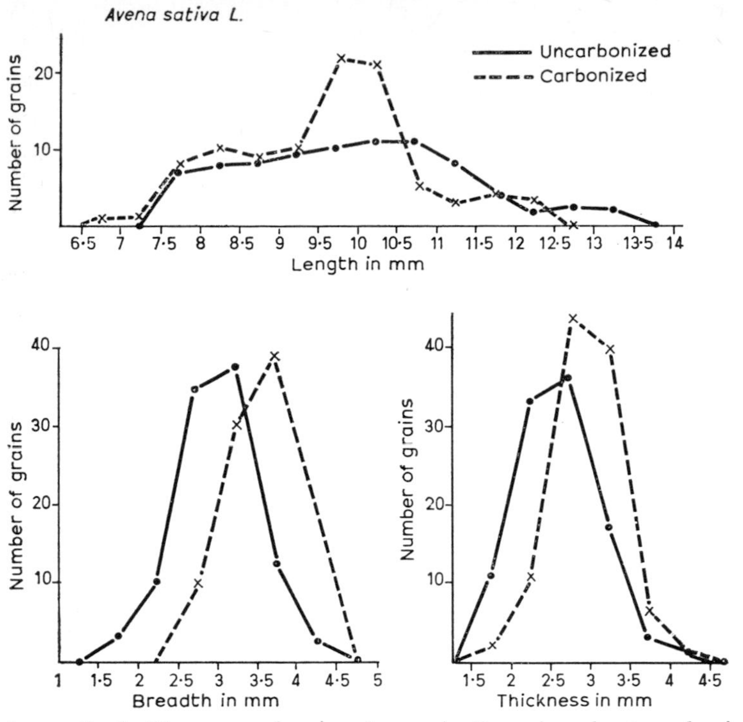

Fig. 3 Avena sativa L. Histograms showing changes in dimensions due to carbonization of oat grains.

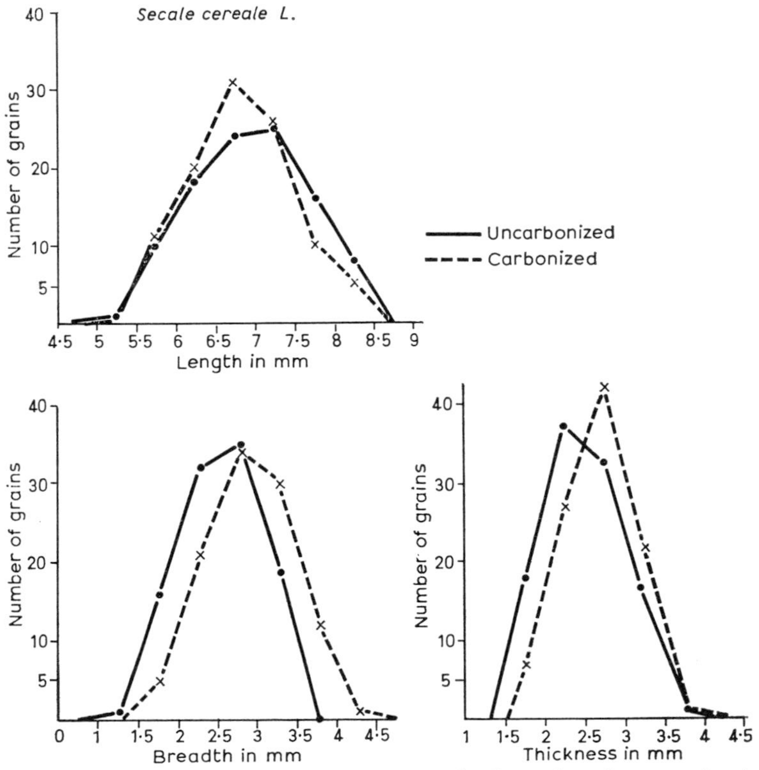

Fig. 4 Secale cereale L. Histograms showing changes in dimensions due to carbonization of rye grains.

The breadth of the grains increases slightly in general – most markedly in oats, but in hulled barley this measurement shows a very slight decrease, since the grain is tightly invested by its lemma. In the case of oats, rye and wheat the thickness of the grain shows an increase, whereas in hulled barley it remains almost unchanged. Exposure to higher temperatures would lead to a greater degree of distortion.

It has been pointed out that prehistoric cereal grains were in fact smaller than their exact counterparts of the present day – even when these are also carbonized. Maria Hopf (1955; 1921; cf. also Hopf 1957, 8 f.), gives this interesting table of measurements comparing the sizes of prehistoric carbonized einkorn, emmer and six-row barley grains with their modern fresh and carbonized counterparts.

Table comparing the dimensions of prehistoric cereal grains with those of their modern counterparts

	Length	*Breadth*	*Thickness*
	mm	mm	mm
Triticum monococcum–einkorn wheat			
prehistoric, carbonized	5·18	2·25	2·34
modern, fresh	7·46	1·89	3·24
modern, carbonized	6·87	2·62	3·01
Triticum dicoccum – emmer wheat			
prehistoric, carbonized	5·71	3·06	2·74
modern, fresh	7·29	2·72	2·77
modern, carbonized	6·67	3·52	2·69
Hordeum vulgare – hulled six-row barley			
prehistoric, carbonized	5·3	3·1	2·5
modern, fresh	7·81	3·55	2·68
modern, carbonized	7·2	3·94	3·11
Hordeum vulgare var. *nudum* – naked six-row barley			
prehistoric, carbonized	5·2	2·58	1·97
modern, fresh	7·3	3·12	2·21
modern, carbonized	6·49	3·41	2·97

This point is again illustrated in pl. 3 where carbonized grains of wheat, barley and oats from the Roman fort at Ambleside, Westmorland, are compared with their modern counterparts.

Little work has yet been done on the carbonization of fruit. Helbaek conducted experiments in order to discover whether the crab apples from Nørre Sandegaarde had been dried before they were carbonized. He carbonized a number of bisected modern crab apples in an electric stove at 190°C for 24 hours, having first measured them in their fresh state. A second batch were left for a week in a central-heated room. They were then carbonized at 190°C in the oven. Those apples which were exposed to direct heat shrank fairly evenly about 12% in length and 16% in diameter. The skin remained smooth, wrinkling only along the margin, due to the vapour pressure as the high water content was driven off. Those apples which were dried collapsed slowly and irregularly; the cut surface became hollow and the skin wrinkled; the diameter decreased by 16–35% (most noticeable in the smaller examples), and the length decreased less due to the stiffness of the endocarp (only 9–19%). On carbonization these dried apple halves shrank further, by 20–28% of their original length and 24–37% of the diameter in their fresh state. The core at this stage gave way, decreasing the length, but not altering the diameter measurements so much. The skin of these carbonized, dried apples was finely creased with the margin along the cut surface involuted and wrinkled (Helbaek 1952C, 111).

Carbonized grain is often insufficiently well preserved for the observation of finer morphological details: but occasionally when it has been protected, a surprising number of delicate features may survive. Sometimes entire spikes or large portions of them are found intact (e.g. the ears of einkorn wheat and six-row barley from Ezerovo II, Bulgaria: pl. 4), at other times the floatation samples may well contain diagnostic fragments of rachis, glumes, lemmas or awns, which would not be readily seen in the soil by the excavator (see pl. 5).

Occasionally the surface of the grain is so well preserved that it is possible to study the cell pattern, and even the structure of the hairs, on the husks and grain shells. These exceptional circumstances are especially useful in any attempt to identify the crushed remains of seeds incorporated into bread, buns or porridge which have become carbonized. Helbaek has successfully identified the ingredients of buns from the iron age village of Glastonbury in this way (Helbaek 1952B, 212).

Unless there are exceptional circumstances, in which the carbonized grain is found in pots or containers, it is best recovered from the soil by flotation – this ensures that the grains are not damaged by clumsy handling in the field, and that the sample is a more or less random representation of the deposit. Thus small weed seeds – not easily visible in the soil – and fragments of glumes and lemmas may be recovered. The technique is comparatively simple. Once specks of charcoal or grains are spotted in the excavation, a sample of soil is taken and sealed in a plastic bag. In the laboratory it is dried at room temperature, and is then ready for flotation. Flotation of carbonized seeds depends on the relationship of their specific gravity to that of the flotation medium, and this is further affected by the amount of enclosed porosity within the complete carbonized seeds. The true specific gravity of charcoal is 1·4–1·7, but because of the high percentage of porosity the apparent specific gravity ranges between 0·3 and 0·6, whereas inorganic material has a higher specific gravity of about 2·5 and will sink to the bottom of the flotation medium.

The technique of flotation consists of pouring the dried carbonized material and soil into a liquid medium such as water, at a slow, steady rate. The surface tension of the liquid together with the lower specific gravity of the carbonized material combine to keep the carbonized seeds floating at, or just below, the surface of the medium, whilst the inorganic material sinks. If the soil is particularly hard and does not readily disintegrate to mud in the water, a small amount of dilute hydrochloric acid may be added to break it up and thus release the carbonized material.

This method of water flotation is not totally satisfactory for collecting all the carbonized material – although it has the advantages of cheapness and ease of processing, and causes the minimum amount of handling of the material to take place. The carbonized seeds can be dried at once after being collected in a fine (0·5 mm) screen and are then ready for examination. There are, however, several disadvantages to the water flotation of carbonized seeds: the small seeds and fragments may not float since the bulk density of large seeds is in general higher than that of smaller seeds. Unless there is closed porosity within the seed which is not available to water during the flotation process the seeds will ultimately sink; this is especially noticeable in the case of the smaller seeds. If the surface of the seed is porous the air spaces quickly become filled with water and this decreases the differences in specific gravity between the carbon and water. Sometimes larger structures, such as acorn cotyledons whose surface is slightly eroded, may actually explode into a cloud of carbon dust on contact with the water. This is caused by the differential wetting of the residual structure along the planes of weakness.

To counteract some of these difficulties the use of a flotation medium with a higher specific gravity than water is recommended. Carbon tetrachloride with a specific gravity of 1·59 or solutions of it with carbon tetraiodide may be used. This latter has the advantage that the specific gravity may be adjusted by varying the proportions of the constituents so that even the smallest seeds and fragments may be floated. But once the material has been collected in this way it is necessary to wash it before drying it out prior to microscopic examination. A number of machines are now in use for the process of floating carbonized seeds from the soil which enable a larger quantity of earth to be sampled, and may lead to an embarrassing quantity of carbonized material being recovered. Once the material has been floated, and washed if necessary, it is collected in a fine-mesh sieve and allowed to dry out slowly, away from direct heat. Rapid drying may result in extensive damage to the seeds, especially to pulse seeds and acorns which split into their two cotyledons. Once the material is dry it is ready to be sorted and examined under a lens or low-powered microscope. When dry the seeds are usually hard enough to be handled gently with paint brushes and tweezers.

Grain impressions

Another common source of information to the palaeoethnobotanist is the impressions of grains and seeds in baked clay (see pl. 6). This is most noticeable in the case of handmade pottery. It seems that the clay was often worked into vessels close to the domestic hearth which was also used for cooking purposes. In consequence stray grains dropped during food preparation, became incorporated in the wet clay and thus found themselves embedded in the wall of the vessel. Occasionally grain impressions were used deliberately for decorating pottery vessels (Hopf 1965, 183). Dry grains absorb a certain amount of water from their moist surroundings, and the passage of water from the clay into the grains embedded in it deposits a layer of fine clay particles around them. This fine mould often reproduces minute morphological details of the grain surface (Helbaek 1955B, 653). Thus the grain impression corresponds to the size and shape of the wet grains. Helbaek has experimented with soaking barley grains and finds that when soaked for 24 hours their length increases by 1–2%, the width and thickness show greater increases: between 10 and 15%.

10 grains of hulled barley

	Dry	*Soaked*
	mm	mm
Length	8·77	8·88
Breadth	3·76	4·30
Thickness	3·09	3·46

(Helbaek 1955B, 654)

During drying and firing the vessel shrinks a certain amount – coarse, gritted pottery such as was frequently made in prehistoric times probably shrank no more than 5%. In the case of grain impressions in mudbrick or daub which was sun-dried and not subjected to firing, the shrinkage is much less, and thus the impressions correspond in size more closely to those for soaked grains. The impressions in the clay thus show slightly different proportions to those of the original grains, and should not be compared directly with those obtained for carbonized grain, or grain preserved by any other means. Grain impressions may most conveniently be studied by pouring a latex solution into the cavity exposed on the surface of the pot, allowing it to set in the air, and then extracting a positive cast of the original grain. For this purpose I have found the *Revultex VRB 949* solution the most successful since it dries in the air and need not be baked.

Grain impressions are a most convenient source of information. Their dates are usually secure provided the dates for the pottery forms, or the stratigraphical context in which they were found, are well established, and the picture they give of prehistoric agriculture in the locality is likely to be less biased than that obtained from a single large sample of carbonized grain. Many deposits of carbonized grain from contemporary levels may, however, give a fuller picture than a few grain impressions. The ideal situation is to have material from the same site preserved in both ways.

Silica skeletons or semi-fossilized grains

Occasionally there occur, associated with impressions of grain in mudbrick or coarse pottery, silica skeletons of small fragments of the glumes or epidermis of the grain which retain the exact form of the cell and the characteristics of the walls and pits (Helbaek 1960C, 540). During its growth the epidermis cells of many grasses undergo a mineralization process whereby the minute interstices in the cellulose cell wall become filled with, amongst other things, deposits of silica. When the plant is burnt or decays this silica skeleton survives, and, when undamaged by subsequent mechanical action, it will represent a true copy of the cells: the differences in shape and dimensions of cells of individual species can then be recognized under a microscope (idem 1963, 183). There is scope for much further research into cell patterns, their outlines and surface relief, in this material.

Finds of silica skeletons appear to be most frequent in the Mediterranean region and the Near East – this may be accounted for by the fact that the silica deposition in the epidermis is greater in arid climates (Helbaek 1963, 183). In temperate regions the formation of ice crystals caused by frost in such delicate structures is likely to destroy them completely. The author has found them in pisé mud at Saliagos in the Cycladic Islands (pl. 7) and also in similar material from neolithic Italy: no doubt they exist elsewhere.

Waterlogged material

In central and northern Europe a great deal of palaeoethnobotanical material survives in peat bogs preserved by the anaerobic conditions and by the slow action of humic acid. In these circumstances stones and seeds of berries and fruits are well preserved, and in some cases whole ears of wheat and barley have been recovered – for example in the marsh-dwelling sites around the Alps (Neuweiler 1924, 254).

In Denmark exceptional finds have been made of the corpses of iron age men whose skin and internal organs are amazingly well preserved. Their stomachs and intestinal canals contain remarkably well-preserved remains of seeds and grain fragments, those from Tollund and Grauballe man have been especially well studied (Helbaek 1950; 1958A). Grauballe man's last meal had consisted of some sort of gruel composed of seeds of sixty-six species of plants, only five of which were cultivated. These plant remains were so well preserved that fragments of the epidermis and sometimes of the inner integument of cereals could be recognized, and identification of the fragmented material was made by studying the cell patterns, and the microscopic hairs which survive on the seed coat.

A stagnant saline solution appears to have been responsible for the preservation of the almonds in the shipwreck off Kyrenia, Cyprus. This environment, unfavourable to bacterial activity, has retarded their decomposition. The nuts are heavily waterlogged, the kernels have disappeared, and the shells have become swollen with water. Typically, they weigh 5·6 gm when wet and only 1·2 gm when dried out slowly in a humid oven. Work is continuing in Sheffield under the direction of Dr E. Rothwell and the author to establish the exact composition of these nuts, and the processes to which they have been subjected since the ship was wrecked.

The comparative importance of different cultivated and wild plants found in the stomachs of Tollund and Grauballe men

Species	*Tollund*	*Grauballe*	
Cultivated			
Hordeum tetrastichum Kcke.	xxx	xxx	
Avena sp.	x	xx	
Triticum dicoccum Schübl.	—	xx	
Triticum spelta L.	—	xx	
Secale cereale L.	—	x	
Linum usitatissimum L.	xxx	x	
Wild			
Echinochloa crus-galli (L.) Beauv.	x	x	
Rumex crispus L.	x	x	
Rumex acetosella L.	x	xx	B
Polygonum lapathifolium agg. / *Polygonum persicaria* L.	xxx	xxx	B
Polygonum convolvulus L.	xx	xx	B
Chenopodium album L.	xx	xx	B
Stellaria media L.	x	x	
Spergula arvensis L.	xx	xx	
Camelina linicola Sch. et Sp.	xx	x	B
Thlaspi arvense L.	x	x	
Capsella bursa-pastoris L.	x	x	
Erysimum cheiranthoides L.	x	x	
Viola arvensis Murr.	xx	xx	
Galeopsis tetrahit agg.	x	x	
Plantago lanceolata L.	x	xx	

(Note: B = also present in the Borremose corpse – Helbaek 1954, 253).
xxx – major component of the deposit – numerically but not necessarily by volume
xx – present in quantity but not the major component
x – present in small quantities in the deposit

Desiccated or 'mummified' grain

Finds of remarkably well-preserved material are known from Egypt where the arid climate has preserved them in very nearly their original state. They have been recovered both from silos in the ground at Fayum (Caton Thompson and Gardiner 1934) and from inside the Pyramids, e.g. Queen Ichetis' tomb in the Saqqarah Pyramid (Helbaek 1953A, 5).

The husk is often a shade darker than that of contemporary grain, but all the details including the hairs and spicules on the kernels, husks and internodes survive, and the cell structure is more easily observed than in fresh cereals (Helbaek 1963, 177). The proteins in the endosperm are, however, completely decomposed, and the embryo is always disorganized and quite incapable of germination (Täckholm, Täckholm and

Drar 1941, 251). From examinations of grains of *Hordeum vulgare* from the Fayum in the University Museum of Archaeology and Ethnology, Cambridge, it was possible to distinguish the falsum (bevelled) grain bases; the presence of spicules on the inner lateral nerves of the lemma, the long-haired long rachillas, and most interestingly, the membranous long-haired collar lodicules were still intact beneath the lemma and overlying the embryo, and could be exposed by carefully removing that part of the lemma which lay over the embryo. Thus if one had a large enough sample it might be possible to distinguish the different agricultural varieties cultivated – much as modern crop inspectors can distinguish cultivars on the basis of a detailed study of the grain.

These are the most common forms of preservation of seeds and grains encountered by the palaeoethnobotanist: others such as their preservation in ice, as in the Scythian tombs in the Altai Mts, are seldom encountered and the present writer has no first-hand knowledge of the distortions which may affect material preserved in this way.

The major part of the palaeoethnobotanist's work is the identification of the plant remains recovered from archaeological contexts; but there are also problems in the interpretation of these results which will now be considered briefly.

Chapter 3

Problems of Sampling and Interpretation

Once the individual seeds in the palaeoethnobotanical samples have been identified, and the range of species present has been established, the palaeoethnobotanist has the challenging task of interpreting the significance of the finds in the context of the prehistoric community being studied. This involves a close scrutiny of the composition of the samples, the means by which they were collected, the archaeological context in which they were found, and, in sites covering more than one period, a comparison between the results for the different phases to establish any changes in agriculture which may have taken place in the vicinity during the occupation of the site. The importance of the different plants from the point of view of their contribution to the balanced diet is discussed in chapter 18; this chapter is concerned with the information the plant remains provide for reconstructing the nature and importance of crop production in the life of the prehistoric community.

Samples and crops

For the purpose of this discussion attention will be focused chiefly on samples of carbonized grain – which is the form in which palaeoethnobotanical material is most commonly preserved. It should be remembered, however, that in the ideal situation evidence preserved in several different forms will give a more comprehensive picture of the true importance of the different species represented. Wherever possible, even on sites rich in carbonized seed material, it is valuable to have the evidence from grain impressions in pottery, or silica skeletons in mudbrick, to confirm the results obtained from an examination of the samples of carbonized grain.

Examination of the samples of carbonized grain should give an indication of the type of agriculture practised. The samples should reveal whether the crops were healthy or diseased, whether they were well filled or poor, or if they were harvested unripe. They should show if the crops were grown mixed as a form of 'maslin', or were heavily infested with weeds or were relatively 'pure' and uncontaminated. They should also indicate the importance of wild plants as supplements and substitutes in the diet, and give some indication of the relative importance of the balance between cereal crops, pulses and oilseeds in the agricultural system.

Before we can start to interpret the material in this way it is necessary to examine the samples themselves to understand the limitations which the method of their collection places upon them. It is already clear that samples of carbonized grain cannot be considered truly 'random'. They have become carbonized due to some accident in antiquity – being scorched in a parching oven, spilled by the hearth or singed in the conflagration of part of a building – and are thus only indicative of their own immediate circumstances, and not even of the entire harvest of that season. It is clear, therefore, that we must try to obtain the maximum information from each sample, and by a comparison of a number of samples build up an outline picture of the agriculture of any one period.

The method of collecting the samples of carbonized grain is of prime importance in evaluating the amount of information which can be obtained from them. If the grains have been hand-picked by the excavator he has probably already unintentionally selected from the deposit only the larger seeds, such as cereal grains which were clearly visible to him in the soil, and the weed seeds will escape his, and thus the palaeoethnobotanist's, attention. Seeds handled in this way also tend to disintegrate before they are examined by the palaeoethnobotanist. Seeds which are recovered from earth samples as a result of sieving to obtain small artefacts and bones also are open to some selection dependent on the size of the mesh of sieves used (it should be remembered that cereal grains for example do not always lie obligingly on their sides but can slip endways on through quite small-meshed sieves). The movement of the sieve and the presence of larger objects within it frequently breaks or crushes the more fragile specimens present.

Samples obtained by flotation, whether carried out in the field or later in the laboratory, will give a more representative indication of the species present in the archaeological deposit; and wherever the excavator has reason to think there might be some seed material he is strongly advised to take a sample of earth to be examined by a competent person. By keeping the seeds in the earth they are less likely to be damaged if they have to travel far, and the sample will be a comprehensive indication of the species present. The techniques of flotation do have limitations as discussed in chapter 2, since some seeds may not always float and others may explode, but at the moment this is the most satisfactory method of dealing with the material from most deposits.

In the case of storage jars and other pots found filled or partly filled with carbonized seeds it is important to examine all the material found, to see if, for example, the jar was filled with the same species from the top to the bottom – sometimes it may happen that the base of a pithos contains a different species which had previously filled it or it may contain smaller seeds which have filtered down to the bottom. Sometimes the grain is stored threshed, at other times careful examination may indicate that whole ears had been placed in the vessel, to be threshed just before use. Occasionally smaller vessels have been found which contain seeds: in this case it is possible that we are confronted by a preparation of food, and one might assume that if there were several species they had been deliberately mixed. It is clear then that the origin and manner in which the sample is collected will govern the amount of information which can usefully be drawn from it.

The most satisfactory approach so far has been to study the material from each deposit separately, and then to compare the results from contemporary levels to obtain a comprehensive picture. The deposits found in cultural contexts, as in a pot or a storage pit, have to be compared with others from less obvious cultural contexts such as the sweepings from the floor caught in the rough area around postholes and those found by flotation of samples from a complete column of earth from all levels at a site (tell sites have produced unsuspected amounts of palaeoethnobotanical material by this means).

The amount of material to be examined by the palaeoethnobotanist may well be considerable, and in this event it is necessary to adopt some form of sampling procedure in order to extract the maximum information from the material without undertaking a great deal of time-consuming and unnecessary work. At Sitagroi, in east Macedonia, examination was made of a complete column of earth approximately 10 cm × 10 cm through the whole of the mound, and in addition a minimum of forty samples of seeds from the deposits elsewhere of each of the major phases of the occupation of the site. Sampling techniques must always be adapted to the particular circumstances of the site.

Not only may the number of samples available for examination be considerable but also the size of the individual samples may be such that only a small proportion of each can be usefully examined: one pithos at Sitagroi contained 6,000 grains of hulled six-row barley mixed with some 200 acorns; their proportions could have been established without counting them all. A random sample of each large deposit should indicate its main components.

When the species have been identified and their proportions established seeds of each species are usually measured to give some idea of their relative size. Again not every seed can possibly be measured for every sample, even when they are perfectly preserved, and palaeoethnobotanists have agreed that it is sufficient to establish the range of size based on the measurements of at least ten and preferably fifty or one hundred complete grains of each species. The size of grains may vary considerably within a single ear, and ear size can depend on the part of the field in which it was cultivated. So it is not possible to draw conclusions from the size of a small number of grains.

The most satisfactory approach to studying the samples has been to regard them in general as crops of the most numerous species present, mixed with other species which are less numerous. It should be pointed out that we cannot be certain that the mixing occurred in the field in every case – in the case of samples in small pots it may well be that they were mixed in the preparation of food, and where acorns are mixed with cereals this clearly is the result of mixing after harvest – but it holds well as a general rule. The less numerous species have to be divided into other species of the same genus, other cultivated plants of the same type – for example cereals or pulses – other cultivated plants of different types, wild plants which were growing as weeds within the cultivated field, other wild plants with edible seeds and obnoxious weeds. Deposits consisting entirely of weed seeds should also be considered: although more than 1,900 seeds of *Polygonum aviculare* L. found together in a phase II deposit at Sitagroi do not fill half a thimble they do represent a deliberate collection of these tiny seeds presumably for use as food.

By comparing contemporary prehistoric deposits on the basis of their chief components, the cultivated and wild species associated with them, in this way, it is possible to see how many different species were cultivated and how many others were known which could have been grown as crops also in the region of the site. The emphasis in interpreting these results thus leans towards the positive presence of a species rather than its absence in a particular deposit. But once a large series of grain samples from contemporary levels at one site, or from contemporary sites in one region, has been studied and can be seen to form a distinct pattern, then the absence of species may also be considered important.

The study of numerous deposits from contemporary levels at a single site has only become possible since the development of flotation techniques for recovering the seeds. Previously the excavator would have been fortunate to recover more than a single deposit from each period of occupation. Although this would yield some information it could not give an adequate picture of the ancient agriculture. In some exceptional cases, however, a good range of contemporary samples have been recovered by the archaeologists without using flotation. For example, at Tell Azmak in Bulgaria the careful work of the excavators has preserved a large range of samples from the Gumelnitsa culture levels, without having floated them – if they had used flotation they would no doubt have obtained even more material. The table, see p. 24, gives the results of analysis of these finds and illustrates well the value of a number of contemporary samples for establishing the main emphasis of crop production. Einkorn and emmer wheat are shown to have been grown as separate minor crops whereas the main cereal crop was hulled six-row barley (the chief component of seven of the twenty-two deposits). Of the pulse crops vetch was the chief component of nine deposits, and lentils of three more, whilst peas were known but do not appear to have been grown in their own right.

It is interesting to compare these results with those obtained from contemporary Gumelnitsa culture sites in the same region, the fertile Maritsa Valley of central Bulgaria (see Map 1). The results are set out in the table, see p. 25, in which the numbers indicate the number of deposits in which the species formed the chief component, those in brackets represent the other samples in which the species was found in smaller quantities.

These figures show several interesting points: first, vetch was clearly an important

Carbonized grains and seeds from the Gumelnitsa culture levels at Tell Azmak, Bulgaria

Sample	*Emmer*	*Einkorn*	*Hulled six-row barley*	*Naked six-row barley*	*Lentils*	*Vetch*	*Peas*
(1)	—	—	—	—	—	xxx	—
(2)	—	—	—	—	xxx	x	—
(3)	x	—	xxx	—	xx	—	x
(4)	—	—	—	—	—	xxx	—
(5)	x	—	—	—	—	xxx	—
(6)	—	—	—	—	—	xxx	—
(7)	—	—	xxx	—	—	x	—
(8)	xx	—	xxx	x	—	—	x
(9)	xx	x	xxx	—	—	—	—
(10)	x	—	xx	—	xxx	x	x
(11)	x	x	—	—	—	—	—
(12)	—	—	—	—	—	xxx	x
(13)	—	—	xxx	—	—	—	—
(14)	x	xx	xxx	—	—	—	x
(15)	—	xx	xxx	—	—	—	—
(16)	—	—	—	—	—	xxx	—
(17)	xx	—	—	—	xxx	—	—
(18)	—	xxx	—	—	—	—	—
(19)	—	—	—	—	—	xxx	—
(20)	—	—	—	—	—	xxx	—
(21)	xxx	xx	—	—	—	—	—
(22)	—	—	—	—	—	xxx	—

xxx – major component of the deposit – numerically but not necessarily by volume
xx – present in quantity but not the major component
x – present in small quantities in the deposit

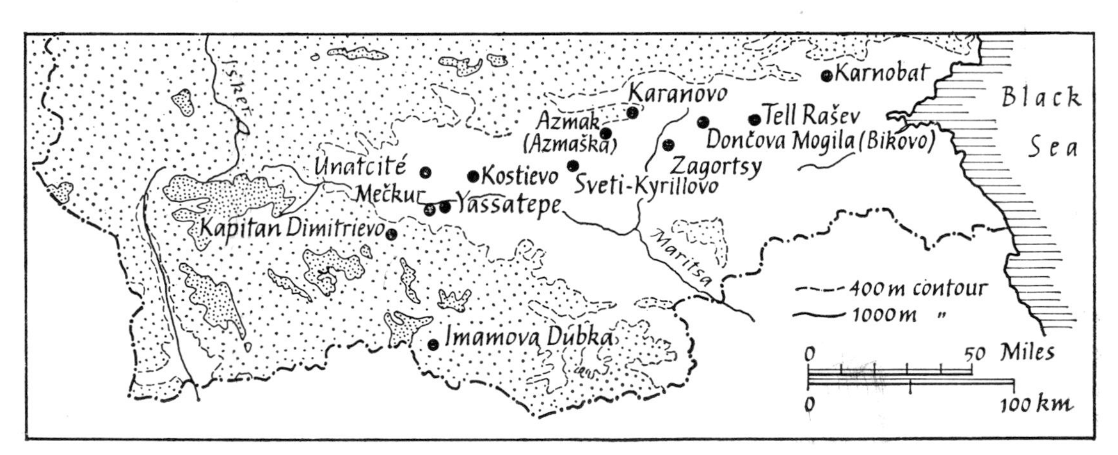

Map 1 Distribution map of finds of cultivated plants from Gumelnitsa culture contexts in the Maritsa Valley, Bulgaria.

Cultivated crops found on Gumelnitsa culture sites in the Maritsa Valley

	Einkorn	*Emmer*	*Club wheat*	*Hulled six-row barley*	*Naked six-row barley*	*Lentils*	*Vetch*	*Peas*
Azmak	1 (5)	1 (8)	—	7 (1)	(1)	3 (1)	9 (3)	(5)
Karanovo VI	1 (1)	1	—	—	—	—	1	—
Dončova mogila	(2)	(1)	(1)	(1)	1 (1)	(1)	(1)	—
Kapitan Dimitrievo	(2)	3 (2)	—	(1)	—	1 (1)	1	—
Unatcité	1	—	—	(1)	(1)	—	—	—
Yassatepe II	—	—	—	—	—	—	1	—
Imamova Dubka	—	—	—	—	—	—	—	—
Totals	3 (10)	5 (11)	(1)	8 (4)	1 (3)	4 (3)	13 (3)	(5)

crop throughout this region at this time, and six-row barley was the chief cereal cultivated, although einkorn and emmer wheat were also quite widely found if not always in large quantities. Thus the finds from Tell Azmak appear to give a fair reflection of the spectrum of crops found in the settlements of this culture in this region, and reflect the emphasis on barley and vetch crops which are characteristic of this area in Gumelnitsa culture times.

Characteristics and problems of subsistence agriculture

In cases where one is fortunate enough to have a series of deposits of grain from several sites in the same geographical region occupied over a long period of time, such as are now available for the fertile plains of Thessaly in central Greece (Renfrew 1966), it is sometimes possible to detect long-term developments in agricultural practice leading up, in some areas, to that final indicator of agricultural stability, the establishment of orchard husbandry.

In Thessaly a series of finds of carbonized grains, seeds and fruits has been recovered from a number of excavations conducted in recent years (see Map 2) and it has been possible to detect the outlines of agricultural development in this region. In order to achieve this the concept of crop purity has been employed in which the principal crop (the chief component of each deposit) is set against the number of other species found in the deposit. Thus a sample of only seventy-one seeds from the earliest neolithic levels at Ghediki contained the following range of species:

44 *Triticum dicoccum*
2 *Triticum monococcum*
9 *Hordeum distichon*
1 *Hordeum distichon* var. *nudum*
5 *Pisum sativum* var. *arvense*
4 *Vicia* sp.

4 *Lens esculenta*
2 *Pistacia atlantica*

Apart from the pistacia nutlets all the other species could have been cultivated in a form of mixed crop or maslin in which *Triticum dicoccum* was the major component. This type of mixed crop which includes both cereals and pulses grown together is still

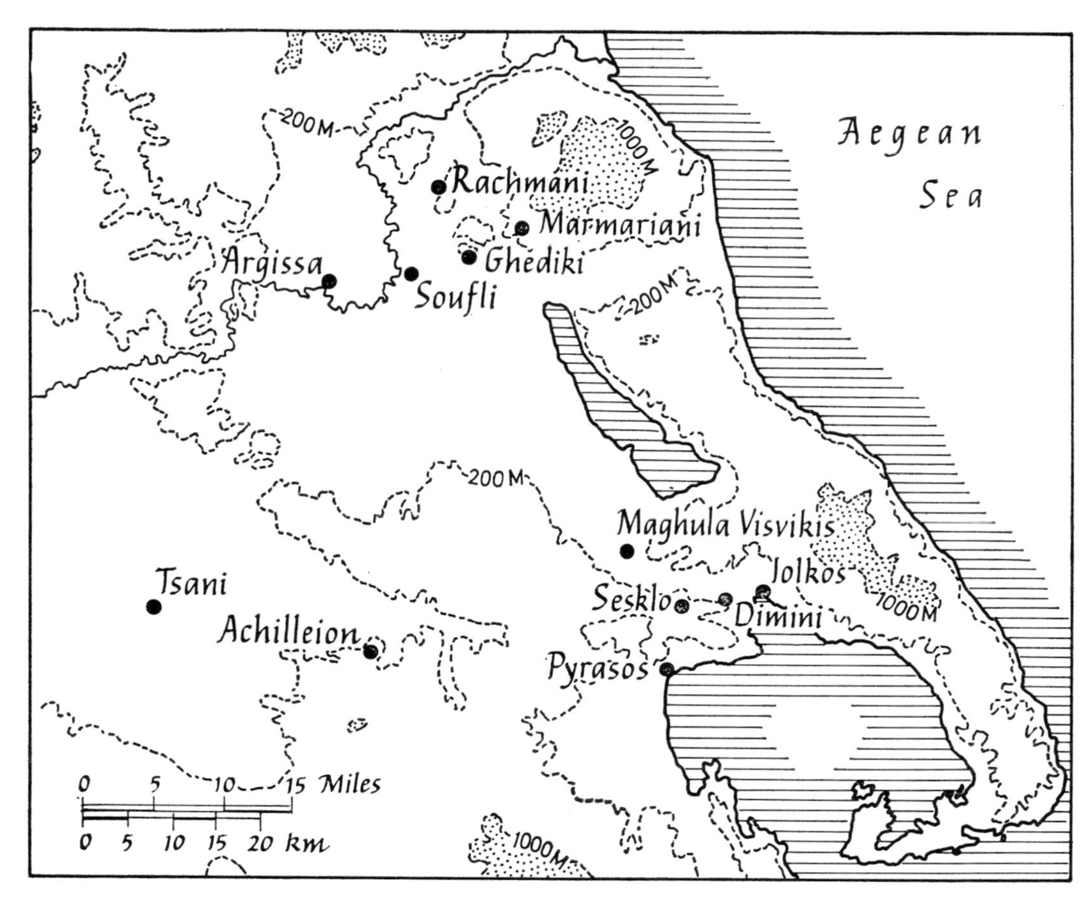

Map 2 Distribution map of palaeoethnobotanical finds in prehistoric Thessaly.

employed in parts of the Cycladic Islands, where the present writer saw a field of wheat, barley and oats mixed with peas and beans as a form of maslin in the summer of 1965. This form of cropping is characteristic of subsistence agriculture and ensures that some return is obtained from the land whatever the climate and soil conditions.

A study of a series of samples from different sites reveals that this sitation of maslin cropping continued in Thessaly throughout the early neolithic period: at Soufli a sample of forty-five seeds comprised chiefly *Triticum dicoccum* (twenty seeds), mixed with *Hordeum distichon*, *Triticum monococcum*, *Pisum* sp. and *Lens esculenta*, giving more or less the same spectrum as in the preceding period. Unfortunately no material has been examined for the middle neolithic period, but when we come to the late neolithic the pattern has altered markedly – at Sesklo, for example, there are now large samples of seeds consisting of over 100, and in one case over 1,000 grains of *Triticum dicoccum* unmixed with any other species. This must, one imagines, represent quite a development in crop husbandry from the mixed crop to the more confident specialization in growing the cereal best suited to the area. Emmer was not, however, the only important cereal at this time: at Pyrasos, on the coast, hulled six-row barley was the chief component of a Dimini culture deposit which also contained some emmer, and very small amounts of einkorn and peas.

The crops found in the palaeoethnobotanical samples from the late bronze age Mycenaean palace at Iolkos show again a change in the agricultural pattern – here hulled six-row barley was found to be the chief cereal, and emmer wheat never accounted for more than 4% of the contents of any of the samples. Apart from the barley crop, we have evidence of the cultivation of peas, and of beans, which are new to the crop spectrum of the area. The evidence for the development of cultivation of the annual crops in prehistoric Thessaly is summarized in the following table.

The development of the cultivation of annual crops in prehistoric Thessaly

(The percentages given are percentages of each species in each deposit examined – further details of the samples are available in Renfrew 1966.)

	Einkorn	*Emmer*	*Two-row barley*	*Six-row barley*	*Peas*	*Beans*	*Lentils*
	%	%	%	%	%	%	%
Late bronze age							
Iolkos 4	—	—	—	80	—	20	—
Iolkos 5	—	4·1	—	76·7	8·2	—	—
Iolkos 6	—	0·9	—	6·1	—	93·0	—
Iolkos 7	—	—	—	0·8	—	98·4	—
Iolkos 8	—	0·7	—	—	—	97·8	—
Iolkos 9	—	—	—	28	72	—	—
Iolkos 10	—	—	—	100	—	—	—
Iolkos 11	—	—	—	50*	—	—	—
Late neolithic							
Pyrasos	4·4	22·5	—	71·7	1·4	—	—
Sesklo 66: 33	—	95	2	—	3	—	—
Sesklo 63: 63	—	100	—	—	—	—	—
Sesklo 63: 24	—	100	—	—	—	—	—
Sesklo 63: 250	—	100	—	—	—	—	—
Early neolithic							
Sesklo 1963	—	100*	—	—	—	—	—
Soufli 1958	9	44·9	22·9	—	18	—	6·1
Earliest neolithic							
Ghediki 1962	2·8	61·9	14·1	—	7	—	5·6
Achilleion 1961	—	85·7	—	—	—	—	—
Sesklo 1962	—	60	10	—	20	—	—

* These figures refer to the find of a single grain of this species.

Apart from the evidence for the cultivation of annual crops the Thessalian finds give an interesting insight into the utilization of wild fruits and nuts, leading up to, in the late bronze age deposits at Iolkos, evidence for the cultivation of grapes and olives. In the earliest neolithic levels acorns (found at Sesklo) and the small nutlets of *Pistacia atlantica* (found at Ghediki and Sesklo) were used to supplement the diet of cultivated plants. These same two species were also recovered from the succeeding early neolithic levels at Sesklo (pistachio nutlets) and Achilleion (acorns – see pl. 39). A much greater interest in wild fruits and nuts seems to be reflected in the finds from the late neolithic period in Thessaly. In House Q at Rachmani Wace found a deposit of forty figs which were remarkably well preserved (see pl. 36). Figs were also reported from Sesklo and Dimini in the late neolithic levels in the excavations of Tsountas (Tsountas 1908, 360). Other fruits include wild pears found by Tsountas at Dimini, and almonds from both Dimini and Sesklo. Acorns made up one deposit in a recent excavation at Sesklo (63: 12) and from the same site came the interesting find of a single nut of *Pistacia vera* (see pl. 42).

It is of interest to note that even in late bronze age times acorns were collected to supplement the food supplies – at Iolkos three samples were entirely composed of acorns. Also interesting from this site is the find of an olive stone and ten pips of cultivated grapes.

The beginnings of orchard husbandry probably go back into late neolithic times. At Sitagroi a series of grape pips have been found from the base of the mound through all the phases of occupation to the top, covering a time span of about 2,500 years. The pips from the earliest three phases of occupation are rounded and squat in shape with very short stalks and the two grooves along the ventral side arranged at an angle to the central 'bridge.' In the fourth phase of occupation at the site they show a distinct change in shape, becoming pyriform in outline with a greatly extended stalk, and with the grooves now running parallel to the 'bridge'. These then must represent the earliest indication we have yet for the cultivation of the vine in the Aegean area and date to about 2500 B.C.

It is clear therefore that the earliest farmers in south-east Europe were concerned with the cultivation of annual crops – cereals, pulses and possibly oilseeds which demanded that the farmer stay settled in a permanent village all the year round to attend to cultivation and harvesting. It was the crops rather than the domestic animals which required this stability of settlement. Permanent settlement had only previously been possible in exceptional circumstances, as for example at Lepenski Vir, Yugoslavia, where possibly the abundance of fish gave the mesolithic inhabitants the same economic security as that enjoyed by the Nootka and Kwakiutl of British Columbia in the last century (Daryll Forde 1966, 69 f.). With the impact of neolithic peasant farming, the hunters' small, temporary camping sites were abandoned in favour of larger permanent villages inhabited all the year round.

Not only was the community settled in a permanent village fully occupied in producing a subsistence for themselves of cereals and pulse crops, they were completely dependent on their success – failure would mean not merely an economic loss, but the 'starvation and destruction or dispersal of the community' (Allan 1965, 3). Recourse to wild plants would seldom counteract complete crop failure. In order to avert such catastrophes there is a tendency in subsistence farming communities to diversify the crops grown – either in separate fields or sown together as a form of maslin. These annual crops were cultivated to provide a storable surplus of large edible seeds which could supply starch, protein and vegetable oil for the diet of the community throughout the year (Gill and Vear 1958, 103 f.).

The cultivation of vines and fruit trees requires a much more stable community than that needed for the production of cereal and pulse crops. Fruit trees represent a much longer-term investment in which the profits are not reaped for a number of years after the first planting and the subsequent grafting, pruning and care of the trees. The trees bearing fruits and nuts provided sources of sweetening and protein respectively for the diet which could be dried and stored for a certain length of time, but unless they were converted to oil or wine, they were distinctly more perishable than dry cereal grains and leguminous seeds.

Thus from these few examples it is clear that the palaeoethnobotanist in interpreting the contents of the seed samples from prehistoric sites, has much information to give not only about the culture level from which they were obtained but about the prehistoric agriculture of the region and the development of crop husbandry on the site and in the general region. This is no easy task, and one not often undertaken, but it is only with full co-operation between archaeologists and palaeoethnobotanists that the beginnings and development of agriculture in prehistoric times can be reconstructed.

Chapter 4

The Cereals: A General Discussion

The cereals, which include some of the earliest to be domesticated by man, consist of annual grasses grown chiefly for their large 'grains' which provide a concentrated carbohydrate food source and may easily be stored until required.

The cereals so far reported from prehistoric contexts in Europe and the Near East belong to the following genera: *Triticum* (wheat), *Hordeum* (barley), *Avena* (oats), *Secale* (rye), *Panicum* (broomcorn millet), *Setaria* (Italian millet) *Echinochloa* (barnyard millet). In the gross morphology of their inflorescences, these genera have an underlying similarity. In this chapter the structure of these cereals will be described in general terms and the major points of difference between them indicated. In this discussion only those points which may be of use in the identification of ears and grains will be mentioned, since it is the identification of these parts which is our main concern. Each plant is considered in greater detail in later chapters.

The grains are borne at the top of a stem (*culm*) either in the form of a more or less

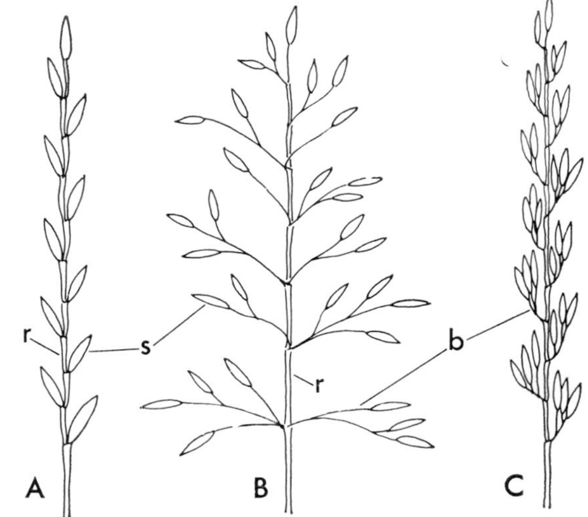

Fig. 5 Comparison of the structure of (A) spike, (B) panicle and (C) spike-like panicle. r = rachis segment or internodes, b = branches, s = spikelets.

dense *spike* (wheat, barley, rye) or of a spreading *panicle* (oats, *Panicum*) or as a spike-like panicle (*Setaria*), or as a panicle of spike-like *racemes* (*Echinochloa*) (see fig. 5). The spike consists of a central axis or rachis which is divided into segments and which bears the spikelets at alternate nodes. One or more spikelets may be found at each node. A panicle consists of a rachis with branches arising at each node. The length of the branches and the internodes may vary considerably. In oats the nodes are distant and the branches are long giving a spreading panicle: in Italian millet the nodes are closely spaced with short branches and tightly packed spikelets forming a dense more or less cylindrical ear known as a 'spike-like panicle' (Gill and Vear 1958, 230–1). In

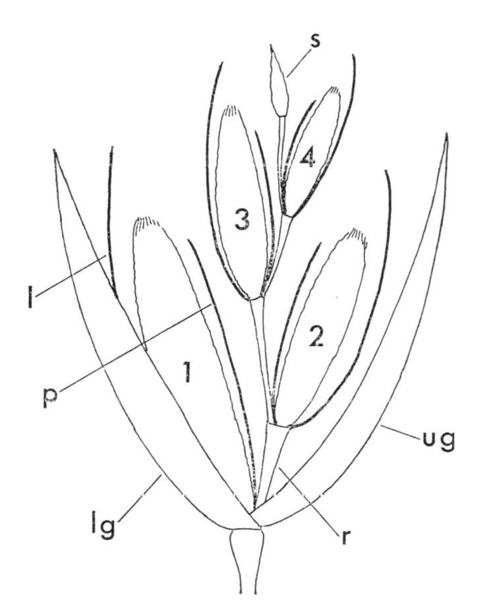

Fig. 6 Diagram to show the structure of a spikelet. Enclosed between the lower glume (lg) and the upper glume (ug) are four fertile florets (1–4) and a sterile floret (s) all borne on segments of the rachilla (r). Each floret consists of a lemma (l) on the dorsal side and a palea (p) on the ventral side of the caryopsis or grain.

barnyard millet the panicle consists of a number of spike-like racemes which bear the spikelets directly on the axis branches.

The spikelet consists of a pair of glumes which enclose a number of florets. The second and subsequent florets are attached successively to the primary floret by a short axis known as the *rachilla*. Each floret is composed of a lemma and palea which enclose the grain (*caryopsis*). The lemma lies over the dorsal side of the grain and overlaps the margins of the palea: the smaller palea covers the ventral side (fig. 6).

The grains are generally ovoid in form, being pointed at the lower, embryo end or base where they were attached to the floret, and blunt at the apex. They are rounded on the dorsal surface, and have a deep, longitudinal groove on the ventral side (fig. 7). The whole surface is covered by the fused pericarp and testa which consist of a few layers of dead crushed cells. At the lower end of the dorsal surface lies the embryo, a small, pointed oval area over which the pericarp and testa are somewhat wrinkled. The interior of the caryopsis consists of the endosperm which is made up of polygonal thin-walled cells containing starch grains and some protein. The outer layer (aleurone) is

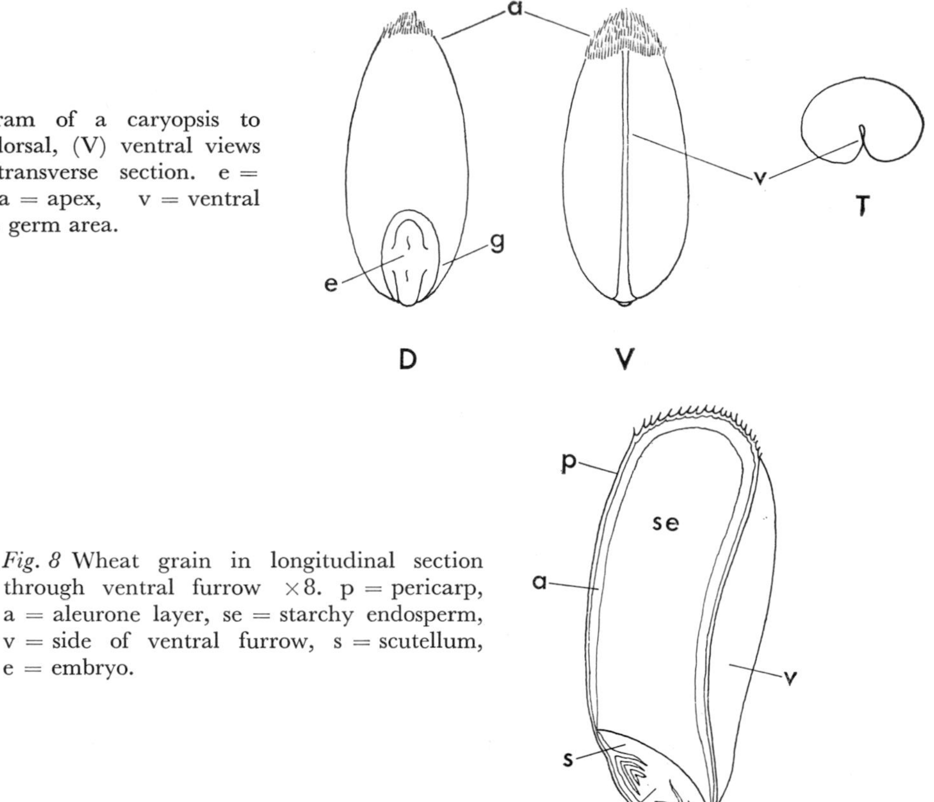

Fig. 7 Diagram of a caryopsis to show (D) dorsal, (V) ventral views and (T) transverse section. e = embryo, a = apex, v = ventral groove, g = germ area.

Fig. 8 Wheat grain in longitudinal section through ventral furrow ×8. p = pericarp, a = aleurone layer, se = starchy endosperm, v = side of ventral furrow, s = scutellum, e = embryo.

formed of more or less cube-shaped cells which contain protein grains. The embryo consists of a flattened oval structure, the *scutellum*, in contact with the endosperm which bears the cylindrical shoot and root regions attached to its outer surface (fig. 8; Gill and Vear 1958, 232).

Triticum

The species of the genus *Triticum* (wheat) share a number of features in common, which serve to distinguish them from the other genera of cereals with which we are concerned. The ear is a typical spike, with a single spikelet borne on each rachis segment. Each spikelet consists of a pair of stiff, keeled, asymmetrical glumes which enclose from one to nine florets carried on a very short rachilla (fig. 9). The lemmas are thinner than the glumes and may be extended at the apex to form a short awn. The palea is papery and less robust than the lemma, and is often forked at the apex. The caryopsis is large and may bear an apical brush of hairs. In some species the caryopsis threshes free from the glumes, lemma and palea – the 'naked' wheats – in others it remains enclosed in the vice-like grip of the glumes – 'hulled' wheats – in the latter case the spike breaks up into rachis segments on threshing.

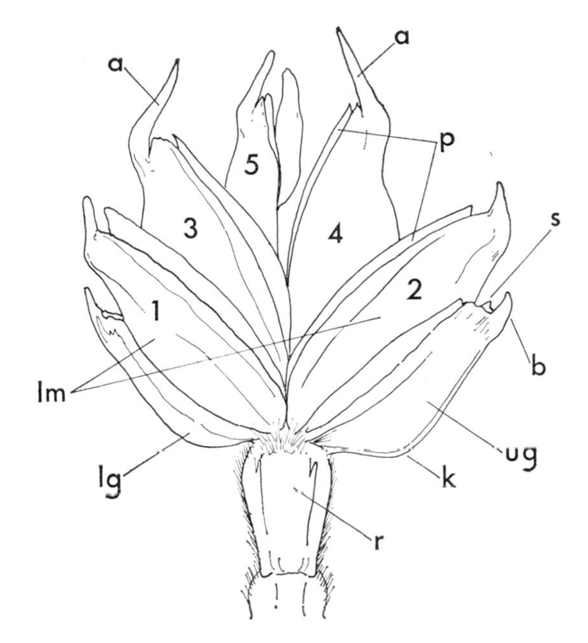

Fig. 9 Spikelet of wheat ×3. ug = upper glume with b = beak, s = shoulder and k = keel; lg = lower glume; lm = lemmae of 1st and 2nd florets with a = short awns; p = paleae; r = rachis segment with hairy margins and patch of hairs below spikelet.

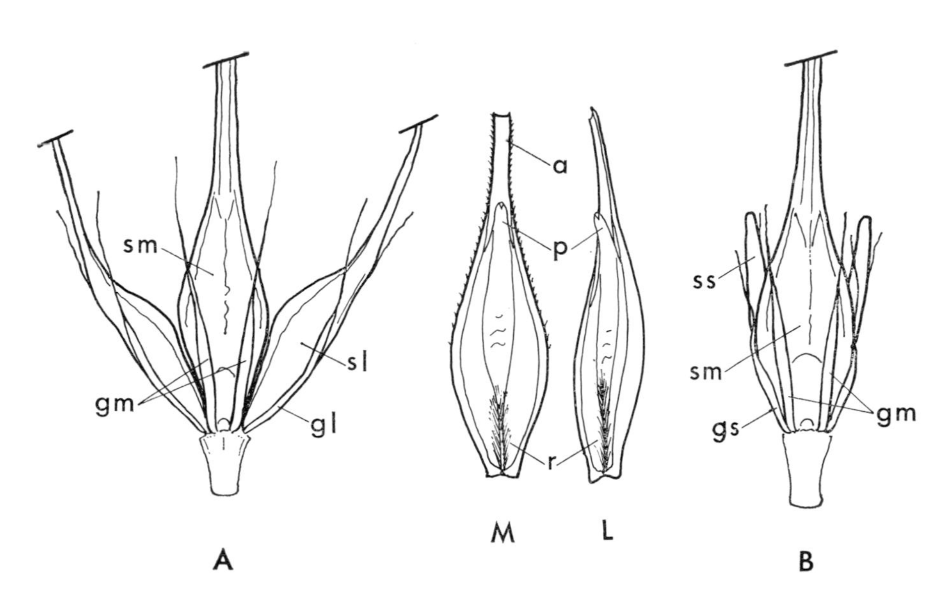

Fig. 10 Comparison of spikelet triplets in (A) six-row barley and (B) two-row barley. The six-row form consists of three single-grained spikelets per rachis internode: a median grain (M), flanked by two fertile, twisted lateral grains (L). The two-row form has a fertile median grain flanked by two sterile florets. gm = glumes of median spikelet, gl = glumes of lateral spikelet, gs = glumes of sterile spikelet, sm = median spikelet, sl = fertile lateral floret, ss = sterile lateral floret, a = lemma awn, p = palea, r = rachilla.

Hordeum

There are two major, distinct types of *Hordeum* (barley), but they both share a similar basic structure (fig. 10). The ear is in the form of a spike, but differs from wheat in that there are three single-flowered spikelets at each internode. The central spikelet is always fertile; the lateral spikelets may be fertile (six-row barley) or sterile (two-row barley). The spikelets consist of two needle-like glumes outside a grain which is invested closely by its lemma and palea. In some species the lemma and palea may actually be fused to the surface of the grain – which thus remains hulled even after the most vigorous threshing. In other forms the grain may be extracted from the lemma and palea – giving naked grains. The apex of the lemma, in most cases, is extended into an awn. The lemma has five prominent nerves extending from the base to the apex of the grain. The palea shows a ventral groove corresponding to that of the caryopsis; in the lower part of the groove lies the slender rachilla. Between the base of the lemma and the embryo of the caryopsis lie two slender structures known as *lodicules* which may be large or small (Gill and Vear 1958, 267). The caryopsis is distinguished from that of wheat by its general shape: the sides are more markedly convex and the lateral profile is much narrower than that of wheat. Wheat and barley are the two cereals most frequently found on prehistoric sites.

Secale

The ear of *Secale* (rye) has features in common with both of the preceding genera. The ear is in the form of a rather lax spike which bears a three-flowered spikelet at each internode. The third floret is almost always abortive. The glumes are short, narrow and acutely keeled: the lemmas are stiff and keeled with a series of stout hairs running up the keel. They taper to form long, stout awns. The lemma and palea of each floret tend to diverge at the tip so that the apex of the mature grain is visible (fig. 11). The caryopsis threshes free from the lemma and palea and is similar in shape to that of

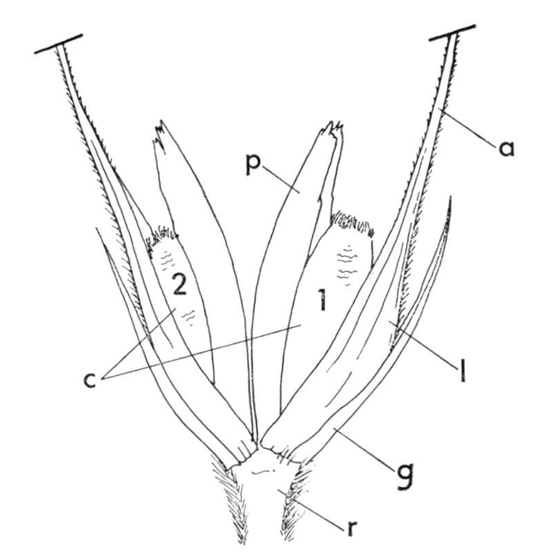

Fig. 11 Spikelet of rye ×3. (a) awn, (c) caryopses of 1st and 2nd florets, (p) palea, (l) lemma with stiff hairs on keel, (g) glume, (r) rachis segment with hairy margins.

wheat but is longer, narrower and more steeply keeled, and the embryo is often asymmetrically placed in the grain (cf. Gill and Vear 1958, 265).

Avena

In *Avena* (oats) the inflorescence is in the form of a spreading panicle of pendulous spikelets. The glumes of each spikelet are extremely large and papery and they enclose from one to three fertile florets (fig. 12). These consist of a hard lemma and softer palea. The lemma often bears a distinctive geniculate awn half or two-thirds of the way to the apex.

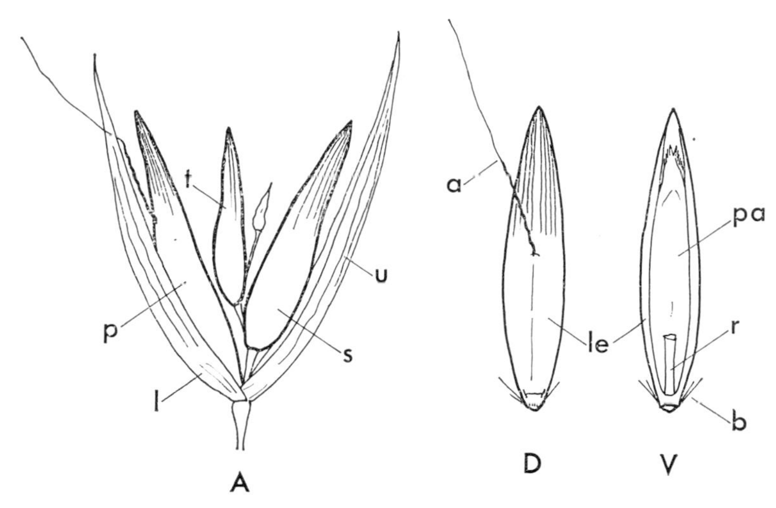

Fig. 12 Spikelet of oats. (A) the (p) primary, (s) secondary and (t) tertiary florets are enclosed between the (l) lower and (u) upper papery glumes. A primary floret is shown in (D) dorsal and (V) ventral views showing the position of the (a) awn, (le) lemma, (pa) palea, (r) rachilla, (b) basal hairs of lemma.

The caryopsis is long and fairly slender – somewhat cylindrical in section, with a very narrow ventral crease. It is distinguished from other cereal grains by the surface of the pericarp which is covered with fine hairs. In most varieties the threshed grain consists of the caryopsis still enclosed by the lemma and palea. During carbonization the lemma and palea become brittle and may disintegrate, leaving the caryopsis naked.

The Millets

The millets are a group of small-grained cereals belonging to a number of different genera. Three types are found in prehistoric contexts in Europe: *Panicum miliaceum* L. (broomcorn millet), *Setaria italica* (L.) Beauv. (Italian millet) and *Echinochloa crus-galli* (barnyard millet).

Panicum miliaceum carries its grains on panicles which arise at successive nodes above the ground. The terminal panicle bears the largest number of grains. The panicles consist of branches terminating in numerous spikelets (fig. 13). Each spikelet consists of a

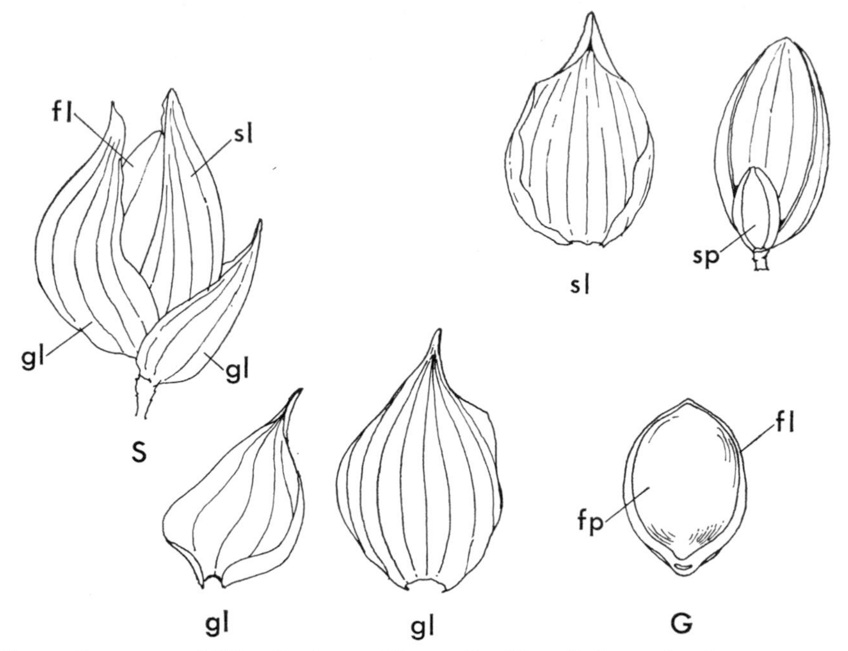

Fig. 13 Broomcorn Millet *Panicum miliaceum* L. (S) spikelet and other parts ×9. (gl) glumes, (sl) lemma of sterile (1st) floret, (sp) palea of sterile floret, (fl) lemma of fertile (2nd) floret, (fp) palea of fertile floret. These latter enclose the caryopsis.

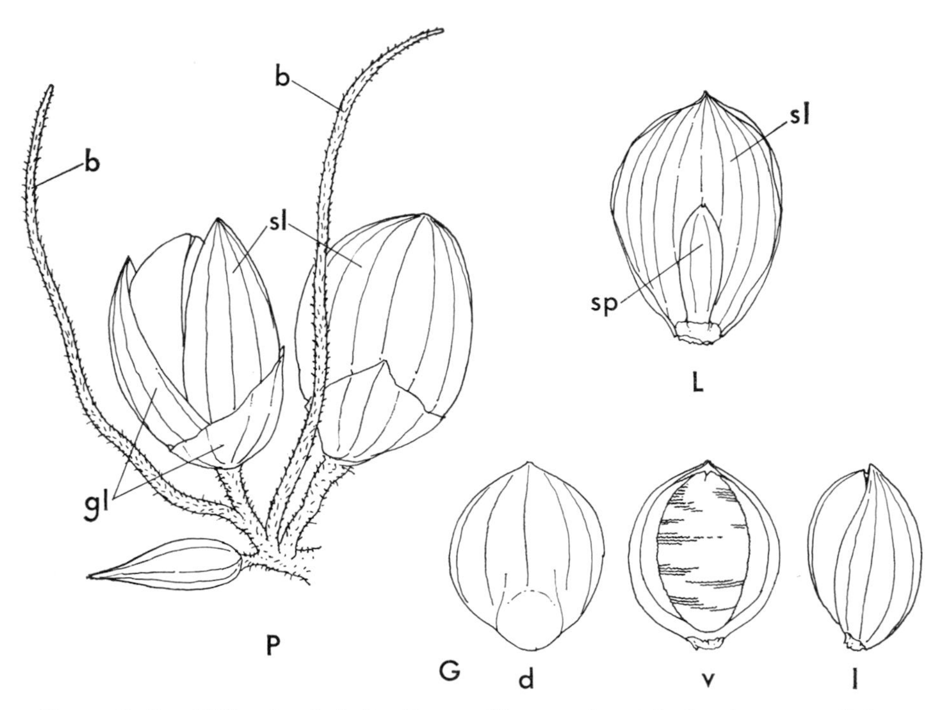

Fig. 14 Italian Millet *Setaria italica* Beauv. (P) part of panicle showing two spikelets and (b) two bristles, (gl) glume, (sl) lemma of sterile floret. (L) showing (sl) lemma and (sp) palea of sterile floret viewed from 'inside'. (G) grain in (d) dorsal, (v) ventral, (l) lateral view. All ×10.

pair of thin membranous glumes, the inner ones being larger than the outer, enclosing two florets: the lower one is sterile. It consists of a lemma similar in size to the inner glume, and a minute palea. The upper fertile floret contains a short, rounded caryopsis about 3 mm long with no ventral furrow enclosed in a thick, shiny lemma and palea (Gill and Vear 1958, 289).

Italian millet (*Setaria italica*) differs from broomcorn millet in having a more compact panicle, often cylindrical and spike-like (fig. 14). The spikelets are similar to those

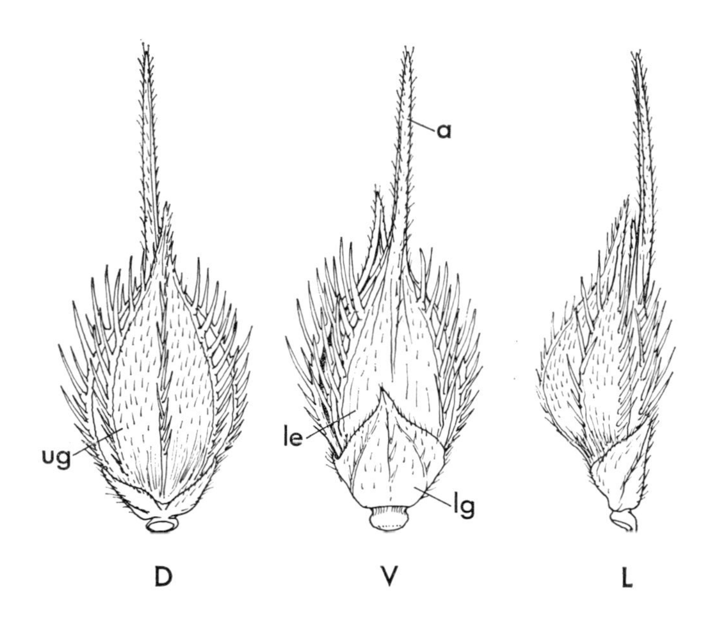

Fig. 15 Cockspur Grass, Barnyard Millet *Echinochloa crus-galli* (L.) Beauv. Spikelet in (D) dorsal, (V) ventral, (L) lateral view ×14. (lg) lower glume, (ug) upper glume, enclosing rounded dorsal side of spikelet and bearing large spines on midrib and four nerves, (le) lower lemma flat, similarly spined and bearing (a) awn; this may vary from 0 to 5 cm in length.

of broomcorn millet but they have one to three bristles representing reduced sterile spikelets below the fertile floret. The grain is smaller and narrower than that of broomcorn millet and is flattened on one side (Gill and Vear 1958, 291).

The spikelets of *Echinochloa crus-galli* are borne directly on the branches which form the upright panicle of spike-like racemes. The spikelets are crowded in pairs or clusters on one side of the axis, from which they fall complete when ripe. Each spikelet contains two florets, only one of which is fertile. The lower lemma bears a short awn at its apex (cf. Hubbard 1959, 336–7). The grain remains tightly enclosed in the hard lemma and palea at maturity (see fig. 15).

Comparison of ears and grains of the major cereals

	Wheat	*Barley*	*Rye*	*Oats*	*Broomcorn millet*	*Italian millet*	*Barnyard millet*
Ear form	spike	spike	spike	panicle	panicle	spike-like panicle	panicle of spike-like racemes
Comparative length of internodes	short	short	short	long	long	short	long or short
Total number of spikelets per node	1	3	1	1 at end of each branch	2 at end of each branch	2 at end of each branch	variable: densely clustered
Number of fertile spikelets per node	1	1 or 3	1	1	1	1	many
Number of sterile spikelets per node	—	2 in two-row form	—	—	—	1–3 in form of bristles below fertile spikelet	—
Number of grains per spikelet	1–9	1 or 3	2	2–3	1	1	1
Nature of caryopsis, naked/hulled	naked and hulled	naked and hulled	naked	naked and hulled	hulled	hulled	hulled

	Wheat	*Barley*	*Rye*	*Oats*	*Broomcorn millet*	*Italian millet*	*Barnyard millet*
Glumes	Broad: stiff keeled, asymmetrical	narrow, lanceolate	short, narrow, keeled	large and papery	thin membranous, inner glume larger	thin, membranous	short, broad, membranous 3–5 nerved with short spiny hairs up the nerves
Lemma	thinner than glumes – extended into awn at apex	fused to surface of caryopsis: 5 strong nerves, usually extended to form awn	sharply keeled with stiff hairs up the keel: extended to form awn. Lemma and palea diverge at tip	hard, strongly nerved, supports geniculate awn arising from central vein in middle of dorsal side, retains caryopsis after threshing	shiny, retains caryopsis after threshing	shiny, retains caryopsis after threshing	tough and smooth with bristles up the nerves: bears awn at apex, retains caryopsis after threshing
Caryopsis	thickness exceeds breadth, ovoid outline, brush of hairs at apex	breadth exceeds thickness, lens-shaped lateral aspect	similar to wheat but steeply keeled and very wrinkled surface. Embryo asymmetrically placed	cylindrical slender, covered with fine hairs	c. 3 mm long, small, rounded, breadth exceeds thickness	small, narrow with pointed base	small 2·7–3·0 mm, rounded on dorsal side, flattened ventral side.

Chapter 5

Wheat

GENERAL DISCUSSION OF GENUS *Triticum*

At the present time two wild and fourteen cultivated species of wheat are widely recognized. They can be divided into three groups on the basis of their numbers of chromosomes and may be further subdivided into those species whose spikelets remain intact after threshing (i.e. hulled grain) and those whose grains thresh free from their glumes (naked grain).

The system of nomenclature is by no means the subject of total agreement among taxonomists: many regard *Triticum boeoticum* and *Triticum monococcum* as a single species, since they may readily be crossed to produce fertile hybrids (Peterson 1965, 3). Similarly among the hexaploids all the six 'species' can be crossed to produce fertile hybrids. This has led to the suggestion by MacKey (1954) and Sears (1956) that all hexaploid wheats should be considered to belong to *Triticum aestivium* L. em. Thell. with six sub-species: *spelta* (L.) Thell.: *macha* (Dek. & Men.) MacKey: *vavilovi* (Tuman) Sears: *vulgare* (Vill.) Host., MacKey: *compactum* (Host.) MacKey and *sphaerococcum* (Perc.) MacKey. Bowden (1959) does not recognize the 'species' of hexaploid wheat even as sub-species but regards them all as cultivars of *Triticum aestivum* L. The tetraploid group is less well understood and Bowden (1959, 668 f.) suggested that they could all be sub-species of *Triticum turgidum*. For this discussion the nomenclature used by Peterson (1965) and set out in the table on p. 41 will be followed.

GENETIC AND ARCHAEOLOGICAL EVIDENCE FOR THE ORIGINS OF WHEAT SPECIES

The first step towards understanding the relationships of the different wheat species came when Sakamura (1918) showed that they form a polyploid series with 14, 28 and 42 chromosomes. In the following year Kihara, studying the pollen grains of pentaploid hybrids derived from crosses between tetraploid and hexaploid wheats, discovered that they contained both single 'univalent' chromosomes and chromosome pairs or 'bivalents'. As a result of this and later work (Kihara 1924) he recognized three

Major divisions of wheat species

Hulled grain	*Naked grain*
1. DIPLOID GROUP (14 chromosomes) *Triticum boeoticum* Boiss. (wild einkorn) *Triticum monococcum* L. (einkorn)	
2. TETRAPLOID GROUP (28 chromosomes) *Triticum dicoccoides* Körn (wild emmer) *Triticum dicoccum* Schübl. (emmer) *Triticum timopheevi* Zukov. (Timopheevi wheat)	*Triticum durum* Desf. (hard wheat) *Triticum turgidum* L. (rivet wheat) *Triticum turanicum* Jakubz. = *Triticum orientale* Perc. (Khorsan wheat) *Triticum polonicum* (Polish wheat) *Triticum carthlicum* (Nevski) = *Triticum persicum* Vav. (Persian wheat)
3. HEXAPLOID GROUP (42 chromosomes) *Triticum spelta* L. (spelt wheat) *Triticum macha* Dek. & Men. (Macha wheat) *Triticum vavilovi* Tuman (Vavilov's wheat)	*Triticum aestivum* L. = *Triticum vulgare* (Vill.) Host. (common or bread wheat) *Triticum compactum* Host. (club wheat) *Triticum sphaerococcum* Perc. (shot wheat)

Based on Peterson (1965, 2)

different genomes, each composed of seven chromosomes, in the genus *Triticum*. These Kihara labelled A, B and D. The diploid wheats have only the A genome; the tetraploid group has both the A and B genomes, and the hexaploids have all three genomes A, B, D.

This situation has evolutionary implications. The diploid form appears to be the most primitive and the tetraploids arose from hybrids between the diploid wheat and another diploid species with the B genome. This hybridization was followed by a doubling in the chromosome number. Subsequently the tetraploid wheats hybridized with another diploid species which donated the D genome to the hexaploid wheats (Riley in Hutchinson 1965, 105–6).

The sources of the B and D genomes appear to be in the closely related genus *Aegilops*. The B genome is believed to have been donated by an ancestor of the present-day *Aegilops speltoides* whose genomes SS appear to be closely similar to the BB genomes of tetraploid wheat (Peterson 1965, 83).

Detailed morphological studies led Kihara in 1944 to suggest that *Aegilops squarrosa* is the donor of the D genome. Independent work in the same year by McFadden and Sears, who artificially synthesized *Triticum spelta* from *Triticum dicoccoides* and *Aegilops squarrosa* (McFadden and Sears 1946), led to the same conclusion.

Thus the evolution of the diploid and tetraploid and hexaploid wheats may be tentatively expressed in the following diagrams.

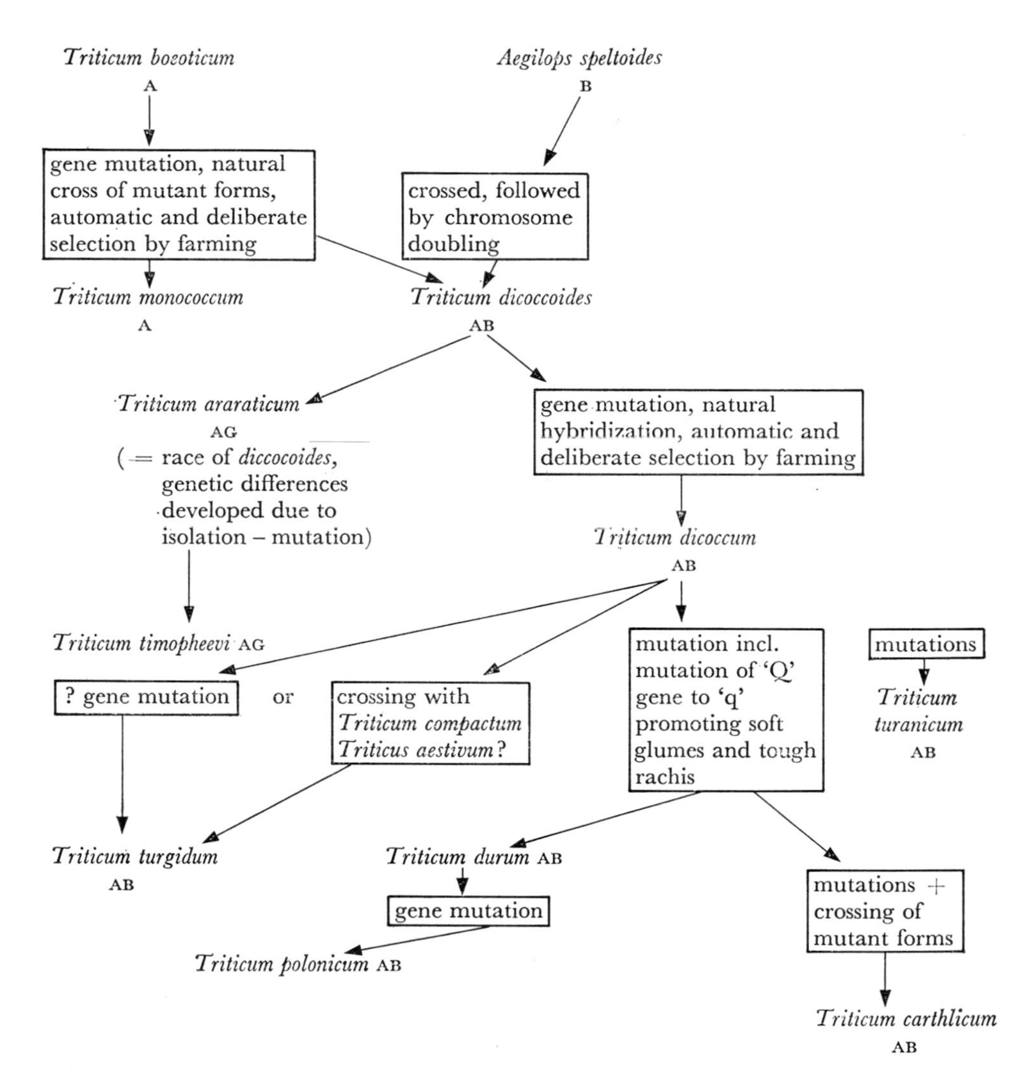

The possible evolution of the cultivated diploid and tetraploid wheats. (Based on Peterson 1965, 80 f., with genomes added after Bell in Hutchinson 1965, 75.)

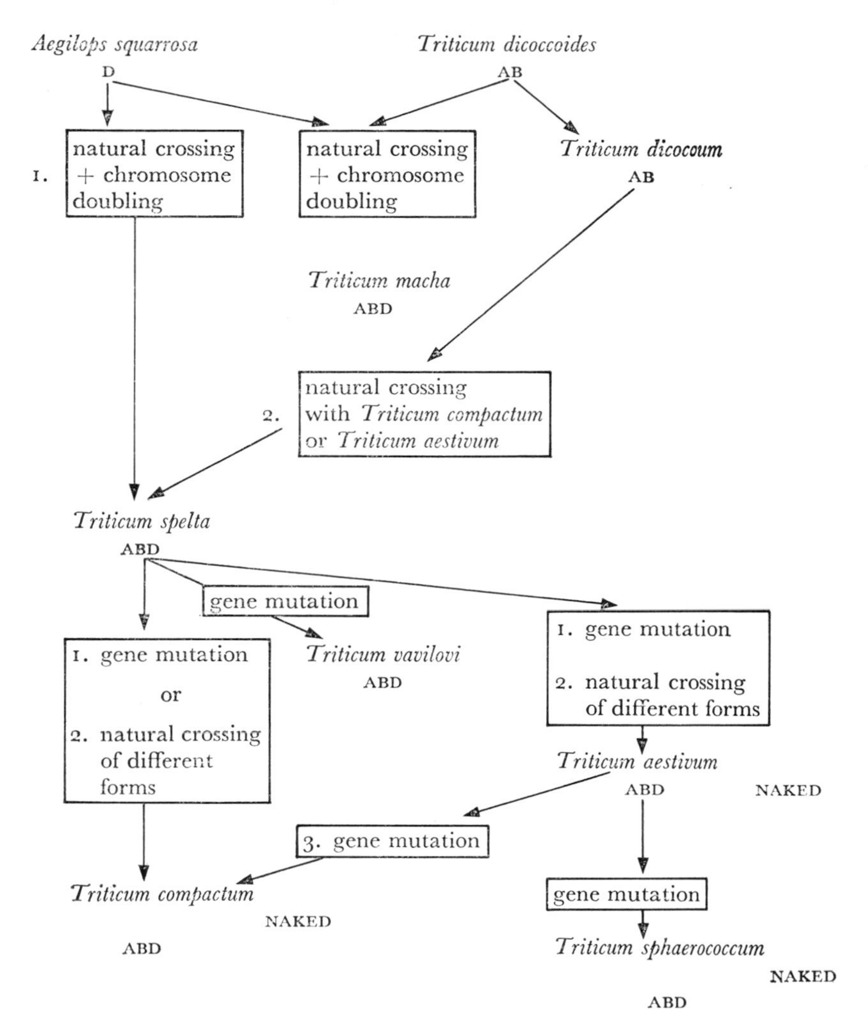

Possible evolution of the cultivated Hexaploid wheats. (After Peterson 1965, 80 f. Genomes added after Bell in Hutchinson 1965, 75.)

The comparative positions of the hulled and naked hexaploid wheats are not totally agreed by leading geneticists as the diagrams on pp. 42–4 illustrate.

It is interesting to compare these conclusions for the origin and evolution of wheat species based on genetic considerations, with those obtained from palaeoethnobotanical evidence in the earliest farming villages yet known in the Near East.

The wild diploid form *Triticum boeoticum* has been found at Ali Kosh – Bus Mordeh phase c. 7500–6750 B.C. (Helbaek 1966B): at Jarmo c. 6750 B.C. (idem 1959B): aceramic Hacilar (idem 1966B): and at Tell Mureybit (see pl. 8) (van Zeist and Casparie 1968). Cultivated einkorn does not occur as widely as the tetraploid emmer wheat on

2. *McFadden and Sears (1946)*

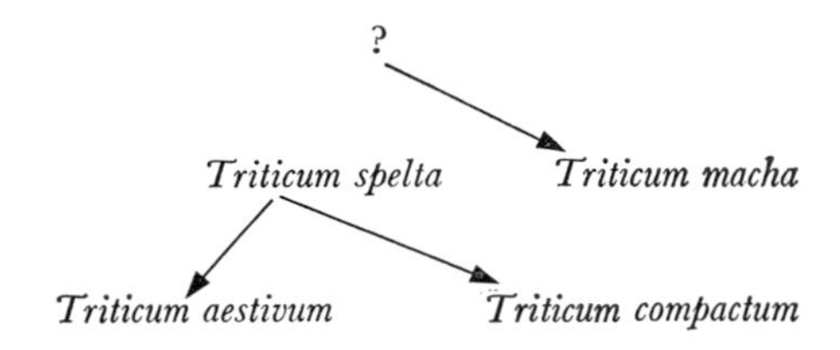

3. *MacKey (1954)*

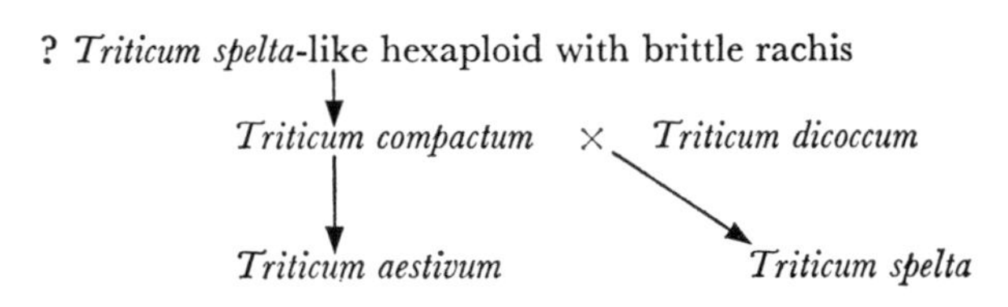

4. *Kihara*

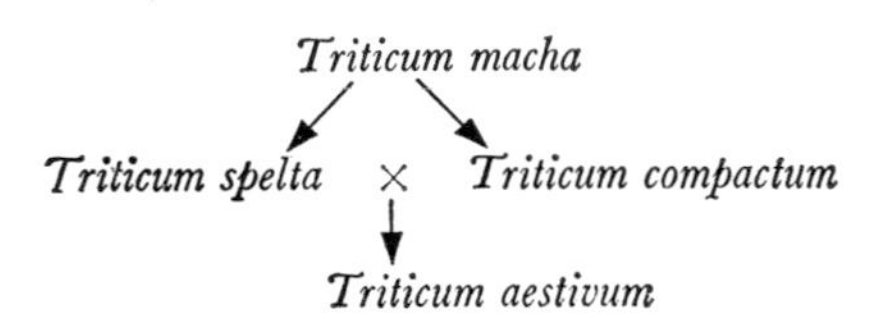

the early sites: it is recorded from Ali Kosh (Bus Mordeh phase: Helbaek 1966B), Tell es-Sawwan (idem 1965, 45), Jarmo (idem 1959B), Çatal Hüyük (idem 1964, 121), late neolithic Hacilar (idem 1966B), Ghediki (Renfrew 1965, 22) and Argissa (Hopf 1962, 101). At Çatal Hüyük, Hacilar and Argissa it is the two-grained form of cultivated einkorn which is found: this is of great interest since the wild form *Triticum boeoticum* ssp. *thaoudar* from which it is derived has a modern distribution from southern Turkey to Iran and the single-grained *Triticum boeoticum* ssp. *aegilopoides* has its modern distribution in the southern Balkans and west Anatolia (fig. 16; Harlan and Zohary 1966). Helbaek (1966B) has postulated that 'west central Anatolia was the primary centre of conscious development and selection took place about 6000 B.C.' This may be true for the twin-grained variety but does not explain how the single-grained form reached Iran nearly a millennium earlier. The real difficulty is that we cannot be sure that the present distribution of wild forms corresponds at all closely to those of the period of the beginnings of plant domestication about 8000 B.C.

The most common tetraploid wheat found on early sites is *Triticum dicoccum* Schübl. (emmer wheat). We have already discussed its origin from *Triticum boeoticum* and *Aegilops speltoides* through *Triticum dicoccoides* and it would seem likely that it arose in the

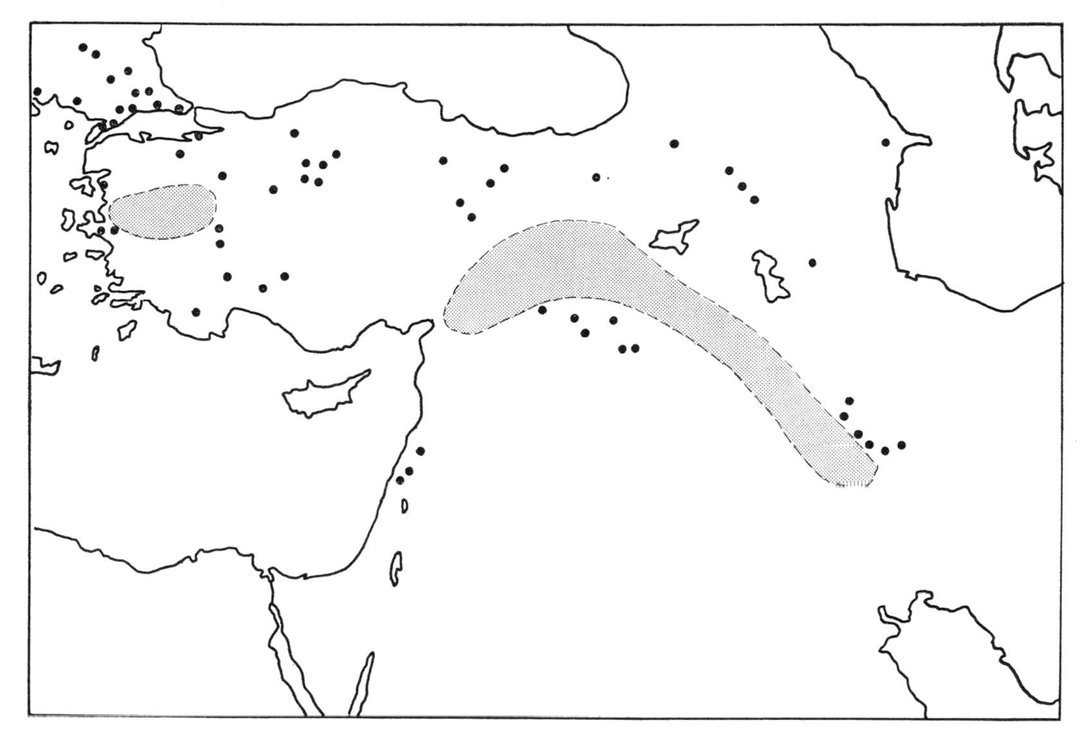

Fig. 16 Modern distribution of *Triticum boeoticum* in the Near East (after Zohary).

area of overlap in the distribution of *Triticum boeoticum* and *Aegilops speltoides* (fig. 17; cf. Peterson 1965, 90), i.e. in eastern Anatolia and Syria – but again we cannot be sure of the exact distributions at this early date. It is likely that *Triticum dicoccoides* arose in the wild from the cross-pollination of its progenitors (Peterson 1965, 98). The race of *Triticum dicoccoides* known as *Triticum araraticum* probably arose in a similar manner, and, being isolated from the main development of *Triticum dicoccoides,* acquired genetic differences which gave rise to its strong sterility barriers with *Triticum dicoccum* and *Triticum durum* (Zohary 1969). So far only the wild *Triticum dicoccoides* has been identified in archaeological contexts and to date only at Jarmo in Iraqi Kurdistan (Helbaek 1960B and 1966B). *Aegilops* grains have been reported from Ali Kosh (Ali Kosh phase) (Hole and Flannery 1967), Beidha (Helbaek 1966A) and Jarmo (idem 1959); and it would be interesting to know if they belonged to either *Aegilops speltoides* or *Aegilops squarrosa* since these species appear to have played such an important role in the origins of the tetraploid and hexaploid wheats.

Emmer wheat is found on almost all the earliest sites in the Near East – it is the commonest of all cereals in the early farming villages of the Near East, Anatolia and southern Europe (Helbaek 1960, 103). Other species of tetraploid wheat have so far rarely been found in archaeological contexts. A mummified ear of *Triticum durum* (macaroni or hard wheat) is reported from a Graeco-Roman context near Fayum in

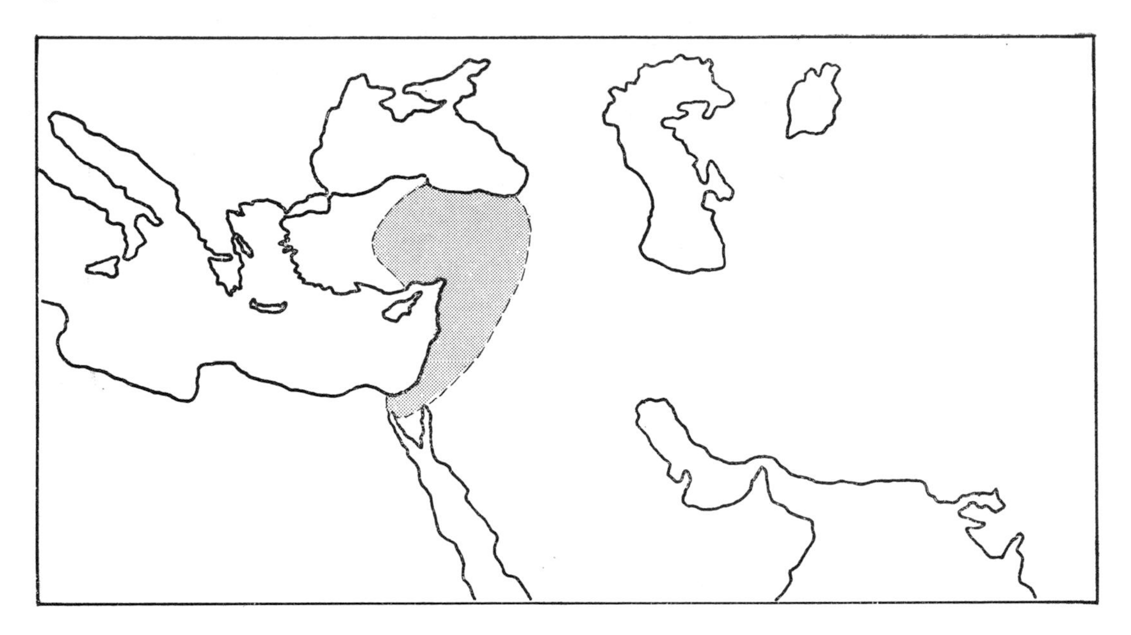

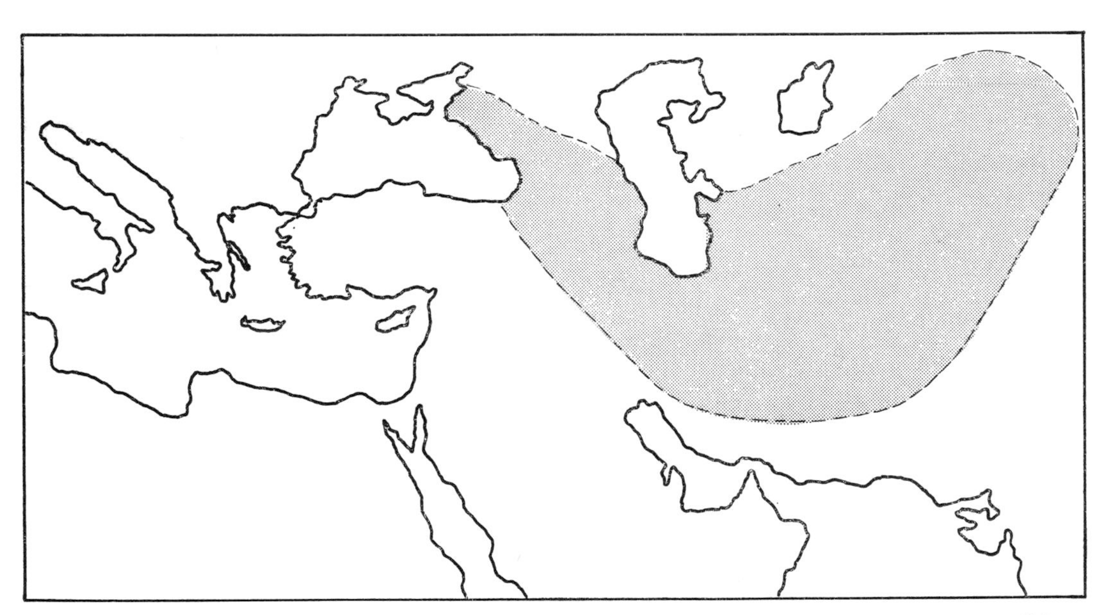

Fig. 17 Distribution of *Aegilops speltoides* (above) and *A. squarrosa* (below) (after Peterson 1965).

Egypt (Täckholm, Täckholm and Drar 1941). Both *Triticum durum* and *Triticum turgidum* are suspected by Helbaek in the finds from Beycesultan (Helbaek 1961, 90). Hopf also suspects *Triticum turgidium* in neolithic contexts in Spain (Hopf 1970, 29).

The question of the origins of the hexaploid wheats is, as we have seen, the subject of controversy, and the archaeological material does not aid us greatly in its clarification. Two hexaploid species are recorded in the Near East from contexts dated before 5000 B.C. These are *Triticum aestivum* and *Triticum compactum*, both naked forms.

At the present time the archaeological evidence suggests that *Triticum aestivum* and *Triticum compactum* are of equal antiquity: *Triticum aestivum* is found at Tepe Sabz (Sabz phase), (Helbaek 1966B), Tell es-Sawwan (idem 1965), Çatal Hüyük (idem 1964 and 1966B) and late neolithic Hacilar (idem 1966B). *Triticum compactum* has recently been recorded in pre-pottery neolithic B levels at Tell Ramad (van Zeist and Botteima 1966). So these naked hexaploid wheats were well established before 5000 B.C. in the Near East. The remaining naked hexaploid form *Triticum sphaerococcum* is found on sites of the Indus civilisation in north-west India dating to the third millennium B.C. (Allchin, in Ucko and Dimbleby 1969, 323 f.).

There is no sign of the hulled *Triticum spelta* in palaeoethnobotanical material until the second millennium B.C. in Central Europe. The earliest find appears to be from the late neolithic settlement of Riedschachen on the Federsee (Bertsch and Bertsch 1949, 40). Its distribution seems to have been restricted in later prehistoric times to Switzerland, southern Germany, southern England, Denmark and southern Sweden (Helbaek 1952, 97; Schultze-Motel and Kruse 1965, 588). The only other hexaploid wheat found in archaeological contexts is the hulled *Triticum macha* which is reported from 'ancient remains in Colchis' (Žukovskij, trans. Hudson 1962, 3). This would tend to support Riley's theory, based on the high frequency of *Aegilops squarrosa* in a sample of modern wheat at Chalus on the Caspian, that the initial hybridization may well have occurred in cultivated fields. This, he says, could have resulted in the development of a free-threshing type of hexaploid wheat resembling *Triticum aestivum* (Riley in Hutchinson 1965).

CRITERIA FOR THE PALAEOETHNOBOTANICAL IDENTIFICATION OF THE EARS AND GRAINS OF WHEAT SPECIES

Triticum boeoticum Boiss. em. Schiem (pl. 9 and fig. 18)

The ear is long, slender, and narrower across the face than across the two-rowed side: the lemma bears awns which are longer than those of *Triticum monococcum*. The rachis is extremely brittle and breaks into segments at maturity: the margins of the rachis are densely pubescent with long hairs. The terminal spikelet is small and aborted. The caryopsis is tightly enclosed in tough glumes after threshing. The glumes have two keels each of which ends in a pointed beak and the shoulder of the glume is notched. The palea splits on maturity. Usually only one fertile floret develops per spikelet (ssp. *aegilopoides*). The caryopsis is small, narrow, laterally flattened and pointed at both ends. In ssp. *aegilopoides* both dorsal and ventral aspects are convex in outline, in ssp. *thaoudar* where two grains mature in the spikelet the ventral side is more or less flat due to pressure from the second caryopsis, the grains are slender, spindle-shaped, with the greatest width in the middle of the grain. Percival (1921) gives the following measurements for fresh caryopses; length 6–7·5 mm, breadth 1–1·5 mm, thickness 2·1–2·6 mm (See Peterson 1965, 9–10; Leonard and Martin 1963 after Flaksberger 1939 and van

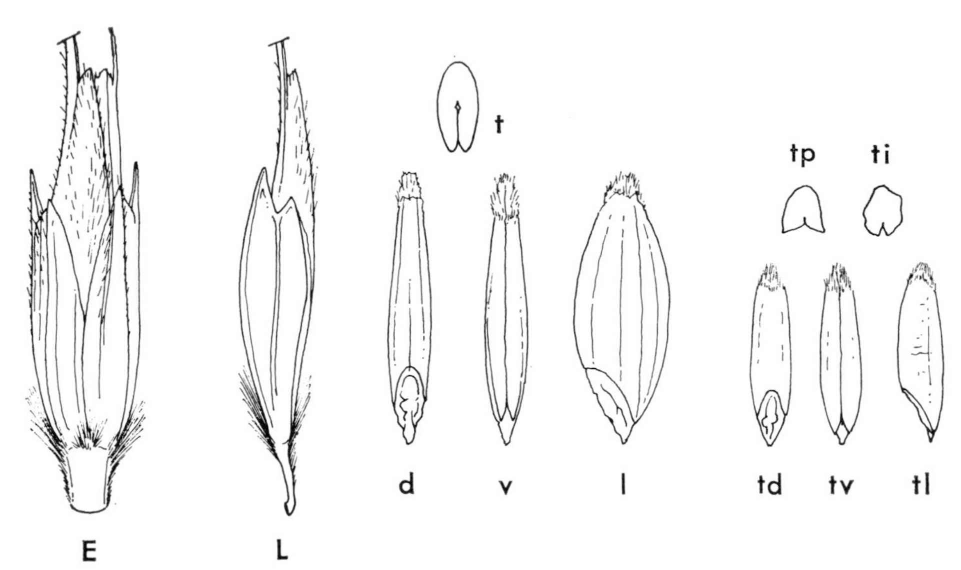

Fig. 18 Wild Einkorn *Triticum boeoticum* Boiss. em. Thiem. Spikelet of ssp. *aegilopoides* in (E) external face view, (L) lateral view, with single large flattened caryopsis in (d) dorsal, (v) ventral, (l) lateral view, (t) transverse section. Caryopsis of ssp. *thaoudar* after van Zeist (1968) in (td) dorsal, (tv) ventral, (tl) lateral view; transverse sections with (tp) protruding and (ti) intruding ventral side. All ×4.

Zeist and Casparie 1968.) *Triticum boeoticum* ssp. *urartu* is found in Armenia and differs from the other two sub-species in the shape of its glumes (Dorofeev 1968, 451 f.).

Triticum monococcum L. (pl. 10, fig. 19)

The ears are narrow, slender and laterally compressed and are shorter than those of *Triticum boeoticum.* The spike is denser, and the rachis segments are shorter than those of *Triticum boeoticum.* The lemmas have short awns. The rachis segments have slightly pubescent margins and the terminal spikelet is aborted. The grains remain enclosed in a tough, tightly fitting outer glume after threshing. The glume has two keels which terminate in two pointed beaks. The shoulder of the glume is notched, and the palea splits into two parts at maturity. Usually one grain ripens per spikelet, but the tendency to form two grains is slightly more common than in *Triticum boeoticum.* The caryopses are small, pointed at both ends but shorter and broader than those in wild einkorn. In single-grained forms the curved dorsal aspect is sharply ridged. The ventral side is curved, steeply ridged and very narrow, and the ventral crease is extremely narrow. When carbonized it cheeks along the ventral furrow and the apex will usually puff. The embryo is relatively large. In the two-grained spikelet the caryopses are usually smaller and narrower but otherwise rather similar to those of emmer. The decisive detail in distinguishing between two-grain einkorn and emmer caryopses is in the relation of breadth and thickness (TB index). In emmer the index range is 75–95, in einkorn 100–179 (average 132). The measurements of modern einkorn caryopses are: length

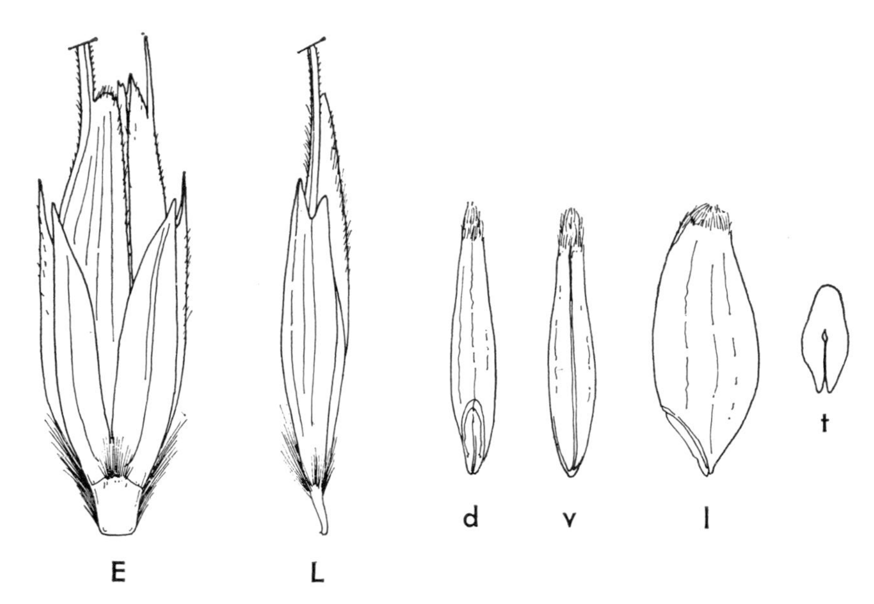

Fig. 19 Einkorn *Triticum monococcum* L. Spikelet in (E) external face view, (L) lateral view. Caryopsis in (d) dorsal, (v) ventral, (l) lateral view, (t) transverse section. All ×4.

7–8·5 mm, breadth 1·8–3·0 mm, thickness 3–3·5 mm. (Percival 1921, 174; see also Peterson 1965, 10; Leonard and Martin 1963, 303; Helbaek 1952B, 219; 1961, 84.)

Another point which may be difficult to establish, if only a few grains are present, is the difference between the single-grained einkorn caryopsis and the grain which may occasionally develop in the terminal spikelet of emmer and may be very similar in form (Schiemann 1951, 316).

The main differences between wild and cultivated einkorn

	Triticum boeoticum	*Triticum monococcum*
Ear	long, slender	shorter, slender
Awns	coarse, long	shorter
Rachis margins	strongly pubescent	slightly pubescent
Glumes	pubescent	slightly pubescent/glabrous
Caryopses	long, slender, spindle-shaped, pointed at both ends	shorter and broader, pointed at both ends
	mm	mm
Measurements of fresh caryopses	L. 6–7·5 B. 1–1·5 Th. 2·1–2·6	7–8·5 1·8–3·0 3–3·5

Triticum dicoccoides Körn (pl. 11; fig. 20)

The ear is laterally compressed. The rachis is very brittle and the rachis margins are densely pubescent. The terminal spikelet is usually sterile, and the caryopses are retained in the tough glumes after threshing. The glumes have a single prominent keel which is extended to form a broad triangular beak. The shoulder of the glume is notched. The lemma bears an awn. The palea is divided at the tip but does not split at maturity. Usually two grains ripen in each spikelet and they are compressed laterally

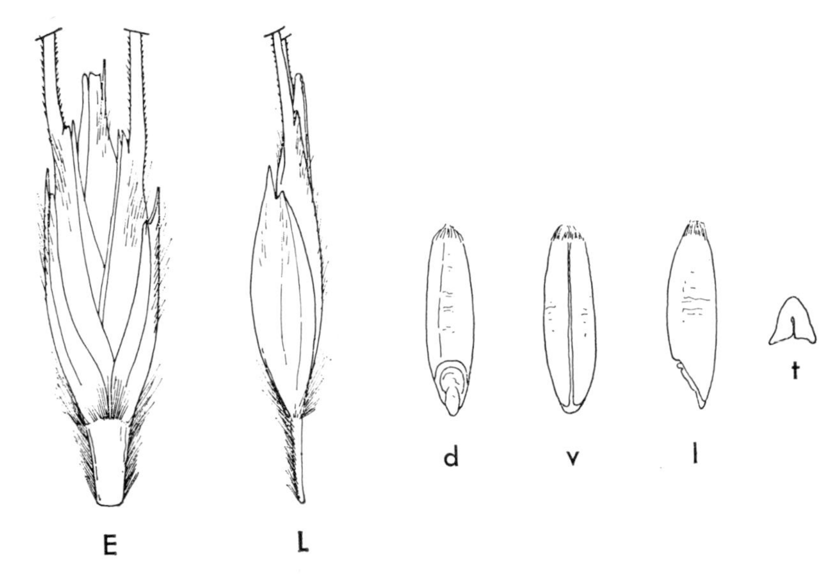

Fig. 20 Wild Emmer *Triticum dicoccoides* Körn. Spikelet in (E) external face view, (L) lateral view. Caryopsis in (d) dorsal, (v) ventral, (l) lateral view, (t) transverse section. All ×2½.

and are pointed at both ends. They appear narrow, sharply keeled and wrinkled. Percival (1921, 182), gives measurements for two forms – the shorter one measured: length 8 mm, breadth 1·7 mm, thickness 1·85 mm. A long-seeded form measured: length 11–12 mm, breadth 2·9 mm, thickness 2·7 mm. (See Peterson 1965, 11; Leonard and Martin 1963 after Flaksberger 1939; Helbaek 1966B, 352.) *Triticum araraticum* Jakubz appears to have developed in Transcaucasia from *Triticum dicoccoides* due to isolation there.

Triticum dicoccum Schübl. (pl. 12 and fig. 21)

The ears are dense, laterally compressed and narrow, and are usually awned. The rachis is brittle but less brittle than in *Triticum dicoccoides* and the rachis margins are less pubescent than those of *Triticum dicoccoides*. The terminal spikelet is fertile. The tough glumes tightly enclose the grain at maturity and the single prominent keel is extended into a beak. The shoulder of the glume is sloping. The palea is broad and does not split

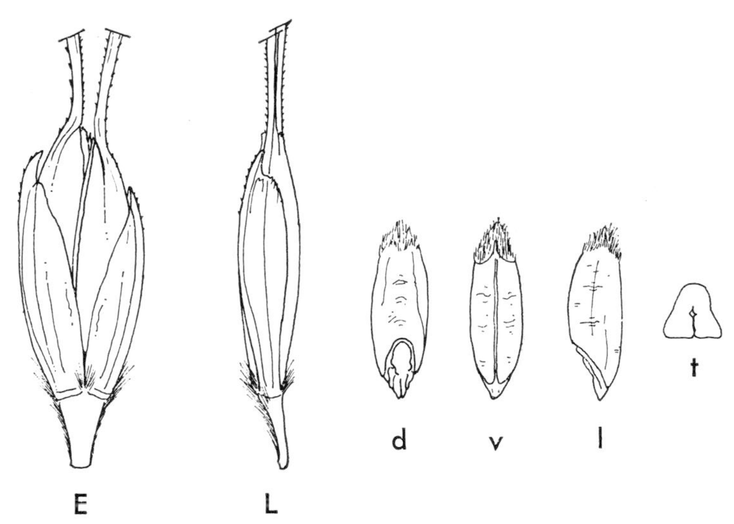

Fig. 21 Emmer *Triticum dicoccum* Schübl. Spikelet in (E) external face view, (L) lateral view. Caryopsis in (d) dorsal, (v) ventral, (l) lateral view, (t) transverse section. All ×3.

at maturity. Two grains develop in each spikelet. The caryopsis is broader and plumper than that of *Triticum dicoccoides.* The grains of emmer are larger than those of einkorn, but taper towards both ends. In cross-section they are triangular and sometimes distinctly asymmetrical. The dorsal side is curved and ridged, the ventral aspect is broad and flat in profile, with a longitudinal concavity along the ventral crease; frequently the ventral margins of the grains are rather sharp. The embryo is rather large. Percival (1921, 191) gives these measurements of modern grains: length 7·2–9 mm., breadth 2·85–3·4 mm, thickness 2·6–3·1 mm. (See Leonard and Martin 1963; Peterson 1965, 12; Helbaek 1952B, 219; 1961, 86.)

The main differences between wild and cultivated emmer

	Triticum dicoccoides	*Triticum dicoccum*
Ear	long, lax	short, dense
Rachis	pubescent margins	slightly pubescent/glabrous margins
Awns	coarse, long	short
Glume shoulder	notched	sloping
Caryopsis	narrow, sharply keeled, wrinkled	broad, dorsal curved ventral groove with longitudinal concavity. Ventral margins sharp
	mm	mm
Fresh caryopsis measurements	L. 9–11/12	7·2–9
	B. 1·7–2·9	2·85–3·4
	Th. 1·85–2·7	2·6–3·1

The main differences between cultivated einkorn and emmer wheat

	Triticum monococcum	*Triticum dioccocum*
	mm	mm
*Spikelet fork articulation width	1·48–2·01	1·79–2·89
*Glume width	0·61–0·87	0·84–1·25
*Glume length	6·04–7·14	6·95–8·78
Glumes	two keels	one keel
Palea	splits at maturity	remains intact at maturity
Caryopsis	steeply keeled, narrow ventral aspect convex ventral furrow narrow	broader, flattened in ventral aspect or slightly concave
Fresh caryopsis measurements	L. 7–8·5 B. 1·8–3·0 Th. 3–3·5	7·2–9 2·85–3·4 2·6–3·1

* Figures from Helbaek 1952B, 202.

Triticum timopheevi Žukov. (fig. 22)

The ear is laterally compressed. It is shorter and broader than that of *Triticum dicoccoides*: it is dense and the rachis segments are short. The rachis is rather brittle at maturity, and the margins are pubescent. The terminal spikelet is in the same orientation as the rest of the ear, and is fertile. The glumes are tough and enclose the grains at

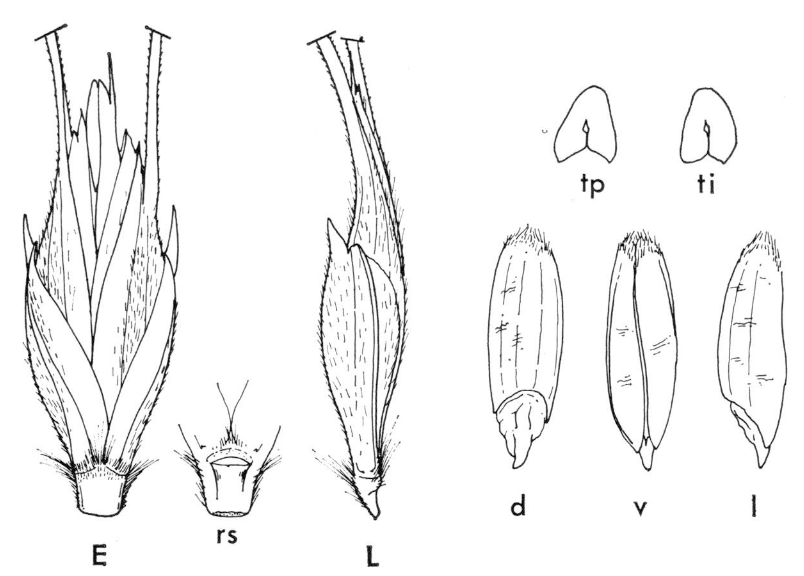

Fig. 22 Timopheevi Wheat *Triticum timopheevi* Žuk. Spikelet in (E) external face view, (L) lateral view, (rs) base of spikelet showing rachis segment in internal face view. Caryopsis in (d) dorsal, (v) ventral, (l) lateral view; transverse sections with (tp) protruding and (ti) intruding ventral side. All $\times 3\frac{1}{2}$.

maturity. They have a single prominent keel, which ends in a broad triangular beak. The glume has a notched shoulder. The palea tends to split at maturity. Normally two grains ripen per spikelet. The caryopsis is usually longer and thicker than that of *Triticum dicoccoides*. The caryopsis is sharply keeled towards the apex, the ventral groove is sharply defined and the ventral cheeks are rounded. The specimens in the author's comparative collection give the following dimensions: length 8·5 mm, breadth 2·3 mm, thickness 2·9 mm. (See Peterson 1965, 11; Leonard and Martin 1963, 297 f.).

Triticum durum Desf. (fig. 23)

The ears are less compressed laterally than those of emmer, the width is often equal across the face and lateral aspect. The ears are lax with long rachis segments. The rachis is tough and the margins are slightly pubescent. The fertile terminal spikelet is placed at 90° to the rest of the ear. The glumes are soft and loose, and have a distinct

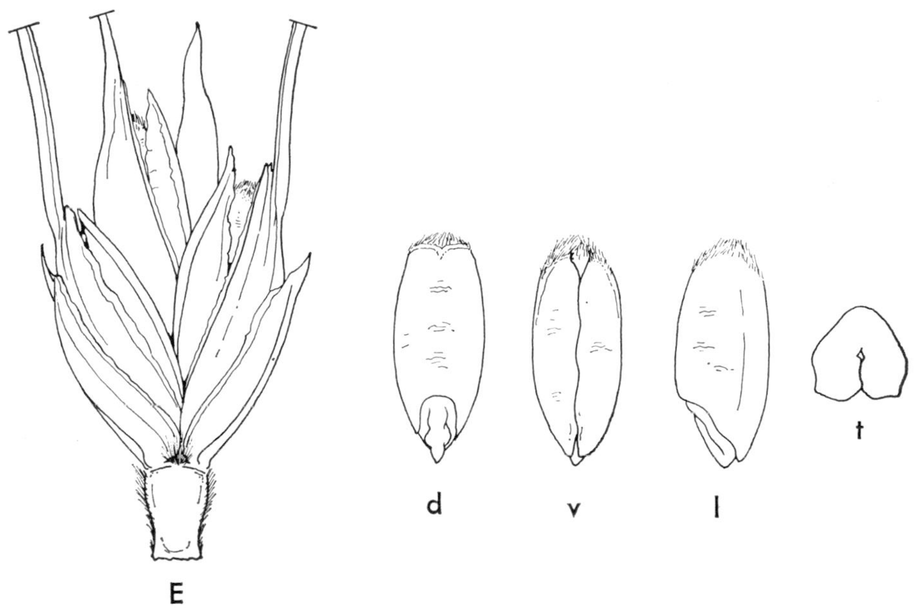

Fig. 23 Marcaroni Wheat *Triticum durum* Desf. Spikelet in (E) external face view. Caryopsis in (d) dorsal, (v) ventral, (l) lateral view, (t) transverse section. All ×3.

keel from the tip to the base of each glume. The glume terminates in a short to medium-length beak. The shoulder of the glume is sloping. The palea remains intact on maturity of the grain. Each spikelet contains two to four free-threshing grains. The caryopsis is large and triangular in cross-section. The ventral furrow is widely v-shaped and the ventral cheeks are sharply angled. The measurements given by Percival 1921, 211, are: length 7·0–9·7 mm (average 8·3 mm), breadth 2·8–4·1 mm (average 3·48 mm), thickness 3·2–4·25 mm (average 3·61 mm). (See Peterson 1965, 12; Leonard and Martin 1963, 297 f.; Gill and Vear 1958, 253.)

Triticum turgidum L. (fig. 24)

The ears are long and occasionally branched. The head is denser than that of *Triticum durum*. The rachis is tough and pubescent along its margins. The fertile terminal spikelet is at 90° to the rest of the ear. The glumes are soft, loose and rather short, and they have a single compact keel terminating in a short beak. The glume shoulder is gently sloping. The palea remains intact at maturity. Three to five grains ripen in each spikelet. The caryopses are shorter and broader than those of *Triticum durum* and

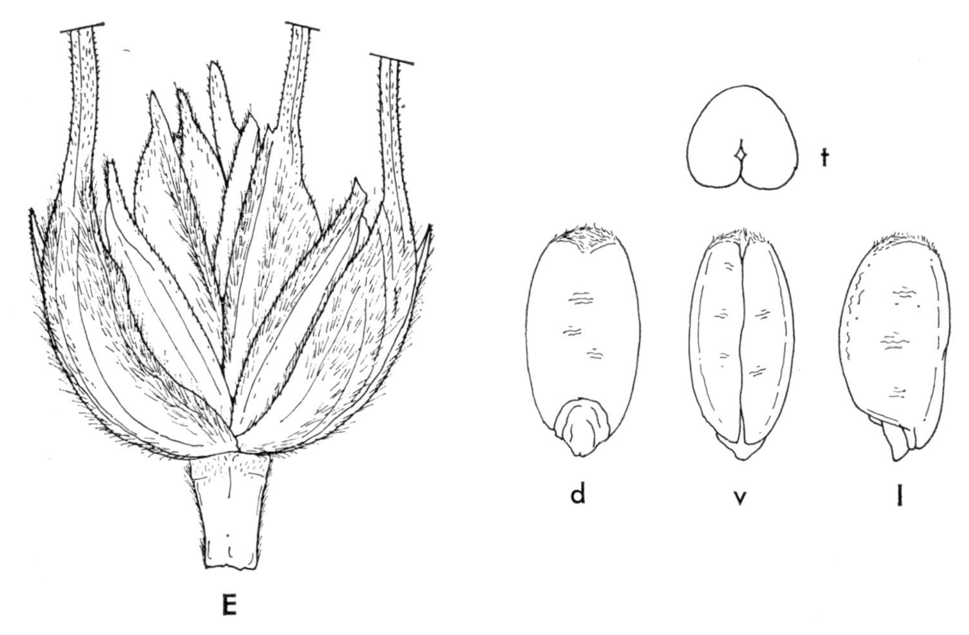

Fig. 24 Rivet (Cone) Wheat *Triticum turgidum* L. Spikelet with four fertile florets in (E) external face view. Caryopsis in (d) dorsal, (v) ventral, (l) lateral views, (t) transverse section. All ×4.

have a characteristic hump on the dorsal side just above the steeply placed embryo. The ventral groove forms a wide v – the ventral cheeks are angled at the lateral margins. Percival (1921) gives the following measurements for grains of this species: length 6·7–8·37 mm (average 7·45 mm), breadth 3·26–4·43 mm (average 3·88 mm), thickness 3·23–4·1 mm (average 3·70 mm)). (See Leonard and Martin 1963, 297–307; Peterson 1965, 13; Percival 1943, 89.)

Triticum turanicum Jakubz. (= *Triticum orientale* Perc.) (fig. 25)

The head is lax. The rachis is tough – the margins being slightly pubescent. The terminal spikelet is set at an oblique angle to the rest of the ear. The glumes are soft, and loose, and often pubescent. Each glume has one main keel and a smaller keel in the broad fold. The main keel is prolonged into an elongated triangular beak with a sharp point. The glume shoulder is notched. The palea remains intact at maturity. The

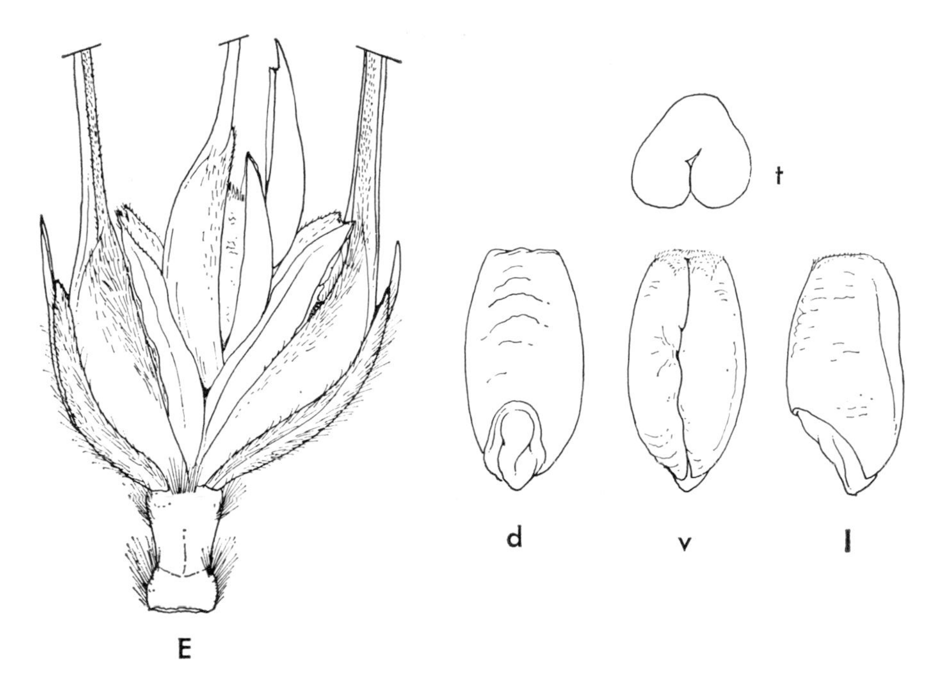

Fig. 25 Khorsan Wheat *Triticum turanicum* Jakubz. Spikelet with three fertile florets in (E) external face view. Caryopsis in (d) dorsal, (v) ventral, (l) lateral views; (t) transverse section. All ×4.

spikelets contain two to three grains. The grains are long and narrow. The widest part is parallel to the upper margin of the embryo. The ventral crease is shallow forming an open v-shape in cross-section. The grains measure: Length 10·5–12 mm, breadth 2·75–3·0 mm, thickness 3·2–3·4 mm (Percival 1921, 205). (See Peterson 1965, 13; Gill and Vear 1958, 253.)

Triticum polonicum L. (fig. 26)

The ears of this species are extremely large and lax. The rachis is rather tough but has a tendency to break: its segments are very long and have pubescent margins. The terminal spikelet is fertile and is placed at 90° to the rest of the ear. The glumes are loose, very long (exceeding the lemmas in length), narrow and papery. They are distinctly nerved but the keel is less marked since the narrow fold of the glume is diminutive. The keel extends to form a tiny acutely pointed beak at the apex of the glume. The shoulder of the glume is notched. The palea splits at maturity. Usually one to three grains ripen per spikelet. The grains are long and slender and sometimes resemble those of rye *Secale cereale*. The caryopses are strongly keeled, the ventral furrow is deep and the ventral cheeks are rounded. Percival (1921, 233) gives the following measurements: length 11–12 mm, breadth 4 mm. (See Peterson 1965, 13; Leonard and Martin 1963, 302.)

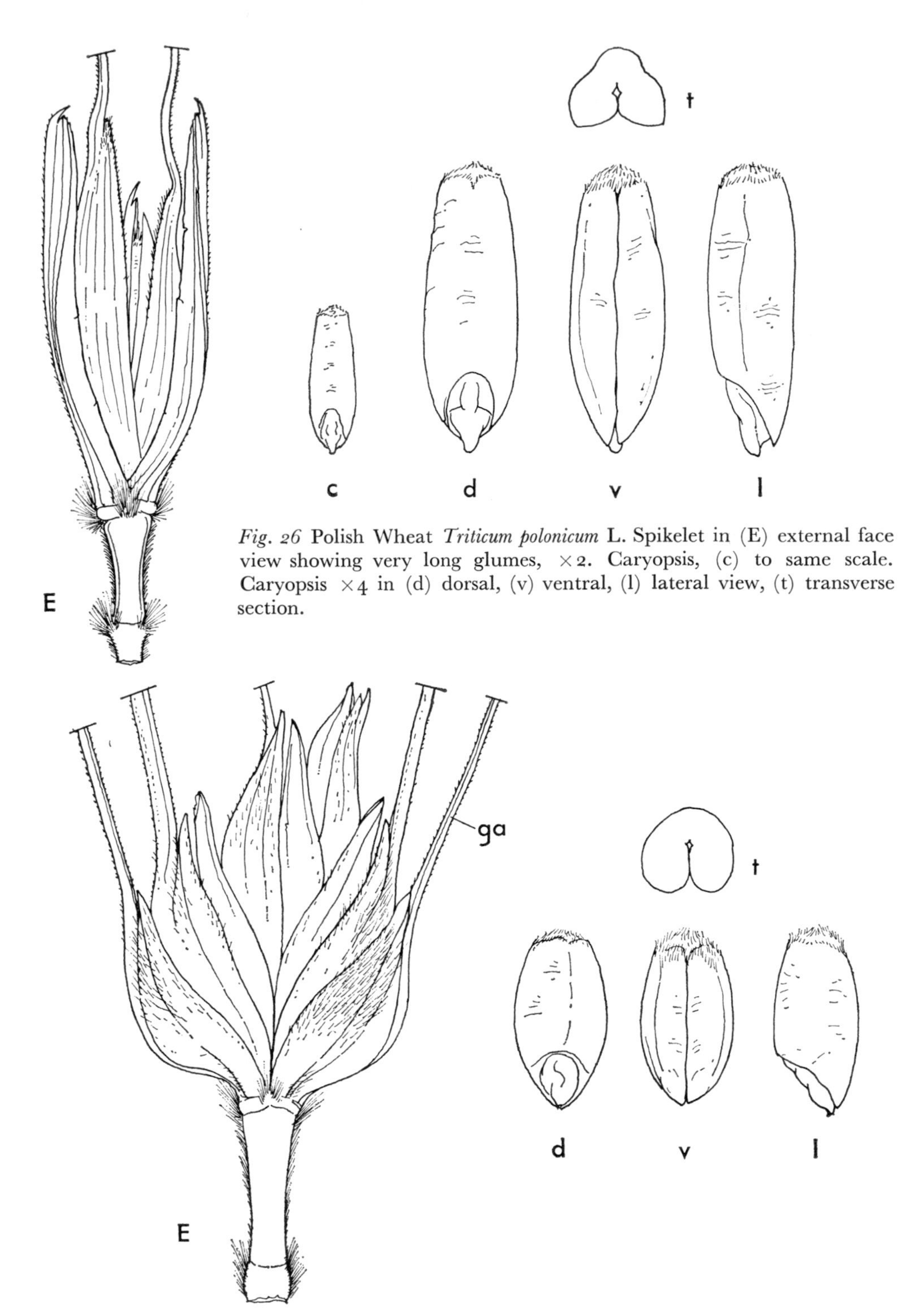

Fig. 26 Polish Wheat *Triticum polonicum* L. Spikelet in (E) external face view showing very long glumes, ×2. Caryopsis, (c) to same scale. Caryopsis ×4 in (d) dorsal, (v) ventral, (l) lateral view, (t) transverse section.

Fig. 27 Persian Wheat *Triticum carthlicum* Nevski. Spikelet in (E) external face view showing long slender rachis segment and (ga) awns on glumes. Caryopsis in (d) dorsal, (v) ventral, (l) lateral view, (t) transverse section. All ×4.

Triticum carthlicum Nevski (= *Triticum persicum* Vav.) (fig. 27)

The lax ear is broad across the face but very narrow across the lateral aspect. The rachis is very narrow, tough and flexible and has pubescent margins. The terminal spikelet is fertile and displaced at 90° to the rest of the ear. The glumes are soft and loose. They are keeled and the narrow keel extending from the base of the glume is prolonged into a long beak which forms an additional awn to those borne on the lemmas (as in all other wheat species). The glume shoulder is nicked sharply. The palea tends to split at the tip on maturity. Two to four grains are produced in each spikelet. The caryopsis is medium-sized and plump. The dorsal side is not sharply keeled: the ventral groove is deep and the cheeks are angular. The specimens in my comparative collection measure: length 7·0 mm, breadth 3·3 mm, thickness 3·5 mm. (See Peterson 1965, 14; Leonard and Martin 1963, 297–8.)

Triticum spelta L. (fig. 28 and pl. 13)

The lax ear is long and narrow. The rachis is brittle, and breaks leaving a segment attached to the face of each spikelet. The rachis segments are broad and strong, the margins are pubescent. The glumes are very tough, holding the grains like a vice. They

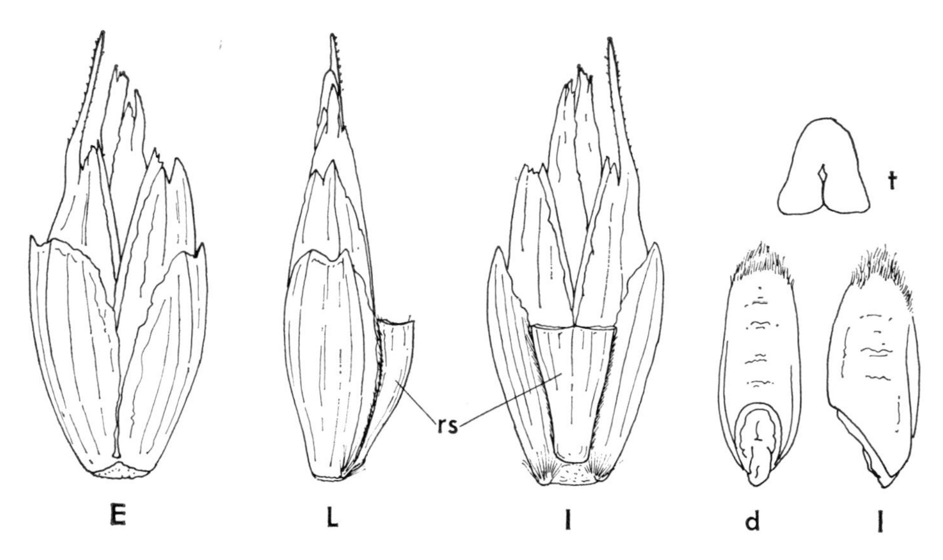

Fig. 28 Spelt *Triticum spelta* L. Spikelet in (E) external face view, (L) lateral view, (I) internal face view. Caryopsis in (d) dorsal, (l) lateral view, (t) transverse section. All ×4. (rs) segment of brittle rachis adhering to inner face of spikelet.

are strongly nerved and have two keels – one more marked than the other, and they do not extend to form beaks. The awns borne by the lemmas have a pronounced S-shaped outline and are much shorter than in the other wheats. Occasionally they survive in archaeological deposits, e.g. at Birknaes, east Jutland (Helbaek 1952A, 101). The breadth of the spikelet fork, measured across the articulation point, ranged between

1·82–2·89 mm in the iron age material from Birknaes, and the breadth of the glume base at the same level: 0·91–1·52 mm. Both these dimensions serve to distinguish spikelet forks of spelt from those of emmer and einkorn (see fig. 29). Usually two grains ripen in the spikelet. The caryopses are slender and taper towards both ends. The dorsal side is evenly curved whereas the ventral side is flat, with a ventral furrow. The apex of the kernel is flatter, shorter and broader than that of emmer. The blunt apex and the flatness

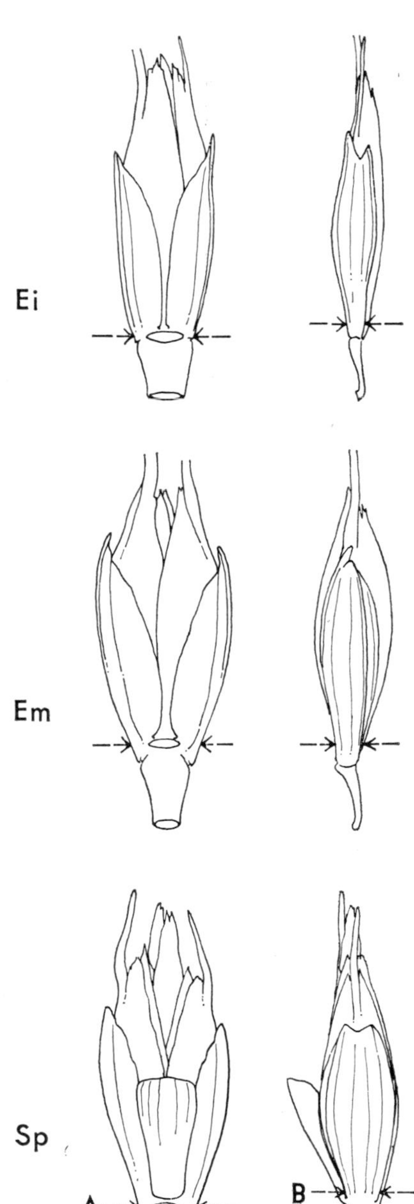

Fig. 29 Comparison of spikelets of (Ei) Einkorn, (Em) Emmer and (Sp) Spelt wheats in inner face view and lateral view to show differences in (A) spikelet articulation width and (B) glume width. All ×3.

of the grain are the chief features distinguishing it from that of emmer. Spelt grains cannot be safely identified, however, unless parts of the spikelets also survive. Percival gives the following measurements for fresh grains of spelt: length 7–10 mm, breadth 3–3·5 mm, thickness 2·7–3 mm. (See Peterson 1965, 14; Jessen and Helbaek 1944, 40; Helbaek 1952A, fig. 1, 103; 1952B, 219.)

Triticum macha Dek. & Men. (fig. 30)

The ears may be dense or lax. The rachis is brittle and the rachis segment may adhere to the face of the spikelet after threshing (as in *Triticum spelta* the rachis often breaks immediately below the spikelet so that the segment which adheres to its face is that

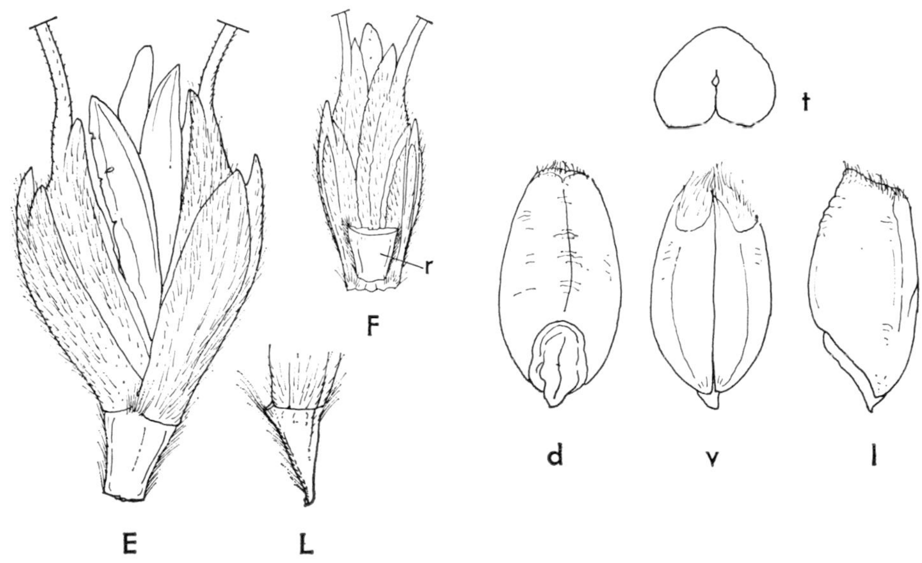

Fig. 30 Macha Wheat *Triticum macha* Dek. et Men. Spikelet in (E) external face view showing characteristic skewing of axis and adherent rachis segment. (F) internal face view with adherent (r) rachis segment, in alternative position $\times 2\frac{1}{2}$. (L) rachis segment in lateral view. Caryopsis in (d) dorsal, (v) ventral, (l) lateral view, (t) transverse section. All $\times 4$.

which supported the spikelet above (see fig. 30F)). The small terminal spikelet is set at 90° to the rest of the ear. The glumes are thick, tough and pubescent, and they have a single keel. The beak is acute and sharply pointed. The palea remains intact at maturity. Two grains ripen per spikelet: they are similar in size and shape to grains of *Triticum spelta*. The grains of the comparative specimen in my collection measure: length 7·5 mm, breadth 3·8 mm, thickness 3·1 mm. (See Leonard and Martin 1963, 303; Gill and Vear 1958, 254; Peterson 1965, 15.)

Triticum vavilovi Tuman (fig. 31)

This species has a large ear with exceptionally long branched spikelets. The rachis has long, wide segments and shows a tendency to break on maturity: the margins are pubescent. The fertile terminal spikelet is placed at 90° to the ear. The grain threshes free from the glumes. The glumes are firm, shorter than the lemma, and covered with bristle-like spines. The glumes are strongly nerved and have one main keel which is prolonged into a short, blunt beak. The shoulder of the glume has a wide notch. The

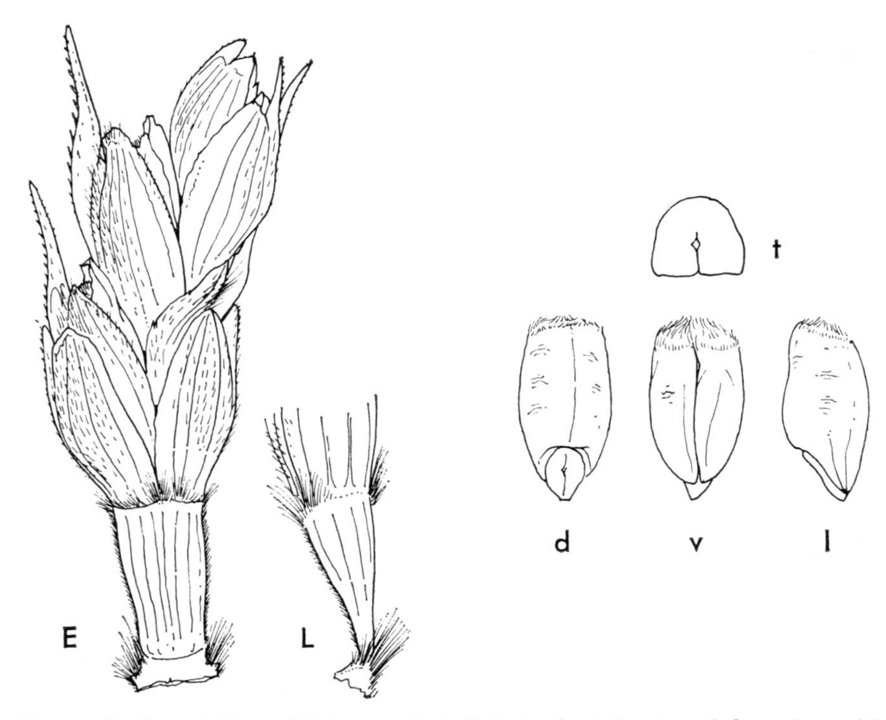

Fig. 31 Vavilov's Wheat *Triticum vavilovi*. Spikelet in (E) external face view with large rachis segment also seen in (L) lateral view. Caryopsis from lower floret in (d) dorsal, (v) ventral, (l) lateral view, (t) transverse section. All ×3.

palea remains intact at maturity. Two to three grains ripen in each spikelet. The caryopses are smaller than those of *Triticum spelta* and *Triticum macha*. They are wide towards the blunt apex and narrow in the region of the embryo. The comparative grains in my collection measure: length 7·0 mm, breadth 3·5 mm, thickness 3·2 mm. (See Peterson 1965, 15; Leonard and Martin 1963, 297 f.).

Triticum aestivum L. (=*Triticum vulgare* Host.) (fig. 32)

The ear is either square in cross-section or broader in face than side view. It may be lax or dense. The rachis is tough and the glumes soft to firm but the grain threshes free. They have a simple keel only in the upper part and are rounder in the lower part. The palea remains intact at maturity. Two to six grains may ripen per spikelet. In length the grains are similar to emmer, but their width is much greater. The thickest point is just above the embryo, the dorsal side is rounded, the cheeks to the ventral furrow are

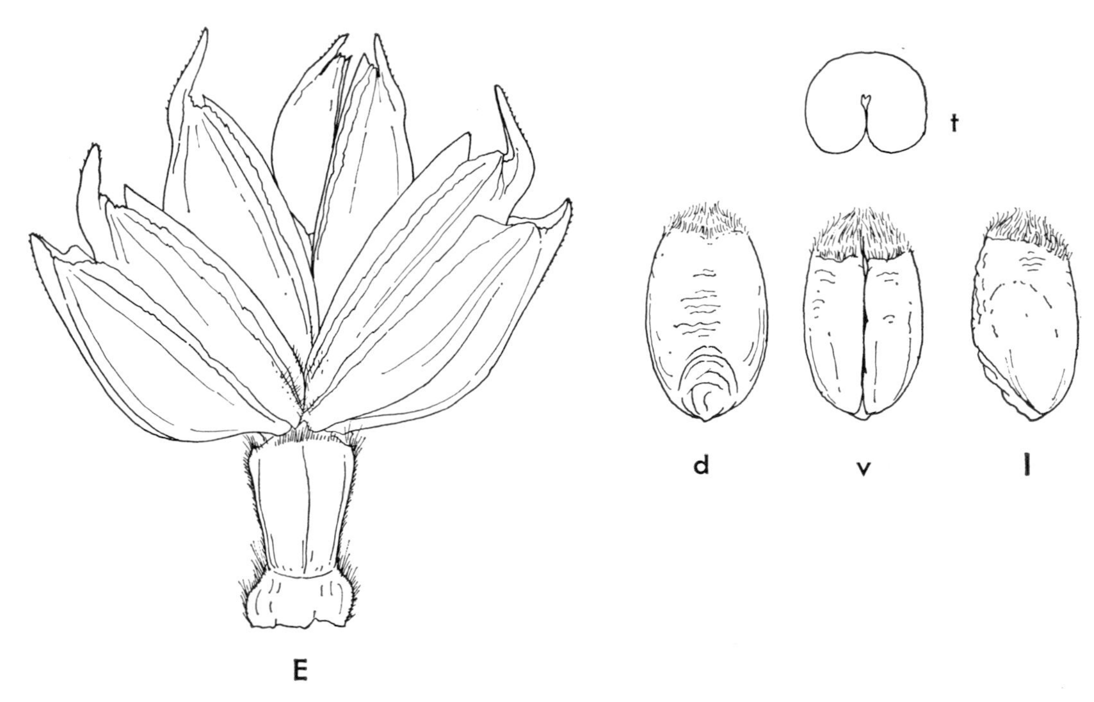

Fig. 32 Bread Wheat *Triticum aestivum* L. Spikelet in (E) external face view. Caryopsis in (d) dorsal, (v) ventral, (l) lateral view, (t) transverse section. All ×4.

also rounded, and the embryo is steeply placed. Percival (1921, 269) gives the following measurements: length 5·2–8·5 mm (average 6·78 mm), breadth 2·9–4·2 mm (average 3.63 mm), thickness 2·6–3·9 mm (average 3.25 mm). (See Peterson 1965, 15; Leonard and Martin 1963, 297; Percival 1943, 95; Helbaek 1961, 88.)

Triticum compactum Host. (fig. 33)

The ears are short, compact and laterally compressed. The rachis is tough and the internodes are extremely short. Those from Store Valby (neolithic Denmark) measured 1·7 mm wide × 1·9–2·7 mm long (Helbaek 1954A, 203). The fertile terminal spikelet is obliquely placed across the ear. The glumes are soft and usually keeled only in the upper part. The keel is extended into a sharp vertical beak, the glume shoulder is sloping. There is a tendency for the apex of the palea to split when mature. Two to five grains develop in each spikelet. The grains are small and laterally compressed due to overcrowding on the compact spikes. There is a narrow, shallow ventral crease, and the embryo is steeply placed.

Fresh specimens in Percival's collection measured: length 5·7–7·0 mm (average 6·19 mm), breadth 2·9–3·75 mm (average 3·31 mm), thickness 2·85–3·65 mm (average 3·13 mm). (Percival 1921, 311). (See Peterson 1965, 15; Leonard and Martin 1963, 307; Helbaek 1952B, 213.)

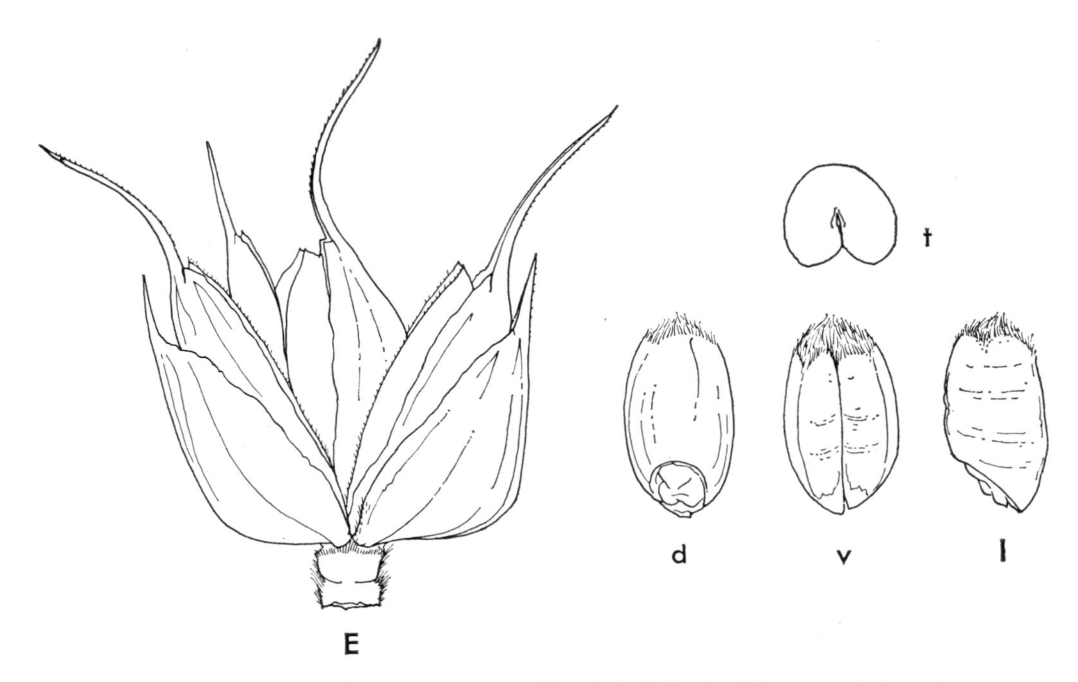

Fig. 33 Club Wheat *Triticum compactum* (Host.) Spikelet in (E) external face view showing glumes with long sharp beaks and very short rachis segment. Caryopsis in (d) dorsal, (v) ventral, (l) lateral views, (t) transverse section. All ×4½

Differences between Triticum aestivum *and* Triticum compactum

	Triticum aestivum	*Triticum compactum*
	mm	mm
Ears	long, lax	short, dense
Carbonized rachis internode length	3·75*	1·0*
Caryopsis	long, plump	short, plump
T: B index	85*	94
B: L index	54*	53
Lengths of carbonized grains*	6·88–4·34 av. 5·48	5·34–2·33 av. 3·67

* Helbaek 1961, 88–9 + Diagram 1.

The difficulty of distinguishing the caryopses of bread and club wheat when no rachis fragments are present has been stressed by several palaeoethnobotanists. Schiemann (1946) has incorporated *Triticum compactum* Host. in *Triticum aestivum* giving it the name *Triticum aestivum-aestivocompactum* Schiem., which overcomes the difficulty of nomenclature, but does not resolve the identification problem (idem 1951, 318).

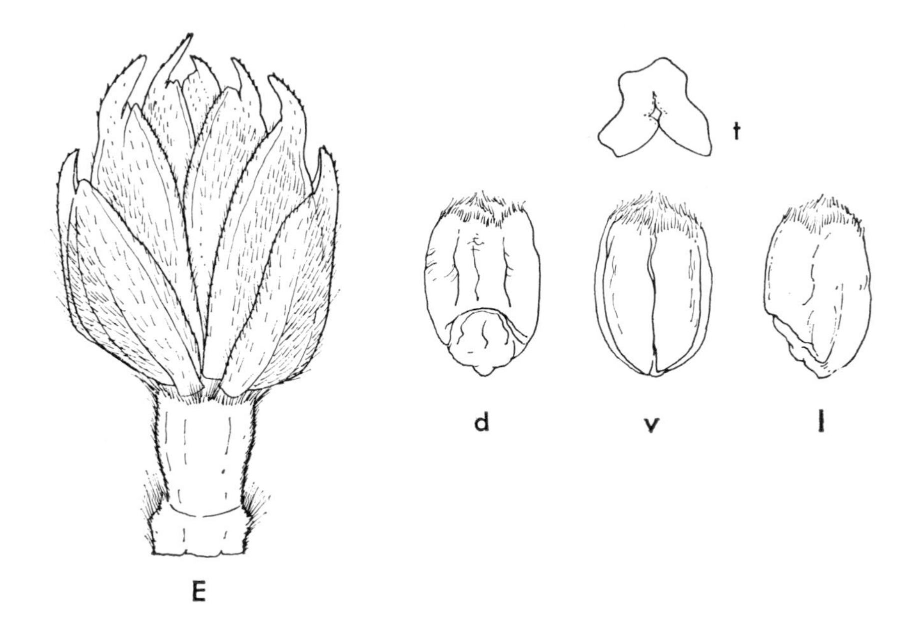

Fig. 34 Indian Dwarf Wheat *Triticum sphaerococcum* Perc. Spikelet in (E) external face view showing glumes and lemmas with hairs and stout beaks. Caryopsis in (d) dorsal, (v) ventral, (l) lateral views, (t) transverse section. All $\times 4\frac{1}{2}$.

Dimensions of fresh grains of different wheat species

	Length	*Breadth*	*Thickness*
	mm	mm	mm
T. sphaerococcum	4–5·5	3–3·7	3–3·7
T. compactum	5·7–7·0	2·9–3·75	2·85–3·65
T. aestivum	5·2–8·5	2·9–4·2	2·6–3·9
T. boeoticum	6–7·5	1–1·5	2·1–2·6
T. turgidum	6·7–8·37	3·26–4·43	3·23–4·1
T. carthlicum	7·0	3·3	3·5
T. vavilovi	7·0	3·5	3·2
T. monococcum	7·0–8·5	1·8–3·0	3–3·5
T. durum	7·0–9·7	2·8–4·1	3·2–4·25
T. spelta	7·0–10	3·0–3·5	2·7–3·0
T. dicoccum	7·2–9	2·85–3·4	2·6–3·1
T. macha	7·5	3·8	3·1
T. timopheevi	8·5	2·3	2·9
T. dicoccoides	9–12	1·7–2·9	1·85–2·7
T. turanicum	10·5–12	2·75–3·0	3·2–3·4
T. polonicum	11–12	4·0	3·0

Triticum sphaerococcum Perc. (fig. 34)

The width of the spike is often less than its thickness, and the whole head is very small. The ears are dense and the rachis is tough. The fertile terminal spikelet is set at 90° to the rest of the ear. The glumes are loose and hemispherical in shape with a single main keel, which terminates in a curved beak. The palea remains intact when mature. Three or more grains ripen in each spikelet. They are small and nearly spherical or hemispherical in shape. Percival gives the following measurements for these small grains: length 4–5·5 mm, breadth 3–3·7 mm, thickness 3–3·7 mm. (See Leonard and Martin 1963, 303; Peterson 1965, 14, 16.)

Aegilops sp.

It is appropriate to examine briefly here three species of the closely related genus *Aegilops* which probably contributed to the development of the tetraploid and hexaploid wheats. The genus is characterized by stalkless spikelets placed broadside on to the rachis, to form compact ovoid or cylindrical spikes. The glumes are tough and

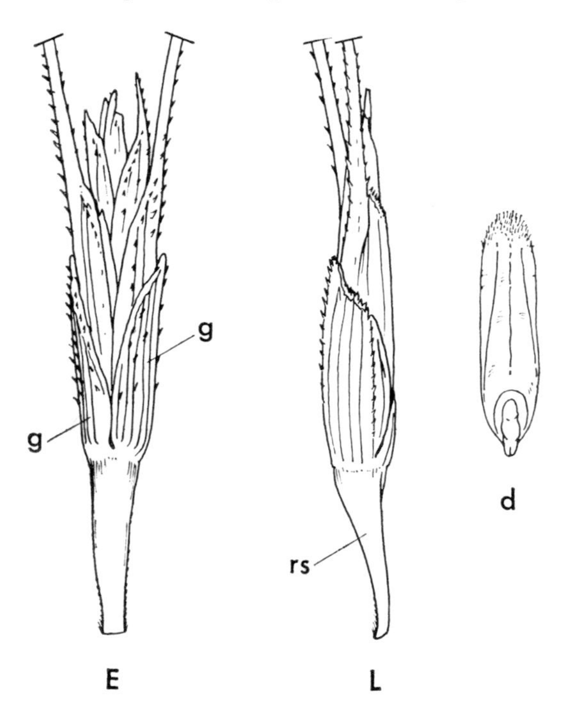

Fig. 35 Aegilops speltoides. Spikelet in (E) face view, (L) lateral view. (g) thick rigid glumes encasing florets, (rs) long slender rachis segment fracturing at its base, ×3. Caryopsis in (d) dorsal view ×4.

strongly veined, and retain the florets even after vigorous threshing. The rachis is brittle and breaks into individual segments on maturity. In *Aegilops speltoides* (fig. 35) the rachis is long and slender, and fractures at the base. The glumes are strongly nerved and the nerves are covered with stiff bristles. The lemma bears an awn at the apex, and like the glumes is strongly nerved, the nerves being covered with bristle-like hairs. Two grains usually ripen per spikelet. The spikelets of *Aegilops squarrosa* consist of thick rigid glumes firmly enclosing the florets. The rachis is brittle but breaks so that the segment bearing the spikelet above remains attached to the face of the one below – in a similar fashion to *T. spelta*. The rachis segments are large and broad. The lemmas bear short awns at their apex and two to three fertile caryopses may develop within the spikelet (see fig. 36).

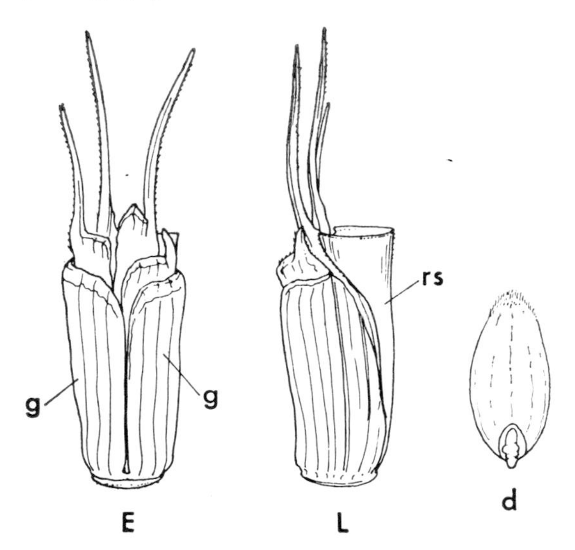

Fig. 36 Aegilops squarrosa. Spikelet in (E) face, (L) lateral view. (g) thick rigid glumes encasing florets, (rs) large rachis segment with fracture above spikelet, ×3. Caryopsis in (d) dorsal view ×4.

Umbellate goat-face grass, *Aegilops umbellulata*, has been identified from the late neolithic levels at Hacilar (Helbaek 1970, 201). The grains are similar to those of *Triticum dicoccum* being long, narrow and fairly flat: they have a length/breadth index of 35 and a thickness/breadth index of 72, both considerably smaller than for emmer grains. Besides the grains, the Hacilar finds also included the basal internodes of the rachis and the large, heavily ribbed, fan-shaped glumes. Many of the glume margins show the basal parts of at least two or three of the four or five stiff awns. The upper rudimentary internodes and glumes were also found, as were the small terminal spikelets. In this species only the two basal spikelets are fertile and may develop two or three grains. On maturity the complete ear breaks off above the culm and acts as a single unit for dispersal.

CONDITIONS REQUIRED FOR THE GROWTH OF WHEAT

Climate

The best wheat crops are harvested in areas which have a comparatively high mean winter temperature and an annual rainfall of 50·8–76·2 cm. The distribution of the rainfall during the growing season is of the greatest importance: moderately heavy rain in early summer when the shoots are in full growth and ears developing is most beneficial, but heavy autumn and winter rains greatly retard the development of the plant and result in small grain yield, as is illustrated by the following figures quoted by Percival (1943, 22) for wheat plots at Rothamsted:

Average rainfall of ten wettest winters (Nov.–Feb.)	33·03 cm
Average yield per acre of three plots	947·6 litres
Average rainfall of ten driest winters (Nov.–Feb.)	15·21 cm
Average yield per acre of three plots	1267·6 litres

The wild einkorn, *Triticum boeoticum*, is distributed widely from the low, warm, foothills of the Euphrates basin to the cool, continental plateaux of Anatolia. It will survive at altitudes between 0–1,600 m in Turkey, and 600–2,000 m in Iraq and Iran. The single-seeded race *Triticum boeoticum* ssp. *aegilopoides* is found chiefly in the cooler Balkans and in west Anatolia, while the larger twin-grained form *Triticum boeoticum* ssp. *thaoudar* occurs in the warmer, drier areas of southern Turkey, Iraq and Iran (Zohary in Ucko and Dimbleby 1969, 49). Wild emmer, *Triticum dicoccoides*, has a more restricted distribution at the present day than the two forms of *Triticum boeoticum*. Robust forms maturing early are found in the winter-warm region around the Sea of Galilee (from 100 m below sea level upwards). Slender and later-maturing varieties occur higher up the mountains in this region, reaching as far as 1,600 m on the east-facing slopes of Mt Hermon (idem 1969, 52).

Soils

Wheat does not thrive well on loose sandy or peaty soils or on wet clays. It gives the best yields on stiff clay loams which are well drained (Percival 1900 (1947). 532). 'A good wheat soil holds together and conserves water supplies and also provides conditions favourable for nitrate formation essential for the production of protein in the wheat grain' (Wilson 1955, 66). Wheat exhausts the land more than any other crop (Theophrastus trans. Hort 1916, II, 199). It tends to lodge when grown on rich, damp 'bottom' land.

The majority of varieties grown in Europe are winter-sown – sown from September to December – and they require a period of cold (vernalization) before they can develop fully. Winter wheat is the chief form cultivated today in Greece and Bulgaria. Spring-sown varieties do also exist in Europe, but the winter forms give heavier yields (Leonard and Martin 1963). It is not possible to ascertain from a morphological examination whether the grains belong to varieties with a winter or spring growth habit. The two habits are often found within the same species.

The two wild forms of wheat *Triticum boeoticum* and *Triticum dicoccoides* show a marked preference for basaltic and hard limestone regions in their present-day distributions. Both species are commonly found as components of oak park-forests and steppe herbaceous formations in the Near East (Zohary in Dimbleby and Ucko 1969, 48–52).

POSSIBLE USES OF WHEAT TO PREHISTORIC MAN

Wheat grains were of importance in the diet of prehistoric man. They were utilized after threshing and grinding into coarse meal, either in the form of gruel or porridge, or were baked into bread. Wheat flour may be of two types, either 'strong' or 'weak'.

'Strong' flour is caused by the high gluten content in the grain which gives it elasticity so that it can be baked into light porous loaves. When the grain is grown in conditions of low rainfall, on rich soil, with a hot, dry and sunny ripening period, strong flour is produced. Weak flours tend to produce compact, hard loaves or brittle biscuits (see pl. 14). They are frequently found in areas with high rainfall and soil moisture, and cool, cloudy weather during the ripening period (Peterson 1965, 311). The hexaploid bread wheat *Triticum aestivum*, as its name suggests, is particularly suited to bread making.

Fragments of emmer and spelt grains were found in the stomach of Grauballe man in Denmark mixed with other cereals and a great number of weed seeds, and they were presumably part of a meal of porridge or gruel. Spelt wheat was also represented by a complete spikelet and a number of glume fragments (suggesting that the grain had not been thoroughly threshed or ground) and the same is true for the emmer wheat fragments (Helbaek 1958, 112).

Finds of buns or bread have also been made in which it was possible to identify complete or fragmentary grains of wheat. Buns from the iron age lake village of Glastonbury, Somerset, were examined by Clement Reid, who describes one cake as being composed of 'whole, unbroken wheat grains with a noticeable proportion of glumes and fragments of awn . . .' (Bulleid and Gray 1916, Vol. 2, 629). Helbaek also examined buns from this site and succeeded in chemically isolating fragments of wheat as well as barley and weed seeds. 'Accordingly it seems evident that buns and bread were made from coarsely ground cereals including impurities of the field' (Helbaek 1952B, 212 and pl. XXIIIa). It is possible, although they have not been studied in detail, that the 'gnocchetti' reported from the bronze age lake-side settlement of Ledro (Battaglia 1943, 27), may have been made from coarsely ground wheat.

Emmer wheat is reported to have been used in preparing beer in ancient Egypt. Täckholm, Täckholm and Drar (1941, 249) describe finds of residues from inside beer jars, and also of dried and exhausted grains left from mashing (i.e. maceration in water). They were examined by J. Grüss, who identified starch grains of emmer wheat, yeast cells (of a variety of wild yeast *Saccharomyces winlocki*), moulds and bacteria. This type of beer appears to be similar to the modern Egyptian 'bouza', made by combining flour and water into a dough with the addition of a little yeast. It is then very lightly baked and the loaf broken up and placed in a jar of water and left to ferment. This process could well have been used in prehistoric Europe also.

Sprouted grains of *Triticum spelta* were found mixed with others of rye and a few grains of bread wheat and hulled barley in the grain from Roman Isca (Caerleon in South Wales). This sprouted grain was probably intended for brewing; for this purpose grains of wheat and rye have the advantage of sprouting more quickly than those of barley (Helbaek 1964B, 158 f.).

Chapter 6

Barley

GENERAL DISCUSSION OF THE GENUS *Hordeum*

It is the section *Cerealia* of the genus which concerns us in this discussion. At the present time two wild species are recognized in this section: the two-row *Hordeum spontaneum* C. Koch. and the six-row *Hordeum agriocrithon* Åberg (including *Hordeum langunculiforme* Bakhteyev).

The cultivated species present a problem of classification since they comprise a homogeneous group of interfertile botanical forms. It has been argued (Atterburg 1899) that they should all be included in a single species, whereas others following Linneaus (1753) have distinguished species on the basis of the fertility of the lateral spikelets at each internode of the rachis, density of the ear, and whether the grains remain invested or thresh free from the lemma and palea at maturity. In this discussion I shall follow Bowden (1959, 680) in his division of the cultivated barleys into three species: six-row barley *Hordeum vulgare* L. emend., two-row barley *Hordeum distichon* L. and irregular barley *Hordeum irregulare* Åberg and Weibe, based on the fertility of the lateral grains, using other features to distinguish sub-groups within each species.

(a) *Hordeum vulgare* L. emend. has the following sub-groups:

var. *nudum* (= *Hordeum vulgare* var. *coeleste* L.)	naked six-row barley
var. *trifurcatum*	hooded six-row barley
var. *hexastichum* (= *Hordeum hexastichum* L.)	dense-eared six-row barley
var. *inerme*	awnless six-row barley

(b) *Hordeum distichon* L. The following sub-groups have been distinguished:

var. *nudum*	naked two-row barley
var. *trifurcatum*	hooded two-row barley
var. *zeocrithon* (= *Hordeum zeocrithon* L.)	dense two-row barley
var. *deficiens* (= *Hordeum deficiens* Steud.)	two-row barley with deficient laterals
var *inerme*	awnless two-row barley

(c) *Hordeum irregulare* Åberg and Weibe. No sub-groups recognized.

GENETIC AND ARCHAEOLOGICAL EVIDENCE FOR THE ORIGINS OF BARLEY SPECIES

Since all cereal barleys have the same number of chromosomes ($2n = 14$) their genetic relationships cannot be established as clearly as they can for the wheats. The major question in setting out to establish the origins of cultivated barley is whether the two- and six-row forms evolved from a single ancestor, and if so, was this of the two- or six-row type; or did they develop separately from the wild two- and six-row forms (Bell in Hutchinson 1965, 81 f.).

At the present time it is accepted that the wild two-row form *Hordeum spontaneum* is the ancestor of all cultivated barleys. The main reason given is that it occupies truly

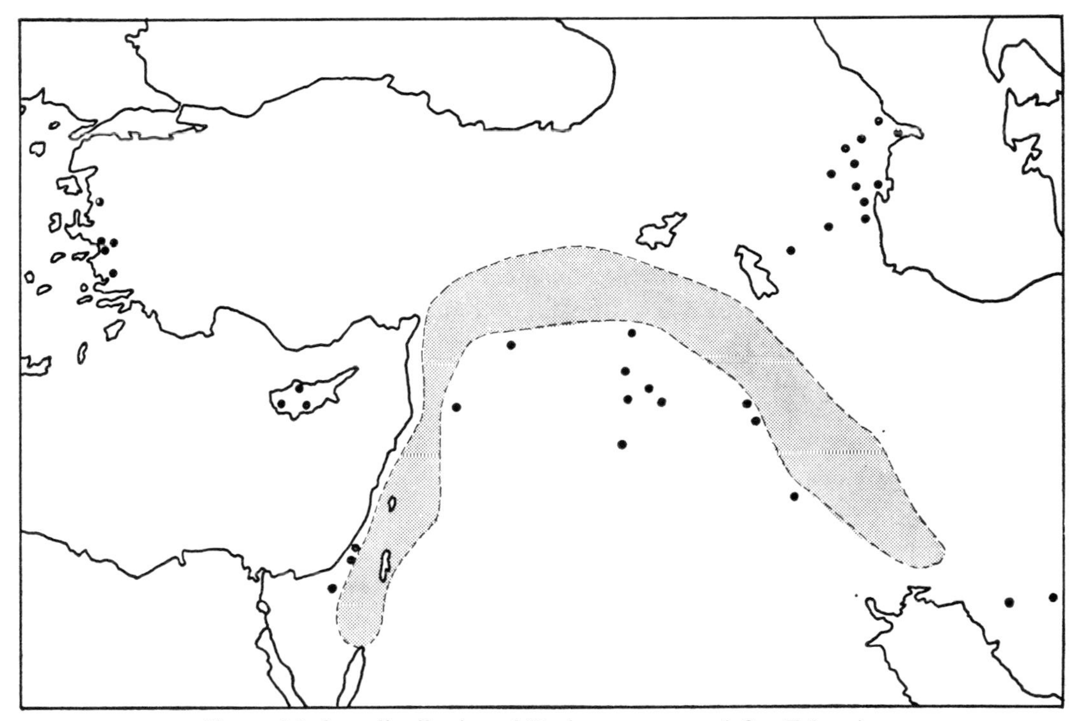

Fig. 37 Modern distribution of *Hordeum spontaneum* (after Zohary).

wild habitats in its modern distribution from the east Mediterranean to Afghanistan (see fig. 37), as well as secondary man-made niches such as abandoned fields and road sides. *Hordeum agriocrithon*, the wild six-row form, on the other hand, has only been found in secondary habitats in widely scattered isolated areas – Tibet, Israel, Transcaucasia and Turkmenia – and is found growing in close proximity to *Hordeum spontaneum* and cultivated six-row barleys (Zohary 1963, 28). Progeny tests carried out by Zohary on naturally occurring *Hordeum agriocrithon* plants revealed 'a whole array of progeny from two- to six-row types, thus verifying their heterozygous and hybrid nature' (idem 1963, 28). *Hordeum spontaneum* has also been made, by X-ray treatment

(Helbaek 1960B, 196), to produce six-row forms which bred true in the second generation. Zohary managed to cross *Hordeum spontaneum* with six-row cultivated forms and obtained the same spectrum of forms between it and *Hordeum agriocrithon* that occur in the natural stands (Zohary 1963, 29).

The palaeoethnobotanical material in the Near East tends to support the case for *Hordeum spontaneum* being the ancestor of all cultivated barleys, as this is the only form of wild barley yet found in the early farming villages. It occurs at Ali Kosh (Bus Mordeh phase) (Helbaek 1966B, 352), Tepe Guran (Meldgaard, Mortensen and Thrane 1963, 112, n. 17), Jarmo (Helbaek 1959A, 187), Tell Mureybit (van Zeist and Casparie 1968, 51; see pl. 18), Beidha (Helbaek 1966A, 61 f.) and Çatal Hüyük (idem 1966B, 357). The Beidha find is of particular interest since many thousand impressions of *Hordeum spontaneum* were found with grains larger than in the truly wild forms and so Helbaek has described them as 'cultivated wild barley' (idem 1966A, 62). At Jarmo, too, the barley showed some signs of domestication: the rachis was no longer brittle as in truly wild *Hordeum spontaneum*: three rachis internodes were found joined together; in other features, such as the pedicellate lateral florets (they are sessile in *Hordeum distichon* L.), and the unwrinkled pales, these barley grains correspond to *Hordeum spontaneum* (idem 1960A, 108).

Fully domesticated *Hordeum distichon* L. occurs at many sites of slightly later date in the Near East: Ali Kosh (Ali Kosh and Mohammad Jaffar phases – Hole and Flannery 1967), Tepe Sabz (ibid.), Tepe Guran (Meldgaard, Mortensen and Thrane 1963, 112, n. 17). Tell es-Sawwan (Helbaek 1965, 45), Matarrah (idem 1960A, 109) and late neolithic Hacilar (idem 1961, 82). Helbaek claims (1966B, 356) that 'there is no evidence for two-rowed naked barley in antiquity', but he reports naked grains from Ali Kosh and Beidha which show no signs of being twisted – as laterals of six-row forms would do – and so could come equally from the two-row species, as be the median grains of six-row form. Only finds of raches with spikelets and grains attached will resolve this problem.

Six-rowed barley also occurs early in the Near East but so far only the cultivated form *Hordeum vulgare* L. emend. has been identified. It occurs at Ali Kosh (Ali Kosh phase – Helbaek 1966B, 356), Tepe Sabz (idem 1966B, 358), Beidha (idem 1966B, 356), Çatal Hüyük (idem 1964, 122; 1966B, 357), aceramic Hacilar (idem 1966B, 356), Tell es-Sawwan (idem 1965, 45), Mersin (idem 1959B, 370), late neolithic Hacilar (idem 1961, 82), Can Hasan (Renfrew 1968) and Argissa Maghula (Hopf 1962). The earliest finds of six-row barley are all of the naked form in *Hordeum vulgare* var. *nudum* and occur chiefly in Anatolia at aceramic Hacilar and Çatal Hüyük. If *Hordeum vulgare* is derived from *Hordeum spontaneum* it appears that three mutations responsible for the firm rachis, the fertile lateral grains, and the naked caryopses, had occurred as a consequence of domestication at least before 7000 B.C. in Anatolia, and had spread to Greece (Argissa Maghula) before c. 6000 B.C. The reason for the development of six-row barley remains obscure: Helbaek (1960B, 190) has suggested that it might be due to 'the violent change of life conditions in . . . being moved from the mountains to the hot plains almost at sea level and deprived of the rain on which it used to thrive up to a short time before maturity'. Finds of hulled six-row barley occur widely in slightly later contexts from Ali Kosh in Khuzistan to Argissa in Thessaly. It is strange that the hulled form occurs later than the naked variety: if it was derived from either *Hordeum spontaneum* or *Hordeum agriocrithon* one would expect the earliest form of cultivated six-row barley to have been hulled, but this does not seem to be the case. Thus there was no significant time lag between the emergence of cultivated two- and six-row barleys; and both occur in their naked and hulled varieties very shortly after domestication (Renfrew, in Ucko and Dimbleby 1969, 153). *Hordeum irregulare* is only reported from the Fayum (Helbaek 1960A, 111), and does not appear to occur in Europe at the present day (Takahashi 1955, 260, table 5).

CRITERIA FOR THE PALAEOETHNOBOTANICAL IDENTIFICATION OF EARS AND GRAINS OF BARLEY SPECIES AND VARIETIES

Hordeum spontaneum Koch. (see fig 38A–C)

This two-row form of wild barley has hulled grains arranged on a brittle rachis. At maturity the spike divides into individual units each comprising one mature seed and two sterile lateral spikelets joined to the rachis segment (Harlan and Zohary 1966,

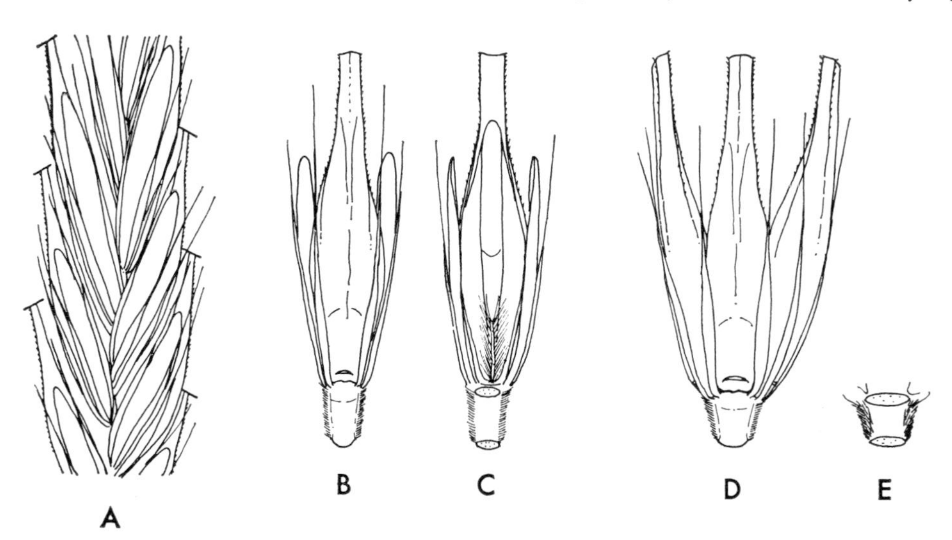

Fig. 38 Wild Barley *Hordeum spontaneum* (two-row). (A) portion of ear in side view, (B) triplet of spikelets in dorsal view, (C) in ventral view. *H. agriocrithon* (six-row), (D) triplet of spikelets in dorsal view, (E) rachis segment in ventral view. All $\times 3$.

1075). Two races have been distinguished: a wadi race, found at present from the Negev desert northwards to the Turkish border and South Afghanistan, which is a rather small, slender plant with small grains – half the size of the second more robust type. The second race is now found in south-west Syria, Jordan and north Israel and is characterized by extremely large seeds, extraordinarily long awns and a generally robust appearance (ibid., 1076). The grains of *Hordeum spontaneum* are flat on the dorsal side and are comparatively thin and angular in cross-section. The maximum width is in the middle of the grain which tapers to a narrow base, and, less sharply, to the blunt apex (see pl. 18) (van Zeist and Casparie 1968, 50). The pales remain unwrinkled at maturity and the lateral florets are pedicellate (Helbaek 1960A, 108). The margins of the rachis segments and the glumes are covered with hairs. The following measurements have been given for modern grains of *Hordeum spontaneum* collected in the mountains west of Damascus: length 8·3–10·05 mm (average 9·20 mm), breadth 2·5–3·2 mm (average 2·84 mm), thickness 1·2–1·8 mm (average 1·55 mm) (van Zeist and Casparie 1968, 51).

Hordeum agriocrithon Åberg (= *Hordeum langunculiforme* Bakhteyev) (see fig. 38D–E)

This wild six-row barley has a brittle rachis with pubescent margins. The grains are hulled and the laterals show the markedly twisted appearance in ventral aspect characteristic of all six-row barleys. The glumes are covered with short hairs. The grains are thin and angular in cross-section. Both median and lateral grains have lemmas extended into stout awns.

Hordeum vulgare L. emend.

This group consists of six-row tough-rachis barleys in which the median grains are symmetrical and the laterals asymmetrical in ventral view (see pl. 20). Two types may be distinguished: those in which the lateral grains are similar in size to the median

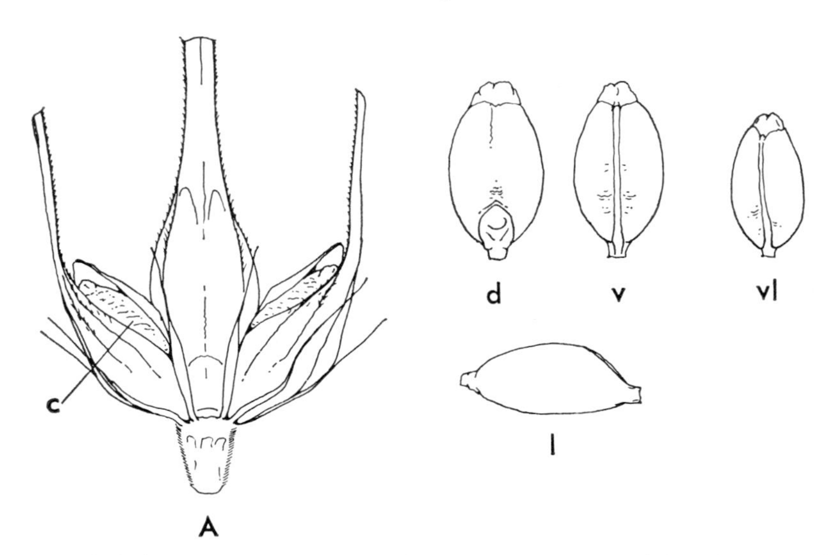

Fig. 39 Naked six-row Barley *Hordeum vulgare* var. *nudum*. (A) triplet of spikelets in face view showing (c) lateral caryopsis visible between lemma and palea. Median caryopses in (d) dorsal view, (v) ventral view and (l) lateral view. Lateral caryopsis in (vl) ventral view showing twisted ventral furrow. All ×3.

grains, and those with laterals which are considerably smaller and lack awns or hoods (Carson and Horne in Cook 1962, 104). Several sub-groups have been distinguished on the basis of major morphological features:

var. *nudum* (= var. *coeleste* L.) has naked grains, with the laterals showing a marked twist in the ventral aspect. The grains of naked barley do not show the distinct lines of the nerves of the lemma on their dorsal side, but have a shallow groove running from the top of the embryo to the apex of the grain. The epidermis of the caryopsis has minute transverse wrinkles (see fig. 39). A marked, narrow ridge is often visible running up the ventral furrow.

var. *trifurcatum*. Instead of the lemma terminating in an awn, it ends in a trifurcate appendage known as a hood (see fig. 40B). Hooded barleys are not widely cultivated,

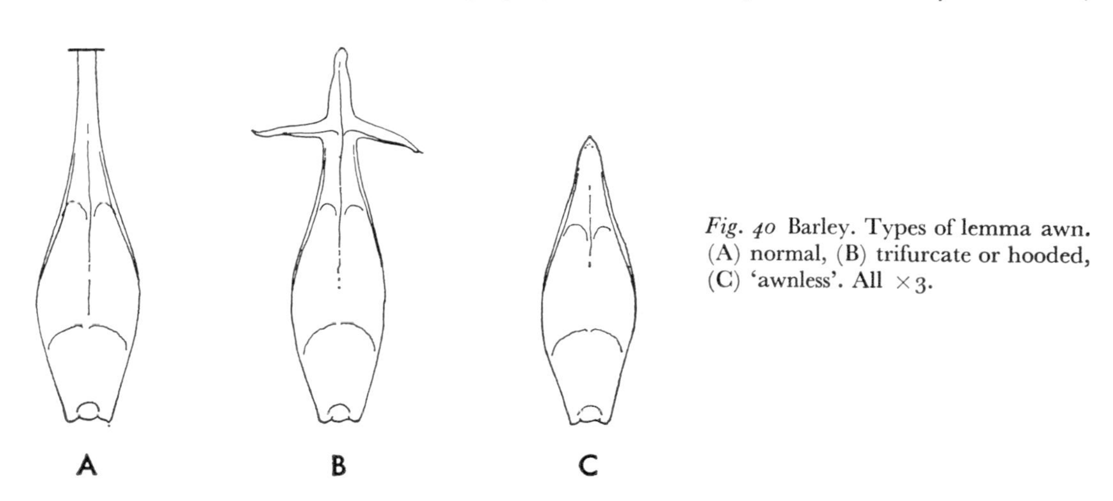

Fig. 40 Barley. Types of lemma awn. (A) normal, (B) trifurcate or hooded, (C) 'awnless'. All ×3.

but are found chiefly in America and Asia (Carson and Horne in Cook 1962, 131; Bowden 1959, 680).

var. *hexastichum* (pl. 17). The dense-eared six-row barley has very short internodes 1·7–2·1 mm long (fresh material) (Hunter 1952, 9). Barley may also appear to be dense when the internodes are longer, and the grains plump, but it is defined for this study on the basis of the average length of the rachis segments. In any single ear the internodes in the central portion of the rachis are longer than those towards the apex and the base (Carson and Horne in Cook 1962, 104, 108).

var. *inerme* (see fig. 40C) lacks awns – the lemmas terminate in a short sharp or blunt point (Carson and Horne in Cook 1962, 113).

Hordeum distichon L. (pl. 18)

This is the tough-rachis, hulled two-row barley in which only the median spikelet of each triplet at the node of the rachis is fertile, flanked by two sterile spikelets. The lemmas of the sterile lateral florets are never awned, but terminate in a rounded apex (Carson and Horne in Cook 1962, 104).

var. *nudum* is the naked form of two-row barley. The grains have a transversely wrinkled epidermis, and a shallow groove runs from the embryo to the apex of the grain on the dorsal side. It differs from the six-row form (fig. 39) in always being symmetrical in the ventral aspect of the fertile grain. In the fertile grains a marked, narrow ridge is often visible running up the ventral furrow.

var. *trifurcatum* is similar to the form of *Hordeum vulgare* var. *trifurcatum* (see fig. 40B) in having a trilobate structure in place of the lemma awn. The lateral grains are sterile.

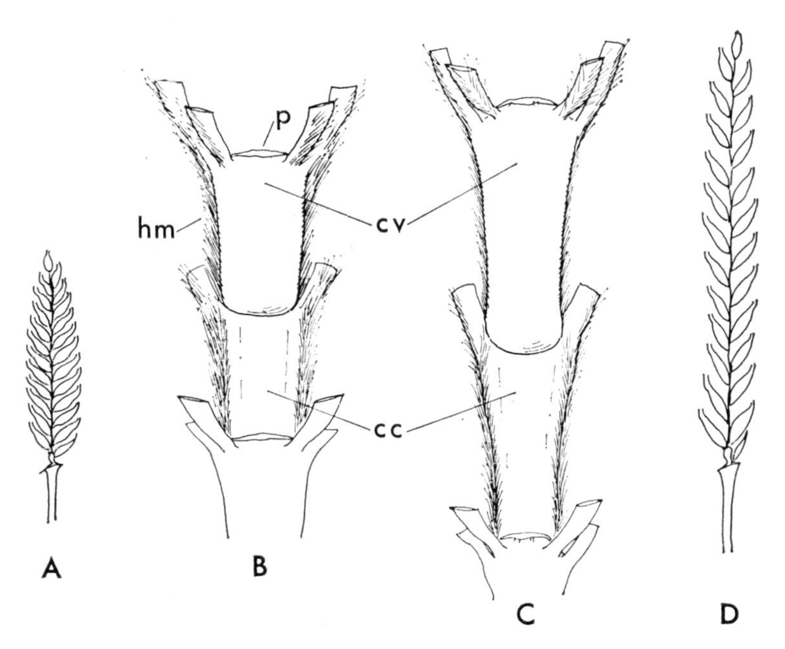

Fig. 41 Barley. Length of rachis segments in dense-eared and lax-eared two-row varieties. (A) dense ear (diagrammatic) $\times\frac{1}{2}$. (B) two rachis segments in face view from central part of dense ear $\times 6$. (C) rachis segments from central part of lax ear $\times 6$. (D) lax ear (diagrammatic) $\times\frac{1}{2}$. (cv) convex face, (cc) concave face, (hm) margin densely clothed with hairs, (p) point of attachment of grain.

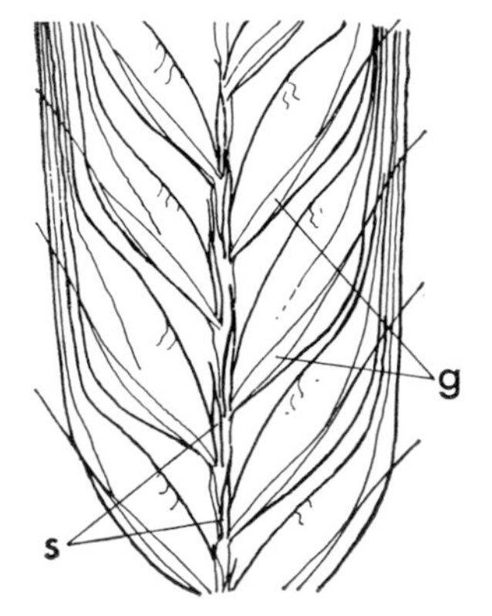

Fig. 42 Hulled two-row barley with greatly reduced lateral spikelets *Hordeum distichon* var. *deficiens*. Portion of ear showing (s) reduced sterile spikelets, (g) glumes of median grains, $\times 3$.

var. *zeocrithon* is the dense-eared form of hulled two-row barley with short rachis segments: 1·7–2·1 mm on fresh material (Hunter 1952, 16), and the grains attached almost at right angles to the rachis – giving rise to the name 'fan barley' (see fig. 41). The 'fan' shape is emphasized when the grains bear awns.

var. *deficiens* (see fig. 42) is a form of hulled two-row barley with deficient lateral spikelets. These are reduced to the glumes, a much-reduced lemma and rarely a palea or rachilla (Carson and Horne in Cook 1962, 104).

var. *inerme* lacks awns on the lemmas as in the sub-group of *Hordeum vulgare* L. (see fig. 40C).

Hordeum irregulare Åberg and Weibe (see fig. 43)

This group is composed of barleys in which the spikes are composed of an irregular

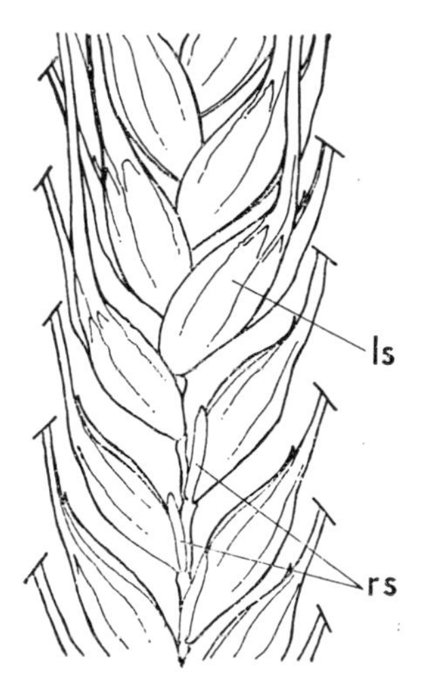

Fig. 43 Portion of ear of *Hordeum irregulare* showing mixed six-row and two-row types of spikelets, (rs) reduced lateral spikelets, (ls) fertile laterals ×3.

mixture of fertile and sterile lateral florets (Bowden 1959, 681), and fertile median grains. The irregular barleys all seem to have originated in Ethiopia (Leonard and Martin 1963, 490).

Within the species and their sub-groups defined above, barley ears and grains can be further attributed to cultivars on the basis of the following characters – which can often be found on palaeoethnobotanical material.

Sterile spikelets

The shape, attitude and hairiness of the sterile spikelets in two-row barley can be used to define cultivars. Their attitude may be straight, lying parallel to the margins of the rachis, or divergent (see fig. 44). The shape of the apex of the sterile spikelet may be round, square or pointed. The pedicel and lemma may be hairy or glabrous (Carson and Horne in Cook 1962, 106).

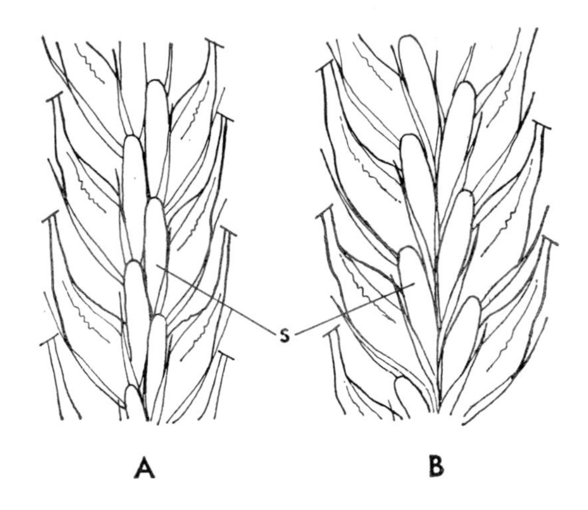

Fig. 44 Barley. Portion of ears of two-row barley to show attitude of sterile spikelets (s) (A) parallel, (B) divergent, ×3.

Glumes

They may be hairy on the outer dorsal surface, and it is noted that the length of hairs on the glume correlates directly with the length of the hairs on the rachilla, and with the presence of barbs on the lemma awn. The hairs may be arranged over the whole surface or in a median band (see fig. 45D–G). The awns on the glumes vary from being twice as long as the glume to only a quarter of the glume's length (cf. fig. 45A–C) (Carson and Horne in Cook 1962, 107).

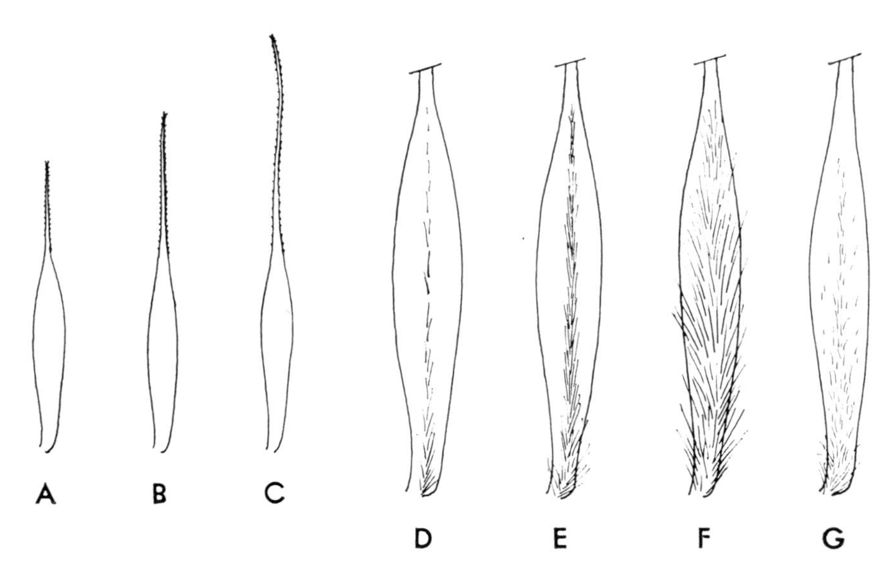

Fig. 45 Barley. Glumes of median (fertile) floret. (A–C) length of glume awn relative to glume, ×6, (A) short, (B) medium, (C) long. (D–G) hairiness of glume, ×12, (D) median line, (E) median band, (F) surface covered, (G) as seen in varieties with short-haired rachilla.

Rachis

(a) The flange-like structure at the junction of the culm and base of the ear (the collar) assumes several distinct forms: they may be closed, v-shaped or open. The v-shaped and open forms of collar only occur in six-row barleys and both forms may be found in a single variety. The closed collar may be either cup- or platform-shaped (see fig. 46) (Carson and Horne in Cook 1962, 107–8).

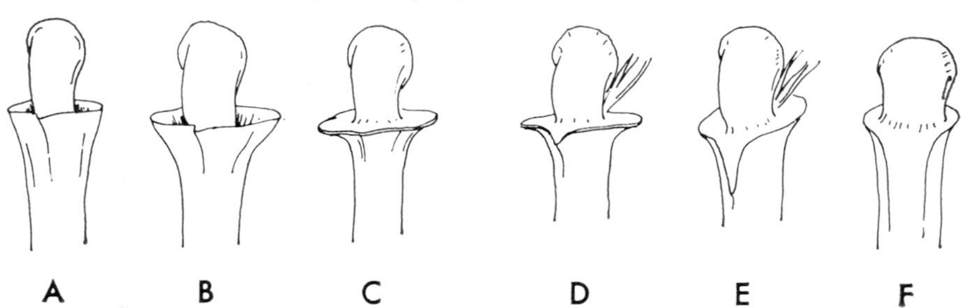

Fig. 46 Barley. Types of collar ×6. (A) cup, (B) shallow cup or 'intermediate', (C) platform. (D, E, F,) platform collars with rim notched, V shaped and decurrent respectively, contrasting with 'entire' rim of (C.)

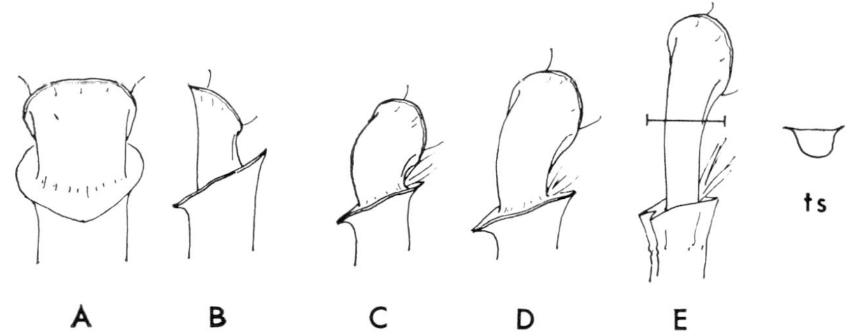

Fig. 47 Barley. Variations in the form of the first rachis segment ×7. Short broad straight type often seen in six-row barley, (A) in front view, (B) in lateral view. (C) short, strongly curved, (D) medium length and curvature, (E) long, slender, weakly curved types as seen in two-row barley. (ts) transverse section of E.

(b) The form of the basal internode of the rachis is distinct from other rachis segments and may belong to various types: it is often shorter than other segments of the rachis, but in some cases it may be much longer and more slender (fig. 47). In dense-eared varieties this segment is invariably short, and may be curved or straight: in lax-eared forms it may be long and curved or of intermediate form (Carson and Horne in Cook 1962, 108).

(c) There may be long or short hairs on the rachis margins, or they may be glabrous. There appears to be a correlation between long or short hairs on the rachis margins and long or short rachilla hairs (see fig. 48) (Carson and Horne in Cook 1962, 114).

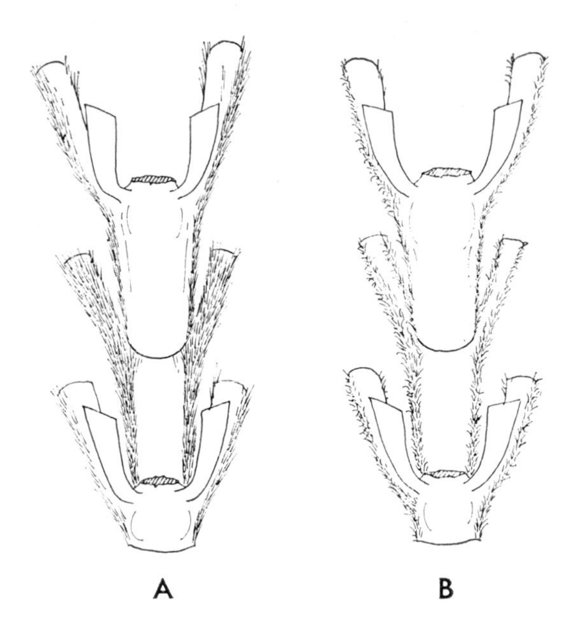

Fig. 48 Barley. Portion of rachis showing hairs on margins of rachis segments in varieties with (A) long straight, (B) short woolly rachis margin and rachilla hairs, ×6. Hairs on glumes of median spikelets have been omitted for clarity.

Grain base

Two major types have been distinguished: the *falsum* base which shows a distinct bevel over the lower part of the embryo on the dorsal side, and the *verum* or nicked base which shows a distinct, sharp, horizontal groove above the attachment to the rachis. A third form in which the base is plain (*spurium* type) is also found (see fig. 49). In the

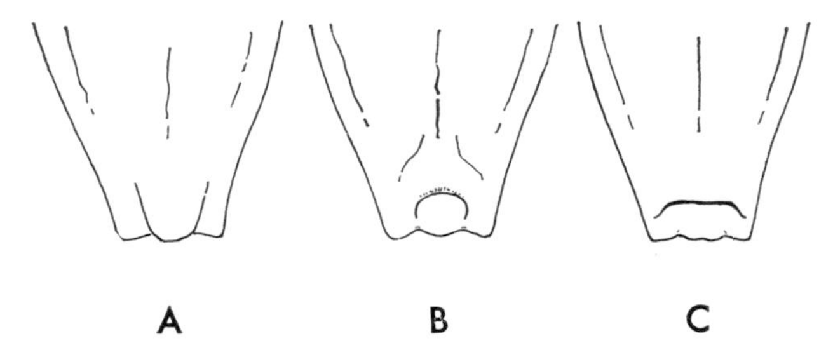

Fig. 49 Barley. Forms of lemma base ×6. (A) plain, (B) bevelled, (C) nicked.

majority of lax-eared barleys bevelled bases occur as the most frequent type – but the cultivar Spratt barley, a very dense form, also exhibits this type of base. In the case of the cultivar Proctor both bevelled and plain bases may occur in a single ear, so that this character can no longer be relied upon to indicate the density of the ear in hulled barleys (Carson and Horne in Cook 1962, 111, 121B).

Rachilla

This lies at the lower end of the ventral groove above the palea. It varies from long to short, and in either case is covered with either long or short hairs (see fig. 50). The

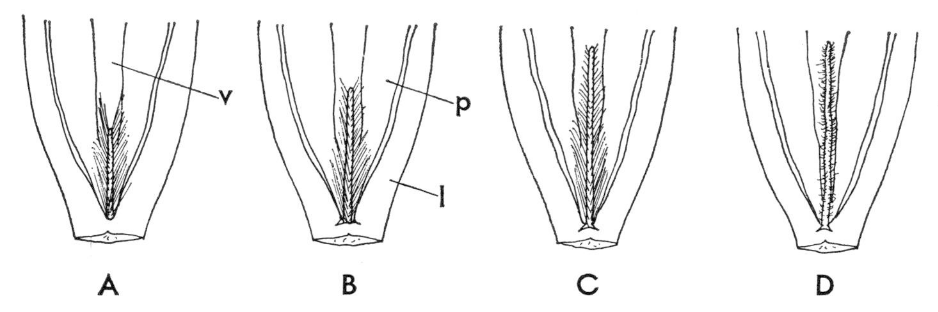

Fig. 50 Barley. Rachilla hair type. (A, B, C) short, medium, long rachillae of 'long silky' hair type. (D) long rachilla of 'short woolly' hair type. (l) lemma, (p) palea, (v) ventral furrow ×6.

length of the rachilla, however, is influenced by the position of the grain in the ear – the longest rachillas occurring in the central portion of the ear, and in six-row forms the laterals have longer rachillas than the median grains (Carson and Horne in Cook 1962, 125–128).

Lemma nerves

The lemma usually shows five distinct nerves running up the dorsal side; these are described as the medial, and the inner and outer pairs of lateral nerves. The presence or absence of teeth (spicules) on the inner lateral pair of nerves combined with the hair type and length of rachilla has been used in describing cultivars (Carson and Horne in Cook 1962, 131; Hunter 1952, 18–19). In six-row barley spicules are typically large, numerous and often present between the nerves as well (see fig. 51).

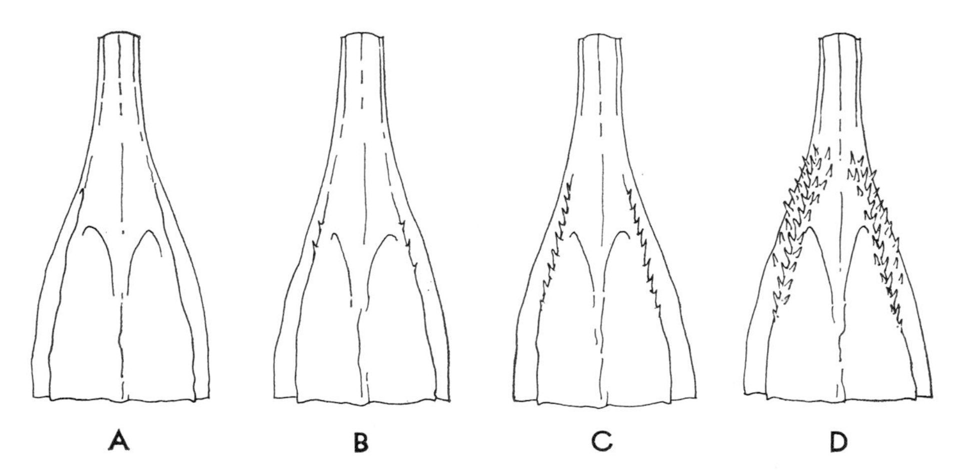

Fig. 51 Spicules on the inner lateral nerves of barley lemmas ×5. (A) spicules absent, (B) weakly, (C) strongly spiculate condition as seen in two-row cultivars, (D) prominent multiseriate and interveinal spicules as seen in most six-row cultivars. Spicules on awn margins and outer lateral nerves are omitted for clarity.

Lodicules

These delicate membranous structures lie at the base of the grain between the lemma and embryo. The shape and size of the lodicules and their position in relation to the embryo vary markedly in different varieties – they may be large (collar) or small (bib) types. Large lodicules are invariably associated with varieties which have a plain or

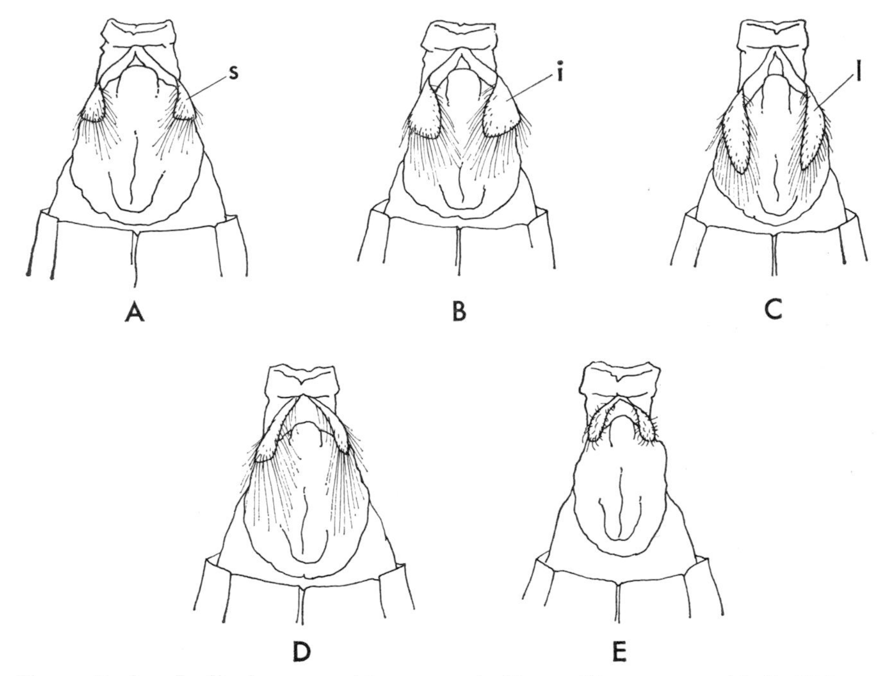

Fig. 52 Barley. Lodicules exposed by removal of base of lemma ×7. (A, B, C) large or 'collar' type, (s) short, (i) intermediate, (l) long. (D, E) small or 'bib' type, from varieties with long silky rachilla hairs and short woolly rachilla hairs respectively.

bevelled grain base, never with the verum grain base. The small lodicules are found in grains with the verum or nicked grain base, or with a plain base; never with bevelled grain bases. The margins of the lodicules are lined with hairs, the lengths of which correspond with the lengths of the rachilla hairs (see fig. 52) (Carson and Horne in Cook 1962, 134–7).

CONDITIONS REQUIRED FOR THE CULTIVATION OF BARLEY

Climate

Barley grows best where the ripening season is comparatively long and cool, and where there is moderate rather than excessive rainfall. As in wheat there are winter- and spring-sown varieties. Some spring barleys may mature in 60–70 days after sowing (earlier than spring varieties of wheat, oats and rye). Some of the naked spring-sown varieties from Tibet show remarkable tolerance of cold as well as early ripening so that these varieties can be sown further north and at higher altitudes than any other cereal.

Barley does best where the annual rainfall does not exceed 89 cm (Leonard and Martin 1963, 500 f.).

Wild barley does not tolerate extreme cold and is only occasionally found at the present day above 1,500 m. On the other hand it penetrates warm steppe and desert areas in the Near East to a greater extent than the wild wheats (Zohary in Ucko and Dimbleby 1969, 53–5).

Soil

Well-drained, fertile deep loam soils suit barley best. It produces low yields on sandy soils, and will grow vigorously and incline to lodge on soils which have a high nitrogen content. Barley is, however, remarkably tolerant in saline and alkaline conditions – more so than other cereals, but it is equally more sensitive to soil acidity (Leonard and Martin 1963, 500 f.). Because of its tolerance of alkalinity, large acreages of barley are found on soils derived from chalk or limestone (Hunter 1951, 15).

USES OF BARLEY

In modern times barley is chiefly grown to produce malt, used in making alcoholic beverages – chiefly beer and whisky (Martin and Leonard 1967, 470). One piece of palaeoethnobotanical evidence for sprouted barley (malt) comes from the fortified village of Eketorp on Öland, Sweden, where Helbaek found sprouted and ground barley grains in a sixth-century A.D. context (Helbaek 1966C, 218). In making beer it is necessary first to allow the grains to sprout until the sprouts reach the whole length of the grain, and in doing so change the composition of the endosperm of the grain from starch to sugar (through the action of the enzyme diastase). When the sprouts reach this length, the grain is roasted just hard enough to kill the sprout, and is then ground up ready to be left suspended in water together with yeast for fermentation. The Eketorp grains had reached this stage (idem 1966, 215). A pot full of sprouted barley grains found in a first-century A.D. context at Østerbølle in Jutland (idem 1938, 216 f.) was also probably intended for brewing.

Barley grains were used for making porridge in classical times: Pliny (*Natural History* XVIII, xiv, 74) describes how the Greeks soaked the barley grain, dried and ground it and then mixed in seeds of flax and coriander and salt. It is interesting to note that hulled six-row barley was the chief cereal found in the stomachs of the Tollund and Grauballe men (Helbaek 1958, 84).

Barley bread was also made. Small buns were found at Wangen, Untersee in Switzerland, about an inch and a half high, and four to five inches in diameter; 'The dough did not consist of meal, but of grains of corn, more or less crushed. In some specimens the halves of grains of barley are plainly discernible' (Keller translated by Lee 1866, 63). Helbaek also found fragments of hulled barley as well as wheat in the buns from Glastonbury (Helbaek 1952B, 212).

Chapter 7

Rye

GENERAL DISCUSSION OF THE GENUS *Secale*

Only one cultivated species is known: *Secale cereale* L. Two wild species *Secale montanum* Guss. and *Secale anatolicum* Boiss. have been considered as the progenitors of the cultivated form (Hector 1936, 229): Žukovskij (trans. Hudson 1962, 5), however, suggests that *Secale cereale* L. arose from selection in the weed species *Secale segetale* Zuk.

GENETIC AND ARCHAEOLOGICAL EVIDENCE FOR THE ORIGINS OF CULTIVATED RYE

Both *Secale cereale* and *Secale montanum* have the same chromosome number ($n = 7$) and may be readily crossed: this has led Hector (1936, 229) to support the case for *Secale montanum* as the ancestor of cultivated barley. The similarity of *Secale montanum* and *Secale anatolicum* makes it difficult to prove which was the ancestral form (Leonard and Martin 1963, 452). They differ from *Secale cereale* in being perennial, having smaller grains and a brittle rachis. An intermediate form which may be annual or biennial and has a brittle rachis occurs as a weed in fields of other cereals in south-west Asia; it has been called *Secale ancestrale* Zuk. and is thought to be a primitive form of *Secale cereale* (Gill and Vear 1958, 265).

Unlike the other cereals rye is cross-pollinated (Hector 1936, 217), and this may account for the 'lack of distinction into discrete botanical forms and agricultural varieties' (Bell in Hutchinson 1965, 98). Not only may *Secale cereale* cross with *Secale montanum* and *Secale anatolicum* (Hector 1936, 222), but successful crosses have also been obtained with diploid, tetraploid and hexaploid wheats (ibid., 223), and with species of *Aegilops* (ibid., 229). It is this close association especially with wheat that prompted Tumanjan to suggest that rye originated from wheat as a result of being moved into areas with short days (Žukovskij, trans. Hudson 1962, 5).

The modern distribution of the wild forms extends from Southern Europe and Anatolia (*Secale montanum*) to Syria, Armenia, Persia, Afghanistan and Turkestan as far as the Kirghiz Steppe (*Secale anatolicum*) (Martin and Leonard 1967, 450).

Finds of wild rye in archaeological contexts are rather rare. An impression of *Secale dalmaticum* is claimed from the Adriatic coast at the neolithic site of Danilo Bitinj (Korošeć 1958, 124, 168). The find of *Secale cereale* (identified by Miss Kozkowska) from a Danubian I context at Ojców in Galicia (Childe 1929, 46) is of great interest, especially if really domesticated and not wild, and may alter the generally accepted theory that rye did not come into cultivation until much later (Bertsch and Bertsch 1949, 61, Abb. 32). Recently, more neolithic finds have come to light: Werneck identified rye at Vösendorf in Austria and four more Polish neolithic finds are now known. It is also present in the bronze age finds of Nitranski Hradok in Czechoslovakia and at Labegg in Austria (Tempir 1964, 88).

Previously it was thought that the earliest certainly domesticated forms belonged to the beginning of the first millennium B.C. in Central Europe, and that rye spread northwards as a separate crop associated with the Hallstatt early iron age (Helbaek 1961, 84). To reconcile the fact that rye was not first grown as a crop within the area of its modern wild distribution it is widely held to have spread into Europe as a weed contaminant of the primary cereals, wheat and barley; subsequently in regions of poor sandy soils and in mountainous districts it gradually ousted the other cereals and became the principal crop (Jessen and Helbaek 1944, 54).

In any case rye was not so successful a crop as wheat and barley and in prehistoric times was chiefly cultivated in Northern Europe: in Britain, Germany and Scandinavia in the iron age, and was subsequently introduced into Hungary, Italy, Greece and Turkey by the Romans (Helbaek 1960A, 113). It reached Mitridat in the eastern Crimea by the third to fourth centuries A.D. as a weed in crops of wheat and barley (Žukovskij, trans. Hudson 1962, 5), and was cultivated at Byzantine Beycesultan in the tenth century A.D. (Helbaek 1961, 83).

CRITERIA FOR THE PALAEOETHNOBOTANICAL IDENTIFICATION OF EARS AND GRAINS OF RYE SPECIES

Secale montanum Guss. (fig. 53B)

It has a perennial habit. The rachis is more or less brittle (Helbaek 1960A, 113), the grains are narrower, and smaller than those of *Secale cereale* (see Gill and Vear 1958, 265; Hector 1936, 222).

Secale anatolicum Boiss. (fig. 53C)

No details of morphology of this grain are available.

Secale ancestrale Žuk. (fig. 53A)

Annual and biennial forms exist (Gill and Vear 1958, 265). The rachis is always brittle and the ear is strongly pubescent (Helbaek 1960, 113). The caryopses are much larger than those of *Secale montanum* and *Secale anatolicum*.

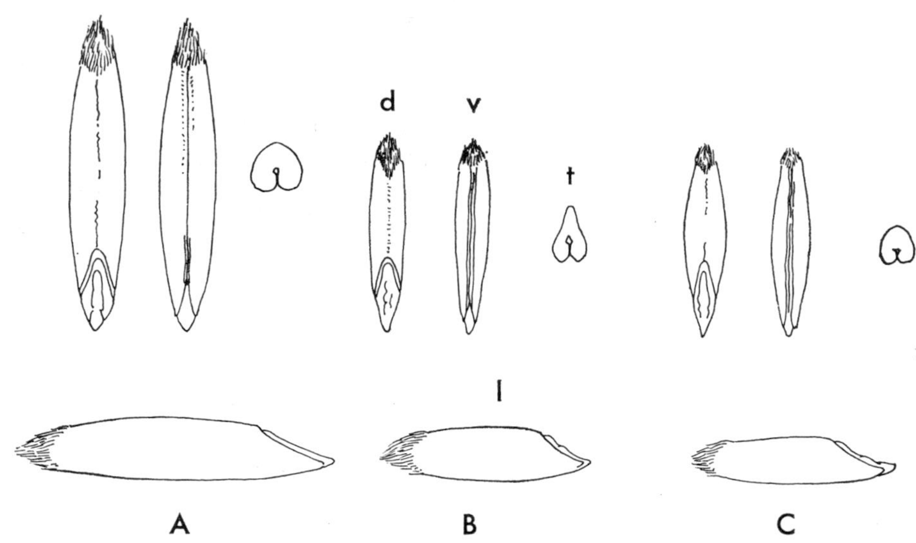

Fig. 53 Caryopses of (A) *Secale ancestrale*; (B) *S. montanum*; (C) *S. anatolicum* ×5 compared in (d) dorsal, (v) ventral, (l) lateral views and in (t) transverse section (after Schiemann 1948).

Secale cereale L. (fig. 54, pl. 21 and cf. fig. 11)

The rachis is tough, and the 20–30 segments have densely pubescent margins. The spike consists of a single spikelet at each node of the rachis. The spikelets are three-flowered but usually only two grains mature – the third floret being abortive and

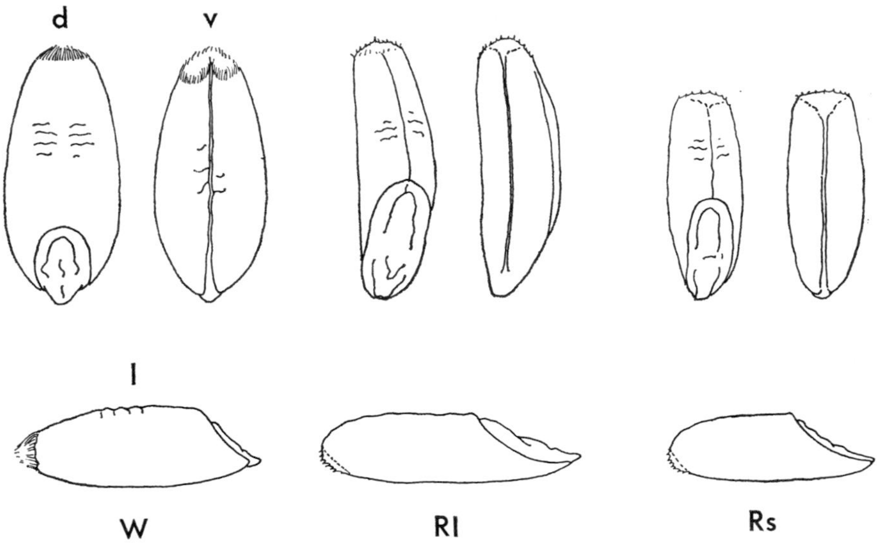

Fig. 54 Caryopses of wheat and rye compared ×5. (W) wheat caryopsis in (d) dorsal, (v) ventral and (l) lateral view. Similar views of rye caryopsis of (Rl) long twisted type, (Rs) short straight type.

minute. Each spikelet consists of a pair of narrow, pointed glumes, with a single nerve. The lemma is broad, keeled and extended into an awn, and has barbs up the keel. The palea is thin and two-keeled. The lemma and palea of each floret tend to diverge at maturity so that the apex of the ripe grain is clearly visible. The grain threshes free, and is longer, more slender and more sharply keeled than wheat caryopses, and has a sharply pointed embryo. The apex of the caryopsis is truncated. The grains of the cultivated species are of two types: some grains are long, slender and curved; others are short, thick and straight.

Buschan (1895, 54) gives measurements for two types of modern rye grains: length 6·0 mm, breadth 3·1 mm, and length 5·4 mm, breadth 2·0 mm. In the Vallhagar deposits fifty carbonized grains measured: length 5·35 mm (3·80–7·42 mm), breadth 1·64 mm (1·04–2·17 mm), thickness 1·61 mm (1·10–2·09 mm) (Helbaek 1955B, 668). The carbonized grains from the Roman period at Castle Cary, Scotland, were similar in length but broader and thicker: length 5·3 mm (6·1–4·7 mm), breadth 2·1 mm (2·6–1·9 mm), thickness 1·9 mm (2·1–1·6 mm) (Jessen and Helbaek 1944, 53). (See Hector 1936, 216 f.; Martin and Leonard 1967, 452; Gill and Vear 1958, 265; Leonard and Martin 1963, 452; Helbaek 1961, 83.)

CONDITIONS REQUIRED FOR CULTIVATION OF RYE

Climate

Rye is chiefly cultivated in cool climates. Winter ryes can be sown where winter temperatures fall as low as −40°F, or where the mean winter temperature is 0°F. Most of the rye in Europe is grown where the average annual rainfall is between 65–70 cms. The crop can withstand all kinds of adverse weather conditions except heat. It sprouts more quickly and grows with more vigour than wheat at low temperatures. Its earliness enables it to escape injury from midsummer droughts and hot winds which may cause the ears to be blasted when the plants are in flower (Leonard and Martin 1963, 452).

Soils

Rye is less exacting in its soil requirements than wheat, barley or oats. In Britain today rye grows chiefly on lighter soils in districts with a limited rainfall. It requires adequate drainage but can tolerate a considerable amount of acidity or alkalinity (Hunter 1951, 115).

Rye has an extensive root system which branches profusely just below the ground surface but also has some roots extending downwards to a depth of 1·6–1·8 m: this development of the roots may account for its better performance compared with wheat in dry climates and on poor soils (Martin and Leonard 1967, 451).

USES PREHISTORIC MAN MAY HAVE MADE OF RYE

Rye is used today in Northern Europe to make bread. Rye bread is inferior to that of wheat, having a bitter taste and the weak flour makes the loaves flat and heavy (Gill and Vear 1958, 266). Rye is subjected to the disease ergot (*Claviceps purpurea*), the sclerotia of which are poisonous to humans and cause the disease known as ergotism or St Anthony's Fire, resulting in either a gangrenous condition or affecting the nervous system to produce convulsive disorders (ibid., 550) in communities dependent on rye bread. No evidence of rye bread has been found in prehistoric contexts in Europe, but it is likely that rye was used for this purpose.

A small amount of rye was found, together with other cereals, in the stomach of Grauballe man (Helbaek 1958, 54), and many sclerotia of ergot were also found – some *in situ* in gross florets of *Holcus* sp., others could possibly have come from rye (ibid., 113).

Rye is largely used for making whisky nowadays, and there is a little evidence that it was used for brewing, but probably not distilling, in antiquity. Sprouting rye grains were found at the Roman fort at Caerleon, Wales, and Helbaek has suggested that they were intentionally induced to germinate in order to make malt (Helbaek 1964B, 163). Other deposits of a few carbonized sprouting rye grains are known from Eketorp, Öland, Sweden (idem 1966C, 218), where it was probably used together with barley as malt; and at Verulamium, England (idem 1952, 213), where it has been suggested that the rye grains germinated due to damp conditions in the granary.

Chapter 8

Oats

GENERAL DISCUSSION OF THE GENUS *Avena*

Oat species may be classified into three groups based on the numbers of chromosomes (Martin and Leonard 1967, 483).

A. DIPLOID GROUP ($2n = 14$)

Avena brevis Roth.	short oat
Avena wiestii Steud.	desert oat
Avena strigosa Schreb.	sand oat
Avena nudibrevis Vav.	small-seeded naked oat

B. TETRAPLOID GROUP ($2n = 28$)

Avena barbata Brot.	slender oat
Avena abyssinica Hochst.	Abyssinian oat

C. HEXAPLOID GROUP ($2n = 42$)

Avena fatua L.	common wild oat
Avena sativa L.	common white oat
Avena nuda L.	large-seeded naked oat
Avena sterilis L.	wild, red oat
Avena byzantina C. Koch.	cultivated, red oat

The chief cultivated forms *Avena nuda*, *Avena sativa* and *Avena byzantina* belong to the hexaploid group. *Avena strigosa*, *Avena brevis*, *Avena nudibrevis* and *Avena abyssinica* have been cultivated to a much lesser extent (Stanton 1955, 47, 104, 59, 98, 101, 53 and 99 respectively).

GENETIC AND ARCHAEOLOGICAL EVIDENCE FOR THE ORIGINS OF CULTIVATED OAT SPECIES

It was previously thought that *Avena fatua* was the progenitor of *Avena sativa* and that *Avena sterilis* was the ancestor of *Avena byzantina*. Coffman (1946) demonstrated, however, that all the hexaploid cultivated oats derived from *Avena sterilis* the wild red oat.

The genetic relationships of the oat species are far from clear. The diploid groups all share the AA genomes (Bell in Hutchinson 1965, 92). The tetraploids also have the AA complex together with a second set – BB – the origin of which is quite unknown. Hybrids between the diploid and tetraploid groups are not easily obtained, despite their close relationship. If the origin of the tetraploid group is not clear, the hexaploids are even more puzzling. They possess the AA genome pair of the diploids and tetraploids, but the CC and DD genomes are quite distinct and of unknown origin. They may have been acquired in one move CCDD or in two separate stages adding CC then DD. There is no evidence for either step and the 'origin and evolution of the hexaploid oats is an unsolved problem' (ibid., 94).

Despite the difficulties it seems likely that the cultivated *Avena strigosa* ssp. *strigosa* was derived from *Avena strigosa* ssp. *hirtula* (= *Avena hirtula*) which is distributed around the shores of the Mediterranean (Hector 1936, 53). This form may also be the ancestor of *Avena brevis* and *Avena nudibrevis* which have been cultivated to a small extent. Bell (in Hutchinson 1965, 92) suggests that these forms may have arisen by differentiation in the cultivated *Avena strigosa* ssp. *strigosa*. *Avena nudibrevis* was made a distinct species by Vavilov, who realized that it was not closely related to *Avena nuda* (previously it was classified as *Avena nuda* L. ssp. *biaristata* (Alef.) (Asch. and Graeb.)), as it crossed readily with *Avena strigosa* and *Avena brevis* but would not hybridize with *Avena nuda* or *Avena sativa* (Stanton 1955, 53).

Avena abyssinica is the only cultivated tetraploid oat, it seems likely that it derived from *Avena wiestii*, the desert oat, which it closely resembles morphologically. Stanton (1955, 99), reports that Trabut found various transitional forms between *Avena wiestii* and *Avena abyssinica* under cultivation. *Avena abyssinica* has a very limited distribution in Abyssinia, Eritrea and Egypt (Bell in Hutchinson 1965, 92).

As mentioned above, *Avena sterilis* is now widely accepted as the wild ancestor of the cultivated forms *Avena sativa* and *Avena byzantina*. The origin of *Avena nuda* is not known but it is closely related to the other hulled, hexaploid, cultivated oats. Its distribution is restricted to the Himalayas, Russia, Tibet and China (Stanton 1955, 48).

Finds of 'wild oat' grains are not unusual in palaeoethnobotanical material from the earliest neolithic villages in the Near East. They are reported from Ali Kosh (Ali Kosh and Mohammad Jaffar phases) (Hole and Flannery 1967, 147f), Beidha (Helbaek 1966A, 63), and in 'Amouq A (idem 1960C, 542). Only in the Beidha find is the species identified – as *Avena ludoviciana* (= *Avena sterilis* ssp. *ludoviciana*).

Avena fatua, *Avena strigosa* and *Avena sativa* have been identified from prehistoric contexts in Europe. The earliest oat grain found in Europe comes from the aceramic neolithic levels at Achilleion, Thessaly, Greece. It unfortunately has lost the diagnostic parts of the lemma so that the species cannot be determined (Renfrew 1966, and see pl. 22). *Avena fatua* has been found in the bronze age lake-side villages at Alpenquai on Lake Zürich and Mörigen on the Bieler See (Bertsch and Bertsch 1949, 79). In the Hallstatt iron age it is known from Lengyel in Hungary (Tempir 1964, 87). Common wild oats are first found in Britain in the early iron age: they occur at Maiden Castle,

Portland 'Beehives', Fifield Bavant, Little Salisbury, Worlebury, Meare Lake Village, Glastonbury Lake Village (Helbaek 1952B, 229) and Aldwick, Barley (Renfrew 1965, 10).

Avena strigosa has also been reported in the Alpine region in the bronze age: at Montellier in Savoy (Heer 1866, 352), and Petersinsel on the Bieler See and at Le Bourget on Lake Annecy (Bertsch and Bertsch 1949, 81). Recent work by Madame Villaret von Rochow (1971, 243 f.), however, suggests that the secondary florets of *Avena ludoviciana* may have been misidentified as *Avena strigosa*. It appears in Britain in the early iron age at Maiden Castle and Fifield Bavant (Jessen and Helbaek 1944, 50). Impressions of oat grains on Corded Ware vessels from Schraplau and Calbe in the Saale valley measuring: length 10 mm, breadth 2–5 mm, and length 8 mm, breadth 2 mm respectively, may possibly be referable – on account of their size – to *Avena sativa* (Matthias and Schultze-Motel 1967, 146).

CRITERIA FOR THE PALAEOETHNOBOTANICAL IDENTIFICATION OF PANICLES AND GRAINS OF OAT SPECIES

Avena brevis Roth. (fig. 55)

The panicle is short and equilateral. The spikelets are usually two-flowered, separating from their pedicels by fracture leaving no distinct basal scars, and the florets separate usually by fracture at the base of the short rachilla segment bearing the second floret.

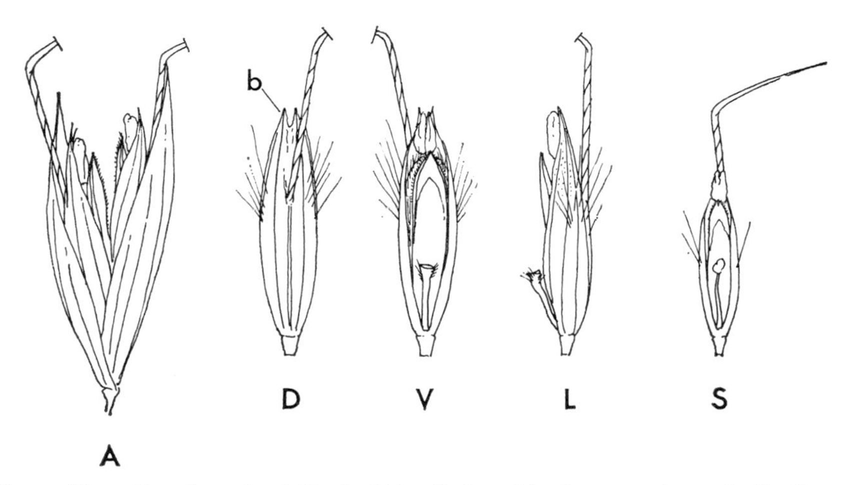

Fig. 55 Short Oat *Avena brevis* Roth. (A) spikelet with glumes and two fertile florets (grains) in side view. Primary grain in (D) dorsal, (V) ventral and (L) lateral view. (S) secondary grain in ventral view. (b) bifid apex of lemma. All ×3.

The glumes are seven-nerved and from 12 to 15 mm long. The lemmas are short (8–11 mm in length in the primary floret, 6–8 mm in the second), and they terminate in two coarse points at the apex. The awns are weakly twisted and geniculate. The caryopses range from 3 to 5 mm in length (see Stanton 1955, 101).

It is morphologically closest to the sand oat *Avena strigosa* from which it is distinguished by a much shorter lemma, with short, coarse apical points, and shorter and weaker awns.

Avena wiestii Steud. (fig. 56)

The equilateral panicles are medium-sized. The spikelets usually have two florets, which separate from their pedicels by abscission leaving small, distinct basal scars on the primary florets: the second and third florets separate by disarticulation of the rachilla segments. The glumes are up to 23 mm long and have seven to nine veins. The lemmas are short, hairy and terminate in two points which may be 3–7 mm long. The

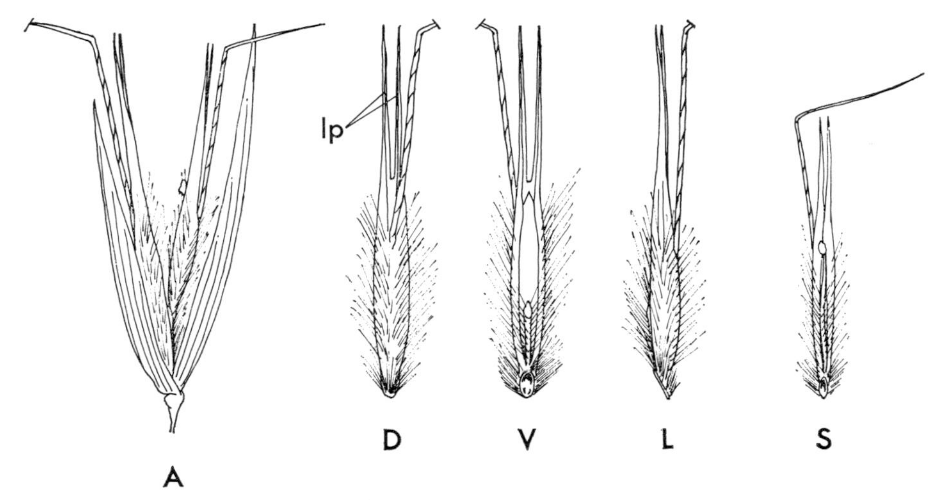

Fig. 56 Desert Oat *Avena wiestii* Steud. (A) spikelet with glumes and two fertile florets (grains) in side view. Primary grain in (D) dorsal, (V) ventral, (L) lateral view. (S) secondary grain in ventral view. (lp) long slender lemma points. All ×3.

basal hairs are short. The lemmas bear twisted geniculate awns. The primary caryopses measure 7–9 mm long and the secondary grains are 6–8 mm in length. The rachillas are flattened and slender with slight cavities at their upper ends and are covered with long hairs which become more concentrated towards the top – forming a 'brush'.

In some details the desert oat resembles *Avena fatua* but it is distinguished by the smaller basal scars, and the long brush-like hairs at the top of the rachilla segments (see Stanton 1955, 95–6).

Avena strigosa Schreb. (fig. 57 and pl. 21)

The panicles are equilateral and small to medium-sized. The spikelets have two florets; the primary floret separates from the pedicel by fracture, and the secondary floret from the primary by disarticulation of the rachilla. The glumes range in length up to 24 mm with seven or eight veins. The lemmas terminate in two long points (5–9 mm long). There are few basal hairs. The awns are twisted and geniculate: 15–30 mm long. The

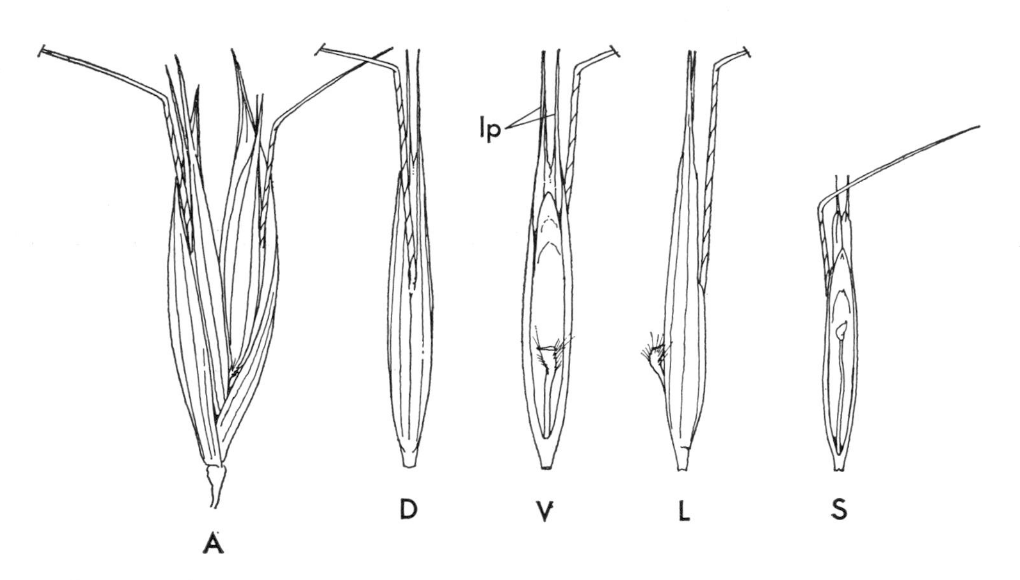

Fig. 57 Bristle-pointed (Sand) Oat *Avena strigosa* Schreb. (A) spikelet with glumes and two fertile florets (grains) in side view. Primary grain in (D) dorsal, (V) ventral, (L) lateral views. (S) secondary grain in ventral view. (lp) long slender lemma points. All ×3.

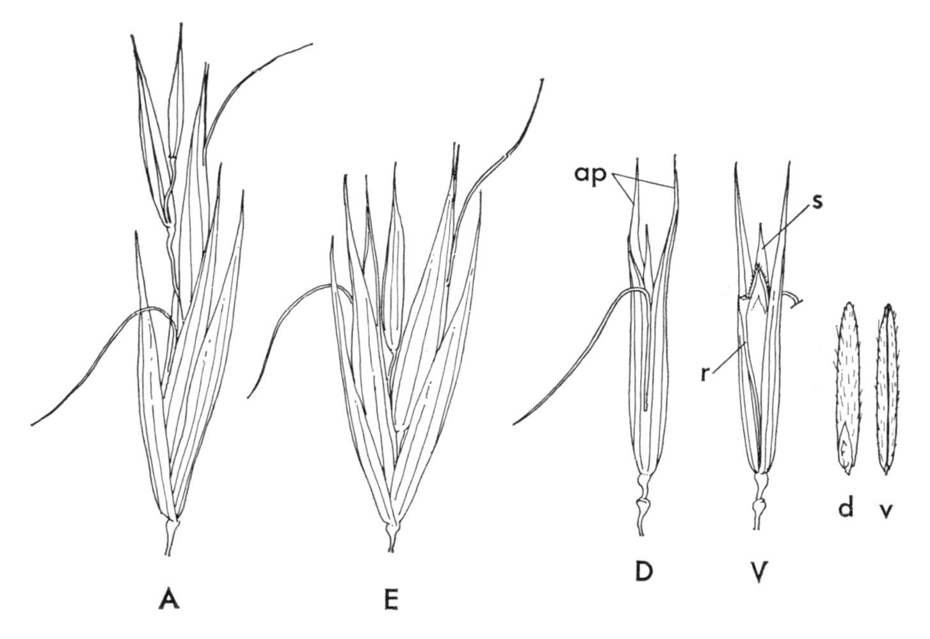

Fig. 58 Small Naked Oat *Avena nudibrevis* Vav. Spikelets with glumes and (A) four, (E) three fertile florets (grains) in side view. Primary grain in (D) dorsal, (V) ventral view showing lemma of glume-like texture with weak awn and two very prominent apical points (ap), with smaller point between (s). Primary grain in (d) dorsal, (v) ventral view. (r) very long rachilla. All ×2.

primary caryopses are 7–9 mm in length, the secondary grains measuring 5–7 mm. The rachilla segments are 3–5 mm long and slender, being glabrous except for two small tufts of hairs on the margins at the apex.

It is distinguished from *Avena brevis* by its greater size, from *Avena abyssinica* by its longer, more distinct lemma points, and in having less hairy rachilla segments; and from *Avena sativa* by these characters and by having smaller caryopses (see Stanton 1955, 98–9).

Avena nudibrevis Vav. (fig. 58)

The small panicles are semi-unilateral. The spikelets have 3–5 florets: the glumes are up to 24 mm long and have nine veins. The lemmas do not invest the caryopses tightly, they are of the same papery texture as the outer glumes and are up to 26 mm long with a distinctive v-notch at the apex. The awns are non-twisted and 12–18 mm long. The first caryopsis is 4–6 mm in length, subsequent grains are progressively shorter. The rachilla segments are glabrous and from 4 to 8 mm long (see Stanton 1955, 53–4).

Avena barbata Brot. (fig. 59)

The panicles are medium- to large-sized and equilateral. The spikelets have two to three florets. The primary florets separate from the pedicel by abscission leaving small, basal, suckermouth scars. The subsequent florets separate by disarticulation of the rachilla segments. The glumes are up to 27 mm long with eight to ten veins. The lemmas

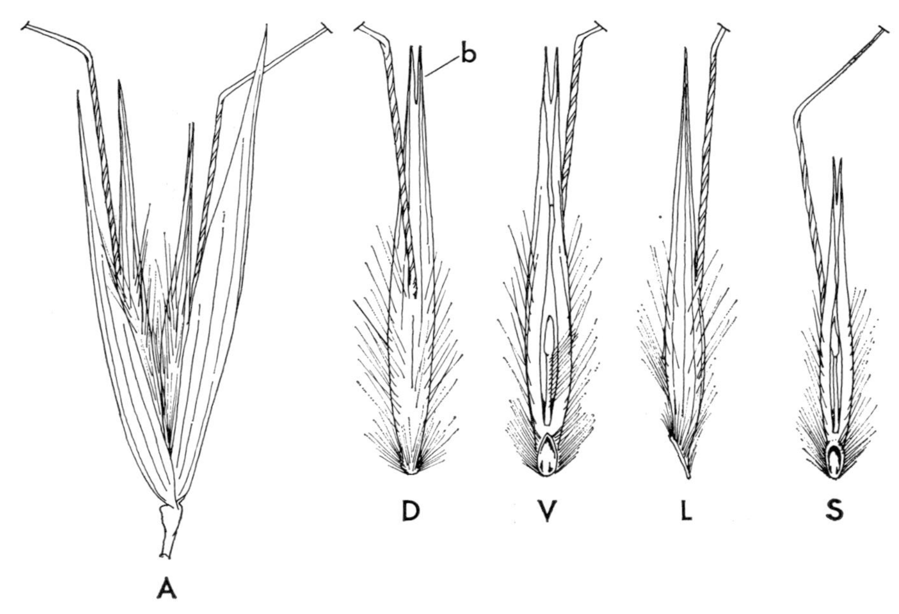

Fig. 59 Slender Oat *Avena barbata* Brot. (A) spikelet with glumes and two fertile florets (grains) both bearing strong awns. Primary grain in (D) dorsal, (V) ventral, (L) lateral view. Secondary grain (S) in ventral view. (b) bifid apex to lemma. All ×3.

are long (on the primary floret they may measure up to 24 mm in length); they are covered from the base to point of insertion of the awn with fine hairs, and terminate in two apical points 4–7 mm long. The awns are twisted and geniculate, and may reach 40 mm in length. The primary caryopses are 6–9 mm long, the secondary ones being 5–7 mm in length. The rachilla segments are flattened with a small cavity at the expanded end. The rachilla margins are pubescent, with long hairs towards the top. The segments may be 3–5 mm long (see Stanton 1955, 96–8).

Avena abyssinica Hochst. (fig. 60)

The panicles are medium-sized and equilateral. The spikelets are usually two-flowered, and they separate from the pedicel by fracture at the base of the primary floret. The second floret separates by disarticulation of the rachilla. The glumes are up to 26 mm long with seven to nine veins. The lemma of the primary floret is up to 20 mm long, glabrous, and terminates in two points (Stanton 1955, 100). However, Hochstetter

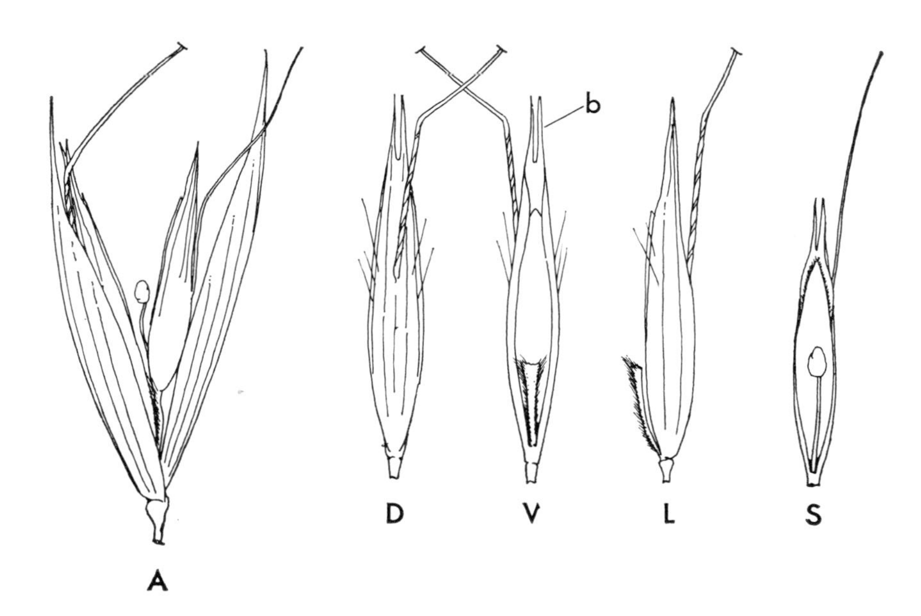

Fig. 60 Abyssinian Oat *Avena abyssinica* Hochst. (A) spikelet with glumes and two fertile florets (grains) both bearing awns. Primary grain in (D) dorsal, (V) ventral, (L) lateral view. (S) secondary grain in ventral view. (b) bifid lemma apex. All ×3.

first described the species as having a lemma which terminated into four awn points or teeth at the apex (cf. ibid., 99), and this feature is visible on the examples in my comparative collection from the Botanisches Institut der Ernst-Mortiz-Arndt Universität, Greifswald. The awns are twisted and geniculate. The primary caryopses measure 7–9 mm in length: the secondary grains being 5–7 mm long. The rachilla segments are slender with numerous hairs on the margins of the upper half. The expanded top of the rachilla is forked (ibid., 100).

Avena fatua L. (fig. 61 and pl. 22)

The panicles are large and equilateral. The spikelets may have two or three florets. The primary floret separates from the pedicel by abscission leaving a distinctive sucker-mouth base. The subsequent florets are separated by disarticulation of the rachilla. The glumes are up to 26 mm long, and usually have nine veins. The lemmas are

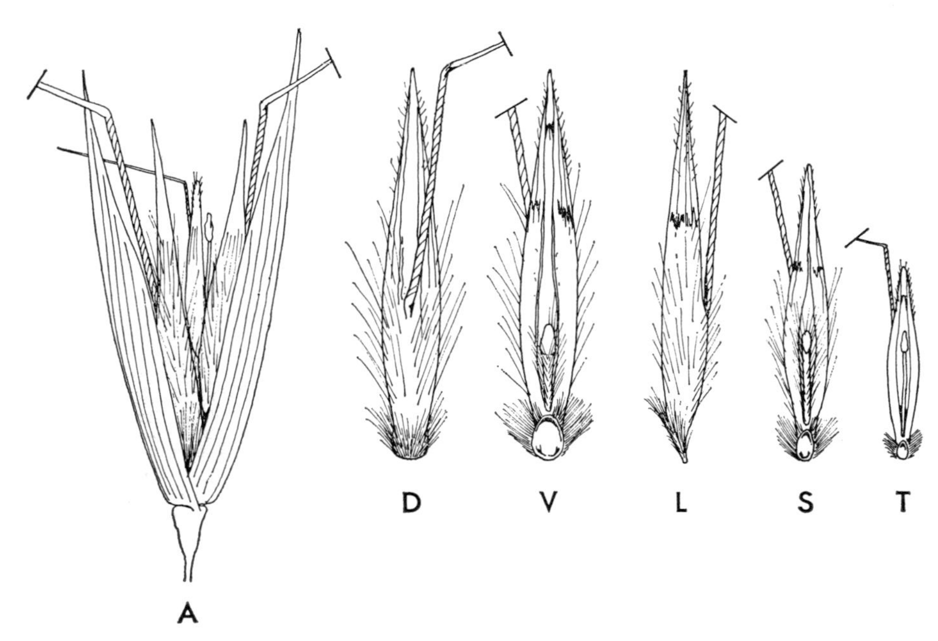

Fig. 61 Common (Spring) Wild Oat *Avena fatua* L. (A) spikelet with glumes and three fertile florets (grains) all bearing awns. Primary grain in (D) dorsal, (V) ventral, (L) lateral view. (S) secondary grain in ventral view. (T) Tertiary grain in ventral view. All ×3.

usually very pubescent, with numerous, long basal hairs, the awns are strong, twisted and geniculate. The primary caryopses may be 9–12 mm long: the secondary grains being 6–9 mm in length. The rachilla segments are usually hairy and up to 3 mm long (Stanton 1955, 102).

It differs from *Avena wiestii*, *Avena barbata* and *Avena strigosa* in not having a forked apex to the lemma, and by its larger florets (Leonard and Martin 1963, 558).

Avena sativa L. (fig. 62 and pl. 23 and 24)

The panicle is equilateral and large; one sub-species, however, has a unilateral habit (*Avena sativa* ssp. *orientalis* Schreb.). Usually three grains ripen in each spikelet (Gill and Vear 1958, 282–3). The primary floret separates from the pedicel by fracture leaving

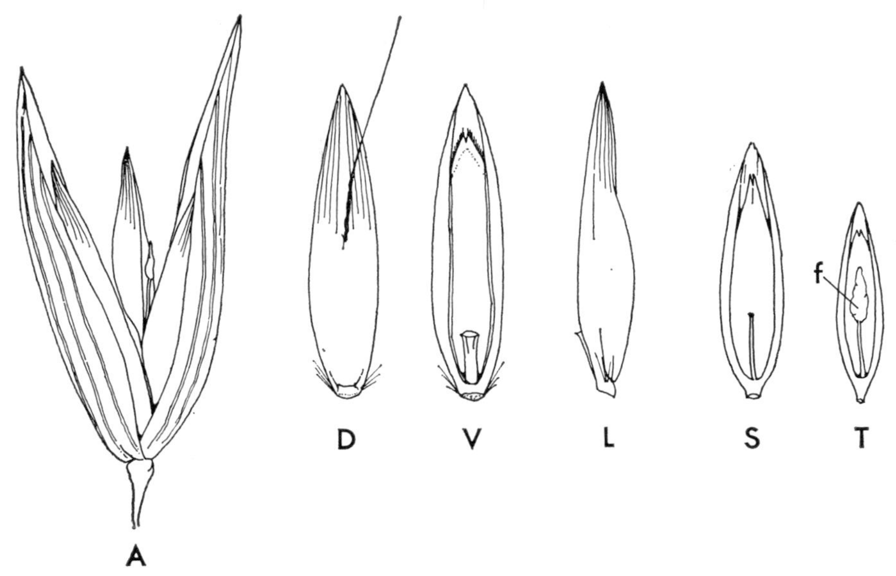

Fig. 62 Cultivated Oat *Avena sativa* L. (A) spikelet with glumes and three fertile florets in side view. Primary grain in (D) dorsal, (V) ventral, (L) lateral view. (S) secondary grain in ventral view. (T) tertiary grain in ventral view, showing sterile fourth floret (f). All $\times 2\frac{1}{2}$.

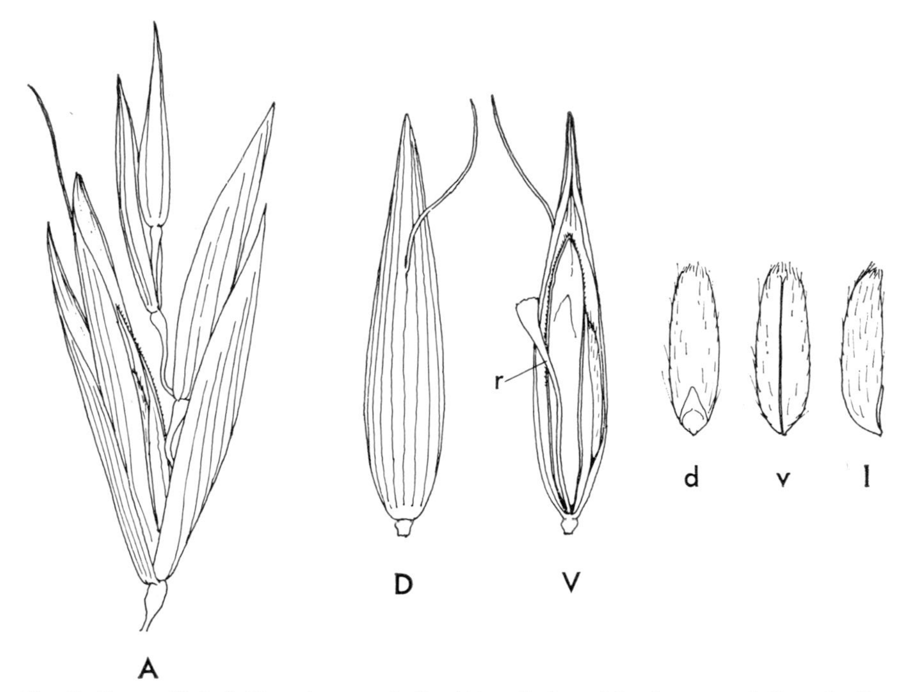

Fig. 63 Large Naked Oat *Avena nuda* L. (A) spikelet with glumes and four fertile florets. Primary floret in (D) dorsal, (V) ventral view, showing glume-like lemma with weak awn. Caryopsis of primary floret in (d) dorsal, (v) ventral and (l) lateral view. (r) very long rachilla. All $\times 2\frac{1}{2}$.

no distinct basal scar (suckermouth) (Stanton 1955, 103). The secondary florets are separated by disarticulation of the rachilla (Hector 1936, 31). The papery glumes are up to 25 mm long (ibid., 57). The lemma is usually glabrous, with occasionally a few hairs at the base, and on the rachilla (Gill and Vear 1958, 284). The lemma of the first floret bears a short, weakly twisted awn, but they almost never occur on the lemmas of the second florets (Leonard and Martin 1963, 555). Modern caryopses of *Avena sativa* range in size: length 12–16 mm, breadth 2·5–3·0 mm (Buschan 1895, 61).

Avena nuda L. (fig. 63)

The equilateral panicles are of medium size. The spikelets consist of four to six florets. The outer glumes are up to 28 mm long and have eight to eleven veins. The lemmas are of the same papery texture as the glumes and do not invest the caryopsis at all closely. They may be 22–30 mm long and terminate in a v-notched apex. The lemmas are glabrous with the occasional exception of a few basal hairs. Small, non-twisted awns are sometimes present on the lemma of the first floret only. The first caryopsis ranges in length from 5 to 10 mm: each successive caryopsis being progressively shorter. The rachilla segments are glabrous, or very slightly hairy and very long (4–8 mm) (Stanton 1955, 47–53).

Avena sterilis L. (fig. 64)

The panicle is medium-sized and equilateral. There are relatively few spikelets, and they are two- to three-flowered. The primary floret separates from the pedicel by abscission leaving a suckermouth basal scar. The second and third florets are separated (with difficulty) by basifracture of the rachilla segments. The glumes are up to 35 mm long and have nine to eleven veins. The lemma of the first floret is extremely long (24–40 mm) and in all cases densely pubescent. The awns are long, coarse, twisted and geniculate. They occur in both first and second florets. The caryopsis of the first floret is 11–14 mm long, the subsequent caryopses are smaller. The rachilla segments are 2–3 mm long and glabrous or slightly hairy. In the sub-species *Ludoviciana* awns may be absent on the second florets, and the lemmas are less densely pubescent than in the form *Maxima*. The variety *Macrocarpa* (= *Avena sterilis sitosissima* Malz) is distinguished by the awns, which are densely covered with short, fine hairs on the lower sections (Stanton 1955, 55–8).

Avena byzantina C. Koch (fig. 65)

The panicles are small, equilateral and have relatively few spikelets. The glumes are usually longer than the lemmas and larger than those of most other species of oats. Both florets in the spikelet often have weak untwisted awns arising from the back of the lemma. The basal hairs on the lemma are more numerous and longer than those of *Avena sativa*. The caryopses are slightly longer than those of *Avena sativa* (cf. Stanton 1955, 59 f.), being about 11 mm long (Hector 1936, 60). The primary florets separate from their pedicels by basifracture as in *Avena sativa* (Stanton 1955, 59). The secondary

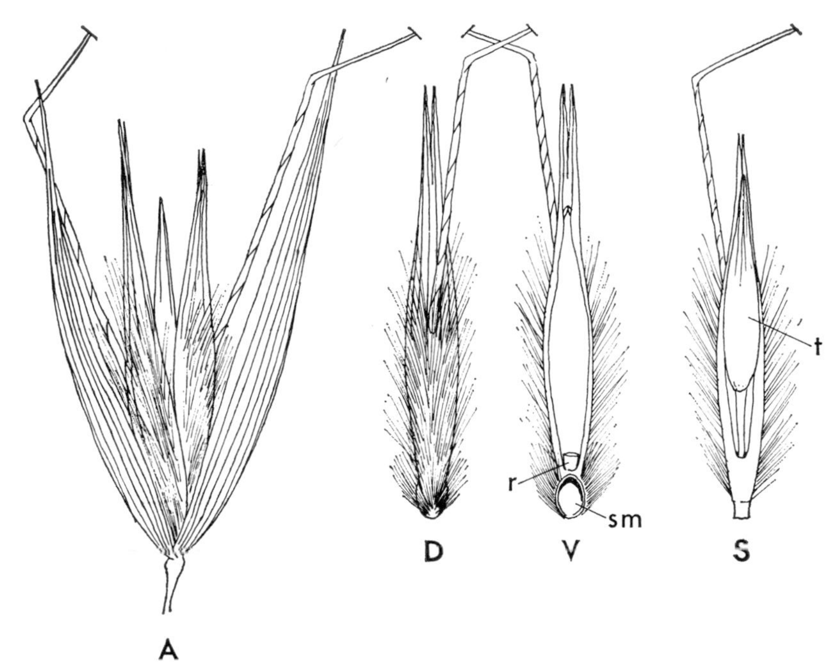

Fig. 64 Wild Red Oat *Avena sterilis* L (A) spikelet with glumes and three fertile florets (grains), the 1st and 2nd with very long strong awns. Primary grain in (D) dorsal, (V) ventral view. (S) secondary grain in ventral view. (r) rachilla of primary grain fractured near base by forcible removal of secondary. (sm) large prominent 'suckermouth' at base of primary grain only, (t) glabrous, awnless tertiary grain. All ×2.

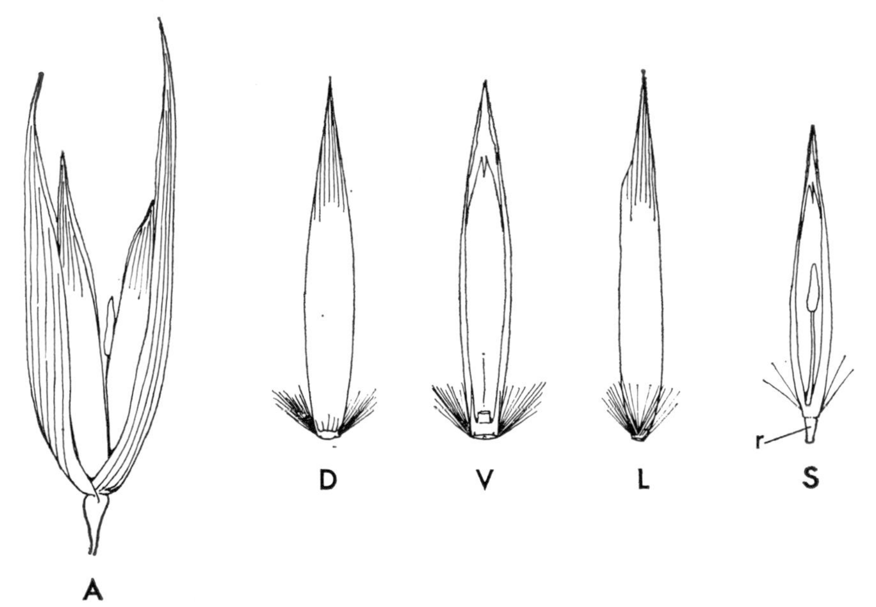

Fig. 65 Red Oat *Avena byzantina* C. Koch. (A) spikelet with glumes and two fertile florets (grains) in side view. Primary grain in (D) dorsal, (V) ventral, (L) lateral view. (S) secondary grain in ventral view showing adherent rachilla (r), separated from primary grain by basifracture. All $\times 2\frac{1}{2}$.

florets are separated by the basifracture of the rachilla segments, which remain attached to the base of the upper florets (Hector 1936, 60).

CONDITIONS REQUIRED FOR THE CULTIVATION OF OATS

Climate

Avena sativa is best adapted to cool temperate regions where the annual rainfall is over 30 in or where the land is irrigated (Martin and Leonard 1967, 79). *Avena byzantina* flourishes best in regions with warm climates as around the Mediterranean (Leonard and Martin 1963, 547).

The oat plant requires more water for its development than do the other temperate cereals, but can survive with less sunshine than the others require (ibid., 548). Hot, dry weather during the development of the grain gives rise to premature ripening: hot, humid weather at this time makes the plant susceptible to disease, resulting in a drop in yield (ibid., 548).

Winter oats are less resistant to cold than are the winter forms of wheat, barley and rye. Where the minimum temperature drops below 10°F winter killing may become serious: it is certainly severe in regions with a minimum temperature 0°F (ibid., 549).

Soil

Satisfactory crops of oats are produced on a wide range of well-drained reasonably fertile soils. They are less sensitive than wheat and barley to soil conditions, particularly to acidity (Leonard and Martin 1963, 549). They do best on silt and clay loams: excessive moisture or nitrogen in the soil causes the crop to lodge (Martin and Leonard 1967, 480).

POSSIBLE USES OF OATS TO PREHISTORIC MAN

Oats are nowadays chiefly used in human diet to make porridge (Leonard and Martin 1963, 582). They were found in the stomach contents of Grauballe man in quantity (Helbaek 1958A, 84), and were also present in Tollund man's last meal (idem 1950, 332).

Oat flour, however, lacks gluten and so it can only be made into flat cakes. It can be used in bread making only if there are four parts of wheat flour to one part of oat flour, to overcome the heaviness of the meal (Leonard and Martin 1963, 582). It is thus not surprising that it is not reported as a chief constituent of any of the finds of prehistoric bread – but Helbaek found traces of wild oats in the carbonized buns from Glastonbury (Helbaek 1952B, 212).

Oats are now widely used for feeding livestock (Martin and Leonard 1967, 490), and may well have been used for winter feeding of stalled animals in prehistoric times. The flour possesses a property that retards the development of rancidity in fat products (ibid., 490), and this may well have been discovered and utilized for keeping products such as butter and curdled milk. Oats were also sometimes used in making malt for beer by the Romans (André 1961, 179).

Chapter 9

Millets

Panicum miliaceum L. Broomcorn Millet

GENETIC AND ARCHAEOLOGICAL EVIDENCE FOR THE ORIGINS OF *Panicum miliaceum* L.

The genetics of *Panicum* are not well understood and the progenitor of *Panicum miliaceum* L. is thought, on the grounds of close morphological similarity, to be the wild Abyssinian species *Panicum callosum* Hochst, (Helbaek 1952C, 108). *Panicum miliaceum* has a diploid chromosome number of 42 (Hector 1936, 317).

The earliest palaeoethnobotanical finds of *Panicum miliaceum* come from the neolithic of Central and Eastern Europe. Broomcorn millet was found in the Aggtelek Cave in Hungary (Danubian I culture) (Bertsch and Bertsch 1949, 90), at the Eisenburg settlement of the same culture in Thuringia (Natho and Rothmaler 1957), at several Tripolye sites in Roumania (ibid., and Gimbutas 1956, 103), in the neolithic Swiss lake-side villages, for example Taubried on the Federsee, Sipplingen, Wangen and Lützelstetten on the Bodensee and Le Bourget on Lake Bourget in Savoy (Bertsch and Bertsch 1949, 90). Further north broomcorn millet has been found in T.R.B. neolithic contexts in Poland (Schultze-Motel 1968, 218). During the bronze age its distribution was extended into Italy (Helbaek 1956, 291), Holland (idem 1960A, 112) and Denmark (idem, 1952C, 108). It is interesting to note that the size of the grains in Italy significantly increased during the early iron age from 2·20 to 2·66 mm long for those from Palafitta di Parma (bronze age) to 2·85 to 3·04 mm (iron age Golasecca, Varese; and pre-urban Rome) (see idem 1956, 291).

Panicum miliaceum is not known from prehistoric contexts in Egypt (Dixon in Dimbleby and Ucko 1969), or India (Allchin in Dimbleby and Ucko 1969), but it is found in neolithic Yang Shao contexts in China (Watson in Dimbleby and Ucko 1969). The earliest find in the Near East at Jemdt Nasr in Mesopotamia dates to c. 3000 B.C. (Helbaek 1959B, 371), and this is followed by a large deposit of the seventh century B.C. at Nimrud (idem 1966D, 615); it is also recorded from Fort Shalmaneser and Hasanlu (ibid.).

THE CRITERIA FOR THE MORPHOLOGICAL IDENTIFICATION OF SEEDS OF *Panicum miliaceum* L. (pl. 25 and fig. 13)

As mentioned above (chapter 4 and fig. 13) the seeds are borne on panicles from 10 to 25 cm long. The spikelets consist of a pair of glumes – the upper one having up to thirteen nerves, the lower being five- to seven-nerved. Within the glumes are two florets: the lower floret is sterile consisting of a lemma similar in size to the lower glume but being seven- to thirteen-nerved; the palea is much reduced in size.

In the fertile floret the caryopsis is enclosed within a broad, smooth, shining seven-nerved lemma, and a similarly smooth three-nerved palea. The ripened caryopsis measures 3 mm long by 2 mm broad. In describing the seed of *Panicum miliaceum* found at Nørre Sandegaard in Bornholm Helbaek mentions traces of the faint grooves along the flanks of the glossy palea left by the revolute margins of the lemma (Helbaek 1952C, 108). The caryopsis has no ventral furrow. (See Hector 1936, 317; Gill and Vear 1958, 289.)

Measurements of prehistoric Panicum miliaceum *grains*

Site		*Length*	*Breadth*	*Thickness*
		mm	mm	mm
Nørre Sandegaard[1]	C	1·90	1·71	1·33
Palafitta di Parma[2]	I	2·20–2·66	—	—
Golasecca, Varese[2]	I	2·85–3·04	—	—
Pre-urban Rome[2]	I	2·93	2·38	—
Ermelo, Holland[3]	C	1·5–1·9	1·2–1·6	1·2–1·4

C = Carbonized I = Impression.

References: 1. Helbaek 1952C, 108.
2. idem. 1956, 290–1.
3. van Zeist 1970, 93.

CONDITIONS REQUIRED FOR CULTIVATION OF BROOMCORN MILLET

Climate

Broomcorn millet can grow further north than any other of the small-grained cereals collectively known as 'millet'. It is adapted to regions where spring-sown crops are fairly successful. It requires only 60–65 days from sowing to maturity, but these must be moderately warm and entirely free from frost. It requires relatively little water, in fact it has the lowest water requirement of any of the cereals, but its shallow root system makes it more susceptible than other cereals to conditions of complete drought (cf. Martin and Leonard 1967, 523).

Soils

Panicum miliaceum grows well on most soils except coarse sands. It requires a better soil than do either pearl millet or finger millet (Leonard and Martin 1963, 743).

THE POSSIBLE USES OF *Panicum miliaceum* TO PREHISTORIC MAN

Broomcorn millet is used as meal for making porridge in the U.S.S.R. at the present day (Leonard and Martin 1963, 742). Pliny tells us that it was made into a white porridge in Campania, and that the Sarmatian tribes lived chiefly on millet porridge (*Natural History* XVIII, xxiv–xxv). Presumably this use was known to the prehistoric farmers who grew it although no physical remains of porridge containing broomcorn millet survive to the present day. No palaeoethnobotanical finds of bread made from *Panicum miliaceum* are known. Pliny also says that millet was specially used for making leaven; if dipped in unfermented wine and kneaded it would keep for a year. This would have been of importance to prehistoric man, ignorant of the cultivation of yeast. (See Pliny *Natural History* XVIII, xxvi.) In Bulgaria in recent times a fermented drink called 'boza' was made from millet (Sanders, 1949, 103); a similar beverage could have been made in prehistoric times.

Setaria italica (L.) Beauv. (*Chaetochloa italica*) Italian Millet

GENETIC AND ARCHAEOLOGICAL EVIDENCE FOR THE ORIGIN OF *Setaria italica*

Setaria italica has a diploid chromosome number of 18 (Hector 1936, 312), and is related morphologically to the wild green millet, *Setaria viridis* Beauv. (fig. 66), which is distributed in Western Asia and the Mediterranean area, and occurs as a weed throughout Europe (Helbaek 1960A, 112). *Setaria viridis* is recorded from the neolithic deposits in the Aggtelek Cave in Hungary and at Schussenthal near Ravensburg (Bertsch and Bertsch 1949, 86). It was also found in the bronze age settlement at Alpenquai on Lake Zürich (ibid.), and in the stomach of Grauballe man (Helbaek 1958, 84).

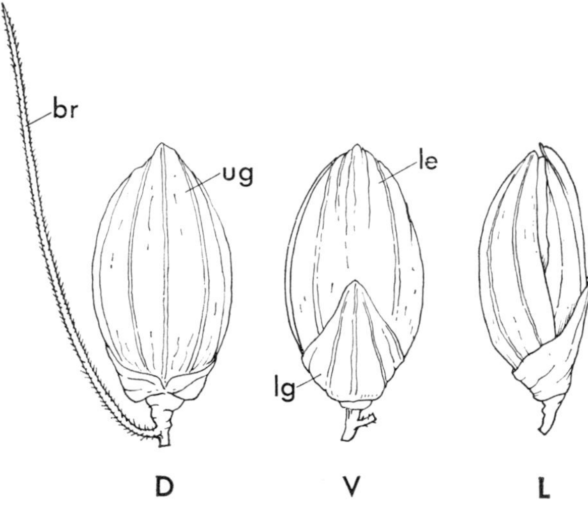

Fig. 66 Green Bristle-grass/Green Foxtail/Bottle-grass *Setaria viridis* (L.) Beauv. Spikelet in (D) dorsal view with a bristle (br), arising below, (V) ventral and (L) lateral view ×17. (lg) lower glume, (ug) upper glume, (le) lower lemma.

Setaria italica is not found as frequently as *Panicum miliaceum* in prehistoric contexts although they do both occur together in some deposits (Helbaek 1953C, 55). Oswald Heer identified *Setaria italica* at the Montellier and Buchs lake-side settlements (Keller, trans. by Lee 1866, 352). Neuweiler records it also from Robenhausen, Irgenhausen, Baden, Nidau, Mörigen and Auvernier in Switzerland (Neuweiler 1905, 48–9 and 111–14). It was found also at Hallstatt in Austria (ibid., 116). It was chiefly cultivated in the bronze age in Alpine Europe.

CRITERIA FOR THE PALAEOETHNOBOTANICAL IDENTIFICATION OF *Setaria italica*

The grains of Italian millet are carried on dense spike-like panicles varying from 5 to 23 cm in length and up to 5 cm wide (see pl. 26). Each short branch carries six to twelve spikelets. The spikelet consists of a thin papery reduced first glume, and a larger second glume. Inside these is a sterile papery lemma, and a fertile floret with hardened, glossy lemma and palea. The spikelet is subtended by from one to three bristles. The spikelet falls free from the bristles so that they may not appear in the harvested grain. The mature caryopsis is firmly enclosed by the lemma and palea. The caryopsis (fig. 14) is flattened on the ventral side, the dorsal aspect is convex with a groove extending from the base for half the length of the grain. The fresh caryopsis measures 2·5–2·75 mm long by 1·5 mm wide. (See Musil 1963, 58; Hector 1936, 309–12; Leonard and Martin 1963, 750.)

The wild green millet (green bristle-grass) *Setaria viridis* (fig. 66) differs from *Setaria italica* in having its lemma roughened by minute tubercles. The caryopsis is 2 mm long and 1–1·25 mm wide (Musil 1963, 57). The seeds are somewhat longer than broad, with the maximum breadth in the middle of the grain, and the greatest thickness towards the apex. The radicle shield forms a depression occupying about three-quarters of the dorsal side. The seeds of *Setaria viridis* tend to increase in size on carbonization: the thickness especially becomes greater (van Zeist 1970, 96).

CONDITIONS REQUIRED FOR CULTIVATION OF *Setaria italica*

Climate

It requires warm weather during the growing season which lasts longer than that of broomcorn millet – for 70–90 days after sowing (Leonard and Martin 1963, 749). It is most productive where there is fairly abundant rainfall, although it can be cultivated in semi-arid regions. It suffers greatly during drought, and lacks the ability to recover afterwards. It can be grown in regions subject to drought if cultivated in periods when acute drought is unlikely to occur. The short growing season is a great advantage from this point of view (Martin and Leonard 1967, 520).

Soil

Italian millet thrives best on a fairly fertile soil: it can grow on poor land, but it will not tolerate waterlogged or arid conditions (Leonard and Martin 1963, 749).

THE POSSIBLE USES OF *Setaria italica* TO PREHISTORIC MAN

As with the other cereals Italian millet can be made into porridge and bread. The wild green millet was incorporated in the gruel eaten by Grauballe man (Helbaek 1958, 84), perhaps by accident since it was present only in small quantities, and is a common weed.

Traces of bread made from *Setaria italica* mixed with a few grains of wheat and some flax seeds were identified by Oswald Heer from Irgenhausen on Lake Pfaffikon, Switzerland (Keller, trans. by Lee 1866, 58 and 338). Netolitzky questioned this identification in 1914 and suggested that this 'bread' may have been made from *Brassica* seeds (Bertsch and Bertsch 1949, 86). Pliny says that millet makes extremely sweet bread (*Natural History* XVIII, xxiv).

Echinochloa crus-galli (L.) Beauv.: Barnyard Millet, or Cockspur Grass

It is appropriate to consider this species with the other small-grained cereals collectively known as 'millet', although the palaeoethnobotanical finds probably were not of deliberately cultivated seeds. Grains of this species have been found at Lengyel, Hungary (Deininger 1890), in the stomach of Grauballe man (Helbaek 1958A), at Valleberga in Skåne, Sweden (Hjelmqvist 1969, 265) and at Elp, Holland (van Zeist 1970, 87).

The spikelets of *Echinochloa crus-galli* (fig. 15) are densely crowded on the branches of the upright panicle. Each spikelet consists of two florets, only one of which is fertile, enclosed in thin glumes which are sparingly covered with minute hairs with long spine-like hairs on the nerves. The fertile floret is broadest midway along its length tapering to the pointed apex and base. The fertile florets are strongly plano-convex and the caryopsis remains tightly enclosed in the hard, smooth, glossy lemma and palea. The sterile lemma of the spikelet may bear a 3·5-cm awn, or a short awn point – both forms may occur even within the same panicle. (See Musil 1963, 51 f.; Polunin 1969, 550 f.; Hubbard 1959, 336f.)

The seeds of *Echinochloa crus-galli* have a flat ventral side, and a domed dorsal surface. The radicle shield forms a depression extending over three-quarters of the dorsal side. The maximum breadth occurs in the middle of the caryopsis and the base and apex are rounded. Seeds from Elp 45 measured 1·7 × 1·2–1·3 × 0·8–0·9 mm (van Zeist 1970, 87).

Chapter 10

The Pulses: a General Discussion

The cultivation of leguminous crops is of equal antiquity to that of the cereals, although for some reason they are often treated as less important. They are valued for their seeds which contain a high percentage of protein, and for their root nodules which are beneficial to the soil. The outgrowths of the roots are caused by the presence of the bacteria *Rhizobium radicicola* in the roots which stimulate the cells of the host to divide, and thus form nodules. *Rhizobium radicicola* is capable of utilizing the free nitrogen in

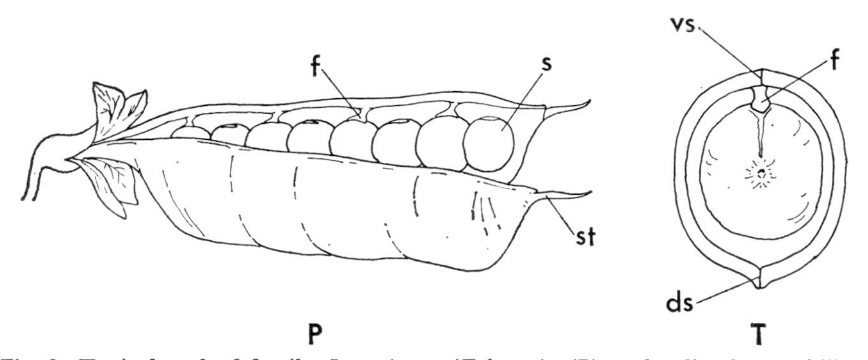

Fig. 67 Typical pod of family *Leguminosae* (*Fabaceae*). (P) pod split along midline showing row of seeds within. Alternate seeds are attached to further half of carpel and detached from nearer half, ×1. (T) transverse section ×2. (f) funicle, (s) seed, (ds) dorsal suture, (vs) ventral suture, (st) style.

the air and feeding it to the host plant in exchange for carbohydrates. This can make the legume independent of the supply of nitrogen compounds from the soil. In time, as the nodules decay more nitrogen is made available in the soil for the use of other plants (Gill and Vear 1958, 149–50). Thus the cultivation of pulse crops enriches the soil, whereas the cereals tend to impoverish the land on which they are grown.

The fruit of pulse crops is known as a *legume*. It consists of a pod derived from a single carpel, usually opening at maturity along both sutures into two halves, revealing a single row of seeds attached to one margin (fig. 67).

The seeds of the different genera and species vary greatly in size and shape, and in the position of the hilum and chalaza. Basically the seeds consist of an embryo but no endosperm, enclosed in a thick, leathery seed coat, the *testa*. The embryo is composed of two round or oval *cotyledons* which contain reserves of protein and starch (and oil in some cases); a *radicle* which grows down to form the root system, and a *plumule* which will form the plant above ground (fig. 68). The radicle is usually indicated on the surface of the seed by a slight depression between it and the cotyledons. The *hilum* lies near the tip of the radicle, and in members of the *Fabaceae* (bean and pea) family it has a fine longitudinal groove down the middle. The position of the *chalaza*, which appears on fresh seeds as a small dark area near the upper end of the cotyledons, is a diagnostic feature in certain species, especially the vetches. Occasionally a raised ridge (the

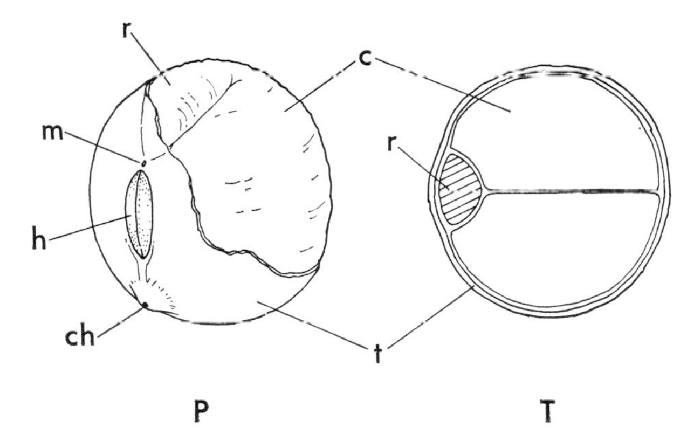

Fig. 68 Typical seed of family *Leguminosae* (*Fabaceae*). (P) in perspective view with testa cut away to show radicle and nearer cotyledon. (T) in diagonal section showing relationship of radicle and two cotyledons. (r) radicle, (c) cotyledon, (m) micropyle, (h) hilum, (ch) chalaza, (t) testa. ×3.

raphe) is found between the hilum and the chalaza in certain species. The micropyle, a minute pore at the opposite end of the hilum from the chalaza, appears not to be a diagnostic feature.

In some species the seed coat (*testa*) may be impermeable to water, and thus, even when the seeds are placed in water, they are unable to germinate unless the testa is cracked by abrasion or frost to permit the water to enter the living embryo.

The edible seeds are valued for food either eaten fresh, or dried for storage. The following species were cultivated in prehistoric times in the Near East and Europe: *Vicia faba* (horsebean), *Pisum sativum* var. *arvense* (field pea), *Lens esculenta* (lentil), *Vicia ervilia* (bitter vetch), *Cicer arietinum* (chickpea) and *Lathyrus sativus* (grass pea). A number of other leguminous plants also yielded seeds which were gathered for food in prehistoric times, including *Lathyrus aphaca*, *Lathyrus nissola*, *Trigonella graecum*, *Vicia angustifolia* and *Vicia cracca* (see chapter 19).

The main differences between different cultivated pulse seeds

Features	*Vicia faba*	*Pisum sativum* var. *arvense*	*Lens esculenta*	*Vicia ervilia*	*Cicer arietinum*	*Lathyrus sativus*
No of seeds per pod	variable	4–9	2	3–4	1–2	2–5
Outline of seed	oblong or oval	circular	circular	angular or rounded	squat, angular	oval, either truncated at both ends, or with one end pointed
Characteristic shape of seed	almost circular in cross-section	spherical	convex in cross-section	triangular in cross-section	rounded and irregular	approximately triangular in cross-section
Surface features	radicle visible in depression at one end	chalaza about 2·5 mm from hilum	—	radicle visible at one end, shallow depressions round hilum	blunt, oblique point formed by radicle	—
Position of hilum	at one end, on short axis	flush with the surface of seed	small, acutely lanceolate hilum on rather sharp margin	on margin of triangular end	large, concave, on the flattened side near pointed end	small, oval hilum at one corner of thick end
Dimensions of the hilum	5–6 mm long 1·5 mm wide	1·5–2 mm long	1 mm long	1·5 mm long	1·5–2 mm long	0·9 mm long
Dimensions of fresh seeds	13–14 mm long 8–9 mm wide	6 mm diameter	3–9 mm diameter	4–6 mm long	8–10 mm long	2·5–5·9 mm long, 2·5–5·0 mm broad
Dimensions of ancient carbonized seeds	rarely exceed 10 mm long 6 mm wide	rarely exceed 5 mm diameter	rarely exceed 4·75 mm diameter	2·0–3·9 mm maximum dimension	3·5–5·4 mm maximum dimension	4·5–5·5 mm long, 3·9–4·0 mm broad

Chapter 11

Horsebean

GENETIC AND ARCHAEOLOGICAL EVIDENCE FOR THE ORIGIN OF *Vicia faba* L. The horsebean is diploid and has 12 chromosomes ($2n = 12$). Sveshnikova has pointed out the similarity between these chromosomes and those of *Vicia narbonensis* and *Vicia serratifolia:* but others such as Senjaninova-Korczagina regard the differences between *Vicia faba* and *Vicia narbonensis* as too great to indicate that the former was derived from the latter; Hector suggests that *Vicia faba* derived from a special branch of the primary *Vicia narbonensis* (see Hector 1936, 651 f.). Bertsch and Bertsch (1949, 161) give a distribution map of the wild *Vicia narbonensis* showing it to be restricted to the Mediterranean region, including the whole of Spain, south-east Europe, Anatolia, and stretching eastwards between the Caspian Sea and the Persian Gulf.

In archaeological contexts *Vicia narbonensis* has been identified from the pre-pottery neolithic B levels at Beidha in Palestine (Helbaek 1966A, 63). The earliest cultivated horsebeans yet found are those reported from the pre-pottery neolithic B levels at Jericho (Hopf, in Ucko and Dimbleby 1969, 355 f.). Their cultivation spread during the neolithic period into Spain: they were identified at El Garcel, Campos near Murcia and Almizaraque, and Pepim in Portugal (Bertsch and Bertsch 1949, 163–4), and they also occur in Eastern Europe, at this time: they were found in the Aggtelek Cave and at Lengyel in Hungary (ibid., 164). Subsequently, they occur in the Aegean at Troy II, Early and Middle Helladic Lerna, at Knossos and Hagia Triada in Crete and late Mycenean Iolkos in Thessaly, and at many of the bronze age lake-side dwellings in Switzerland (Corcelettes and Concise on Lake Neuenburger: Petersinsel and Mörigen on the Bieler See and Alpenquai on Lake Zürich, for example), and are found in the north Italian terramare at this time (e.g. Castione, Parma and Ambrogio). (For detailed references to these finds see Bertsch and Bertsch 1949, 164; Hopf 1964; Renfrew 1966.) The bronze age contexts of the terramare finds has, however, been questioned by Helbaek (1956, 292), who claims that there is 'no good evidence for Horsebean in Italy until the Iron Age'. *Vicia faba* is reported also from the Hal Tarxien cemetery in Malta (Helbaek 1962, 176), c. 2100 B.C. The earliest finds in Britain date from the early iron age at Glastonbury, Meare and Worlebury Camp in Somerset (Jessen and Helbaek 1944, 59). Although *Vicia faba* had already reached Pinnacle, Jersey, in the Channel Islands during the bronze age (Helbaek 1962, 117, n. 1), it does not appear to have reached Britain until the iron age.

CRITERIA FOR THE PALAEOETHNOBOTANICAL IDENTIFICATION OF *Vicia faba*

The seeds occur in large, thick pods curved at the base, generally straight-sided terminating in a pointed and curved/hooked style rudiment. The pods are fleshy when young with a downy inner surface, at maturity they become tough and hard (Hector 1936, 651; Gill and Vear 1958, 177).

The seeds are very variable in size and shape ranging from almost spherical to flattened and broadly oval. Typically they are oblong in outline with an almost circular cross-section and have a radicle visible in the slight depression at one end (fig. 69 and

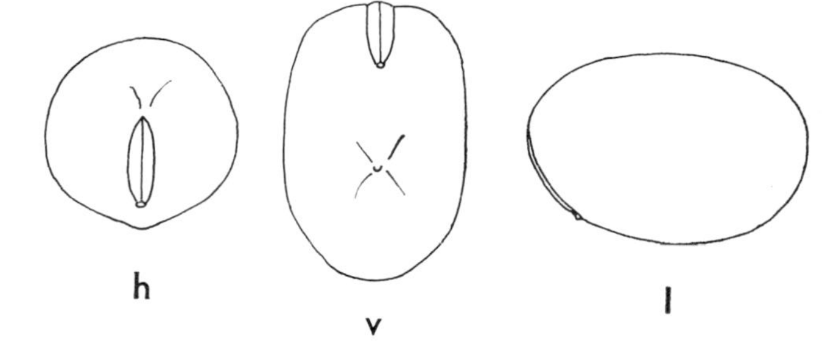

Fig. 69 Horsebean *Vicia faba* L. Seed in (h) hilum, (v) ventral, (l) lateral view ×3.

pl. 27). The hilum is 5–6 mm long and 1·5 mm wide in fresh seeds (Musil 1963, 92), and lies at one end of the short axis of the seed. In palaeoethnobotanical material the hilum is frequently missing, but the characteristic radicle is visible situated above the depressed end from which the hilum has fallen. (See Helbaek 1953B, 157; 1952B, 220; Hector 1936, 651.)

CONDITIONS REQUIRED FOR THE CULTIVATION OF *Vicia faba*

Climate

Beans do not have the same resistance to low temperatures as do some of the cereals. Some varieties are more tolerant than others but in general severe winters can cause almost total loss of the crop. Where the extremes of winter temperature are not too low the plants may recover sufficiently, with ample moisture, to produce a reasonable crop. Both winter- and spring-sown varieties exist. Spring beans tend to be smaller and more spherical in form (see Hunter 1951, 123 f.). In Britain beans do not grow as a field crop north of the Midlands because the climate is too severe. Winter-sown beans are harvested in July and August: spring-sown varieties are cut in September (Porter, in Paterson 1925, 275).

Measurements of Vicia faba *seeds from archaeological contexts*

Site	*Period*	*Length*		*Width*
		mm		mm
Jericho[1]	P.P. neolithic B	av. 5·3	—	—
Pepim[2]	Neolithic	—	(7·1–9·8)	—
Troy II[2+3]	Early bronze age	av. 5·6	(4·8–6·4)	4·4
Knossos[2]	,, ,, ,,	av. 5·4	(4·8–6·8)	4·2
Hal Tarxien[2]	,, ,, ,,	—	(3·48–7·95)	—
Federsee[3]	Bronze age	—	(6·1–10·2)	—
Cezavy Blucina[4]	,, ,,	6·5	—	5·2
Frienwalde[3]	Late bronze age	4·8	—	4·0
Dobenek[3]	,, ,, ,,	5·8	—	4·2
Tupá Skalka[4]	Hallstatt iron age	8·64	—	6·53
Sághegy[4]	,, ,, ,,	6·85	—	5·06
Stans bei Schwaz[4]	La Tène iron age	—	(6·9–10·0)	(3·9–6·6)
Glastonbury[2+5]	Early iron age	6·65	(4·7–7·87)	5·13
Pre-urban Rome[6]	,, ,, ,,	—	(6·4–10·25)	6·69–5·31
Aquileja[2]	Roman	—	(8·0–11·6)	—
Caerleon (Isca)[7]	,,	5·75		4·17
Valkenburg, Holland[8]	,,	5·5–5·6		4·5–5·5
Ouddorp, Holland[8]	,,	6·8		5·1

1. Hopf in Ucko and Dimbleby 1969, 355F.
2. Helbaek 1962A, 117.
3. Bertsch and Bertsch 1949, 162–3.
4. Tempír 1964, 89.
5. Helbaek 1952B, 220–1.
6. idem. 1956, 292.
7. idem 1964B, 161.
8. van Zeist 1970, 165.

Soil

They flourish best on stiff, strong clays, which are well drained. On light soils the yields are low, and on humus-rich soils they tend to run to leaf. They can be grown on light soils provided these have been adequately manured to bring the moisture-retaining capacity up to the required level. (See Hunter 1951, 123; Mortimer 1707, 106; Percival 1900, 423.)

POSSIBLE USES OF *Vicia faba* TO PREHISTORIC MAN

Vicia faba seeds have not been found in food residues in prehistoric Europe. Pliny describes how bean meal was used in making bread by the Romans: the bean meal, *lomentum*, was added to the wheat or Italian millet flour to increase the weight of loaves made for sale (*Natural History* XVIII, xxx). Beans were also used by the Romans for making porridge, in sacrifice to the gods, and were much esteemed as a delicacy. They were, however, thought to have a dulling effect on the senses, and to cause sleeplessness (ibid.). Bean flour was obtained by pounding the seeds in a mortar with a pestle. The Romans made a special purée of bean meal, mixed with fish stock, such herbs as cumin and coriander, and a little oil as described by Apicius (André 1961, 36).

Chapter 12

Field Peas

GENETIC AND ARCHAEOLOGICAL EVIDENCE FOR THE ORIGIN OF *Pisum sativum* var. *arvense*

The field pea has 14 chromosomes. Within the genus *Pisum* extensive hybridization has been carried out over a long period of time. Most of the crosses occurred between varieties of the four species once recognized, but now included as varieties of *Pisum sativum.* Two main varieties are acknowledged today, *Pisum sativum* in the strict sense: the cultivated garden pea, and *Pisum sativum* var. *arvense*: the field pea (see Gill and Vear 1958, 179–80, and Hector 1936, 647). The wild ancestor of *Pisum sativum* is not definitely known: Žukovskij (1950 (1962) 23) suggests that it may be *Pisum elatius* and that *Pisum arvense* (= *Pisum sativum* var. *arvense*) may be an intermediate form. Another possibility he suggests is that *Pisum sativum* arose from the hybridization of *Pisum elatius* and *Pisum arvense* followed by back crossing.

The earliest archaeological finds of peas are those from Jericho: pre-pottery neolithic B levels (Hopf, in Ucko and Dimbleby 1969, 355); Jarmo (Helbaek 1960A, 115), and Can Hasan (Renfrew 1968). These all appear to be *Pisum sativum* var. *arvense,* but at Hacilar in the aceramic neolithic levels *Pisum elatius,* the rough-seeded purple pea, was predominant (Helbaek 1964, 122), and it also occurred occasionally at Çatal Hüyük (ibid.). It was probably *Pisum sativum* var. *arvense* which was found by Dörpfeld in the second city of Troy (it is reported that visitors to the Troy excavations in 1893 'supped on pease from Priam's larder' and that 'one large jar alone contained more than 440 lbs of these pease' (Tsountas and Manatt 1897, 263).

In Europe peas occur in the earliest neolithic deposits in Greece – at Ghediki and Sesklo (Renfrew 1966, 86) and are found from Danubian I neolithic contexts at Heilbronn on the Neckar (Bertsch and Bertsch 1949, 168), also at Westeregeln (Natho and Rothmaler 1957), Eisenberg (ibid.), Zwenkau (ibid.), Tröbsdörf (ibid.), and the Aggtelek Cave (Neuweiler 1905, 116). They occur in some of the Swiss neolithic and bronze age lake-side villages, for example Sipplingen, Bodensee; Mondsee; Robenhausen, Pfaffikersee; Utoquai on Lake Zürich; Thun; Lüscherz on Bielersee (Bertsch and Bertsch 1949, 168). In the bronze age they occur in the villages around the Federsee, at Buchau, Dullenried and Ödenbühl (ibid., 169), as well as in the lake villages of Alpenquai, Sumpf near Zug, Mörigen, Petersinsel and Concise to name a few of the finds from Switzerland (ibid., 169). In the bronze age site of Barca, Czechoslovakia, Tempír (1964, 90) found two types of peas, and Werneck records the same phenomenon from the La Tène site of Stans, Austria (ibid., 90). There are many finds from the iron age in Germany but the cultivation of peas does not appear to have reached Holland or Britain in prehistoric times.

CRITERIA FOR THE PALAEOETHNOBOTANICAL IDENTIFICATION OF *Pisum sativum* var. *arvense*

The pods are about 7·5 cm long and contain four to nine seeds (Martin and Leonard 1967, 706). The seeds are usually spherical and measure about 6 mm in diameter when fresh. The hilum is ovate, 1–2 mm long, and flush with the surface of the seed; the chalaza is about 2 mm from the hilum in fresh seeds (see fig. 70 and pl. 28; Musil 1963, 84). In the palaeoethnobotanical material from Barca and Stans, Tempír and Werneck have noticed two sizes of seeds – both hemispherical in outline (Tempír 1964, 90).

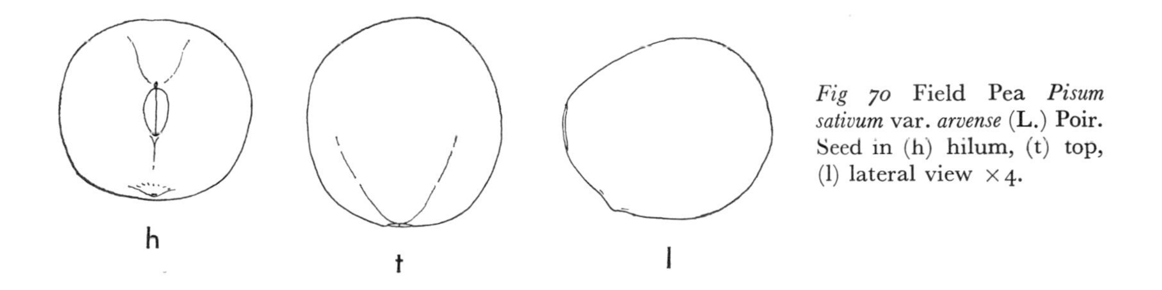

Fig 70 Field Pea *Pisum sativum* var. *arvense* (L.) Poir. Seed in (h) hilum, (t) top, (l) lateral view ×4.

CONDITIONS REQUIRED FOR THE CULTIVATION OF *Pisum sativum* var. *arvense*

Climate

Field peas require a cool growing season; winter-sown peas thrive only in mild climates away from severe frosts, spring-sown peas are more seriously injured by high temperatures and lack of moisture during the ripening season than by slight frosts which are only damaging when the plants are in flower. They flourish best with a fairly abundant rainfall (Martin and Leonard 1967, 706). Pliny (*Natural History* XVIII, xxxi) says that peas must be sown in sunny places, as they stand cold very badly. They thrive best in a relatively humid climate with temperatures of 55° to 65°F (Purseglove 1968, 312).

Soils

Peas give their best yields on medium or slightly improverished soils such as well-drained clay loams of limestone origin. On rich or peaty soils the plants tend to run to

Measurement of peas from archaeological contexts

Site	*Period*	*Length*	*Breadth*	*Thickness*
		mm	mm	mm
Jericho[1]	P.P. neolithic B	3·0–5·5	—	—
Merimde, Egypt[2]	Neolithic	3·0–4·5	—	—
Ghediki[6]	Earliest neolithic	3·5–4·1	3·4–3·8	3·0–3·5
Sesklo[6]	,, ,,	3·7–4·2	2·9–3·5	2·9–3·8
Soufli[6]	Early neolithic	2·8–4·0	3·1–4·1	3·0–4·0
Heilbronn[3]	Neolithic	3·0–2·5	—	—
Westereglen[3]	,,	2·8–4·2	—	—
Zwenkau[3]	,,	2·8–3·8	—	—
Aggtelek[3]	,,	2·5	—	—
Can Hasan[4]	,,	3·9–5·0 (4·4)	—	—
Heilbronn[3]	Late neolithic	3·2–4·8	—	—
Troy II[2]	Early bronze age	3·7–4·9	—	—
Barca[5]	Bronze age	3·9–4·4	4·1–4·3	4·1–4·2
Barca[a5]	,, ,,	2·5–3·3	3·9–4·5	3·8–4·1
Stans[5]	Iron age	3·8–4·8	3·4–4·6	3·8–4·9
Stans[a5]	,, ,,	3·3–4·0	4·1–4·6	4·1–4·6

Note: a = angular seeds.

1. Hopf, in Dimbleby and Ucko 1969, 355 f.
2. Bertsch and Bertsch 1949, 166–7.
3. Natho and Rothmaler 1957, 73 f.
4. Renfrew 1968
5. Tempír 1964, 90.
6. Renfrew 1966, 21 f.

leaf. Like beans, peas require alkaline soil conditions. (See Hunter 1951, 127; Percival 1900 (1947), 421.)

USES OF *Pisum sativum* var. *arvense* TO PREHISTORIC MAN

As in the case of beans, we have no evidence from prehistoric Europe for the way in which peas were prepared for food. Presumably they were eaten in a kind of soup similar to that which the Romans made from beans, but we have no evidence as to how even the Romans used them (André 1963, 37).

Their properties of maintaining and restoring nitrogen to the soil were probably also appreciated.

Chapter 13

Lentils

GENETIC AND ARCHAEOLOGICAL EVIDENCE FOR THE ORIGINS OF *Lens esculenta*

The lentil is diploid, having 14 chromosomes (Hector 1936, 648). *Lens esculenta* is thought to have derived from the wild *Lens nigricans* (Bertsch and Bertsch 1949, 170), which is native to south-east Europe and Western Asia.

Lentils are widely found on the early farming sites in the Near East. They occur at Jericho (Hopf, in Ucko and Dimbleby 1969, 355 f.), Jarmo (Helbaek 1960A, 115), Tepe Sabz (Hole and Flannery 1967, 147 f.), Ali Kosh (Helbaek 1966E, 122), aceramic and later neolithic Hacilar (ibid.) in the Near East. They are found in Europe from the earliest neolithic period, occurring at Argissa (Hopf 1962, 104 f.), Ghediki (Renfrew 1966, 22 f.) in Greece, and in the early neolithic levels at Tell Azmak, Bulgaria (Renfrew in Ucko and Dimbleby 1969, 166). They were also found in Danubian I settlements at Heilbronn and Böckingen near Heilbronn (Bertsch and Bertsch 1949, 172), at the Aggtelek Cave in Hungary (ibid.), at Lengyel (Helbaek 1962A, 179), Butmir and Ripać in Bosnia (Neuweiler 1905, 116), at lake-side sites on the Federsee and Waldsee in southern Germany (Bertsch and Bertsch 1949, 172). Lentils were popular with the bronze age lake-side villagers in Switzerland, being found at Alpenquai, Wollishofen, Sumpf near Zug, Riesi near Seengen, Petersinsel, Mörigen, Lüscherz, St Blaise and Le Bourget in Savoy (ibid., 173), and were widely cultivated in the iron age in Germany and Hungary. However, they are not present in Italy before the iron age and first occur in Holland and Britain in the Roman period (Helbaek 1962A, 179, 1964B, 161).

CRITERIA FOR THE PALAEOETHNOBOTANICAL IDENTIFICATION OF *Lens esculenta*

Lens esculenta Moench is the only cultivated species of lentil. It has been divided by Barulina into two sub-species:

(a) *Macrospermae* (Baumg.) Barul. The large-seeded form with large, flat pods and seeds ranging from 6 to 9 mm in diameter;

(b) *Microspermae* Barul. which has small to medium-sized convex pods, and small seeds 3–6 mm in diameter (Hector 1936, 648). Most prehistoric finds belong to the latter group.

The seeds themselves are circular and flattened with a convex cross-section, and a small hilum on the rather sharp margin (see fig. 71). Usually two seeds ripen per pod (Helbaek 1961, 82). The hilum is acutely lanceolate (idem 1964B, 161). Often in palaeoethnobotanical material the seed coat has been rubbed off revealing the two cotyledons (pl. 29).

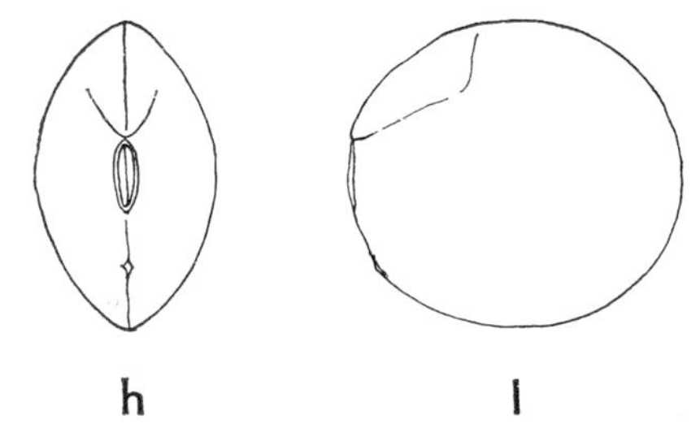

Fig. 71 Lentil *Lens esculenta* Seed in (h) hilum, (l) lateral view × 5.

In the table below, only the finds from Late Assyrian Nimrud and from Saqqarah (fourth century A.D.) may belong to *Lens esculenta* var. *macrosperma*. The rest must belong to the small-seeded variety or to the wild *Lens nigricans*. It should be remembered that the lentil seeds shrink in diameter on carbonization, even though much less consistently than cereal grains (Helbaek 1966E, 122).

CONDITIONS REQUIRED FOR THE CULTIVATION OF *Lens esculenta*

Climate

They require a warm climate, and may be spring-sown to avoid night frosts, to which they are particularly susceptible. Winter-sown varieties do also occur (Martin and Leonard 1967, 682–3).

Soils

Lentils prefer a light, warm, sandy soil – on rich soil they run to leaf and produce few pods. 'The lentil likes a thin soil better than a rich one, and in any case a dry climate' (Pliny, *Natural History* XVIII, xxxi).

POSSIBLE USES OF LENTILS TO PREHISTORIC MAN

As with the other pulses lentils have so far not been found in food residues from prehistoric Europe. A paste of cooked lentils was found in one of the Twelfth Dynasty tombs at Thebes (Brothwell and Brothwell 1969, 105).

In Roman times they were made into soups: prepared first by roasting the seeds, then pounding them with bran in a mortar (André 1963, 39). They were no doubt valued for their high protein content. Lentils were cooked with chestnuts or with mussels, or were added to barley broth together with peas and chickpeas (Brothwell and Brothwell 1969, 107).

Measurements of lentils from archaeological contexts

Site		*Period*	*Diameter*	*Thickness*
			mm	mm
Hacilar[1]		Aceramic neolithic	2·64 (2·08–3·08)	1·33 (1·00–1·75)
Ghediki[10]		Earliest neolithic	3·2–4·1	—
Argissa[11]		” ”	av. 2·48–2·66	—
Soufli[10]		Early neolithic	2·2–3·0	1·5
Tell Azmak[8]		” ”	2·4–2·7	—
Yassatepe I[8]		Middle neolithic	2·5–3·0 (av. 2·9)	—
Heilbronn[2]		Neolithic	2·0–2·5	—
Eisenberg[2]		”	2·9 (2·4–3·2)	1·7 (1·6–2·0)
Aggtelek[2]		”	2·0–2·5	—
Butmir[3]		Late neolithic	3·0–3·5	—
Gornje Polje-Obre[8]		” ”		
Troy II[4]		Early bronze age	2·4	—
Lachish[4]		” ” ”	2·56–4·76	—
Tell Chragh[9]		” ” ”	2·93	—
Barca[3]		Bronze age	3·2–4·0	1·8–2·4
Petersinsel[7]		” ”	3–4	
Mistalbach[3]		” ”	2·5–3·5	1·5–1·9
Mörigen[7]		” ”	3	—
Nimrud[1]		Late Assyrian	4·94	—
Beycesultan[4]		Bronze age	3·26 (1·67–4·00)	—
Apliki[1]		” ”	3·42 (2·00–4·66)	1·98 (1·16–3·00)
Kalopsidha[5]	I	” ”	4·42 (3·00–5·17)	2·50–2·83
Cerveny Hradek[3]		Early iron age	2·8–3·4	1·6
Isca[6]		Roman	2·50–3·84	—
Saqqarah[8]		4th century A.D.	5·3 (5·0–5·9)	2·6 (2·3–3·0)

I = impression: all other material is carbonized.

1. Helbaek 1970, 224.
2. Natho and Rothmaler 1957.
3. Tempír 1964, 90.
4. Helbaek 1961, 82.
5. idem. 1966E, 122.
6. idem 1964B, 161.
7. Bertsch and Bertsch 1949, 171.
8. Renfrew, unpublished.
9. Helbaek 1960E, 79.
10. Renfrew 1966, 21 f.
11. Hopf 1962, 110.

Chapter 14

Miscellaneous Legumes

Bitter Vetch: *Vicia ervilia* Wild

GENETIC AND ARCHAEOLOGICAL EVIDENCE FOR THE ORIGIN OF *Vicia ervilia*

Vicia ervilia is a diploid species of vetch with 14 chromosomes (Hector 1936, 655). At the present day it is cultivated for fodder in north-west Africa, south-east Europe, Asia Minor and eastwards via Iran to Afghanistan and India (Helbaek 1961, 81).

Finds on early village sites in the Near East suggest that it was cultivated in this region for human food from early in the neolithic period. The earliest find so far known is that from Çatal Hüyük (idem 1964A, 122); it occurs at Tell Qurtass in the Shahrzoor valley in Iraq at the end of the third millennium B.C. (idem 1960E, 80), at late bronze age Beycesultan in Anatolia (idem 1961, 81), at Troy II (ibid.), and Nimrud (idem 1966D, 615). In south-east Europe bitter vetch has been found in large quantities on sites of the Gumelnitsa culture in Bulgaria – at Tell Azmak, Karanovo VI, Kapitan Dimitrievo III and in the contemporary levels at Sitagroi in northern Greece (see pl. 30; Renfrew, unpublished). Elsewhere it is seldom found in Europe: Tempír (1964, 91) reports it from the bronze age tell at Toszeg, and from Cervený Hrádek (ibid.), of the Hallstatt iron age.

CRITERIA FOR THE IDENTIFICATION OF *Vicia ervilia*

The seeds are angular – rounded with a triangular plane at one end where the radicle stretches towards the small oval hilum. Shallow depressions are sometimes conspicuous

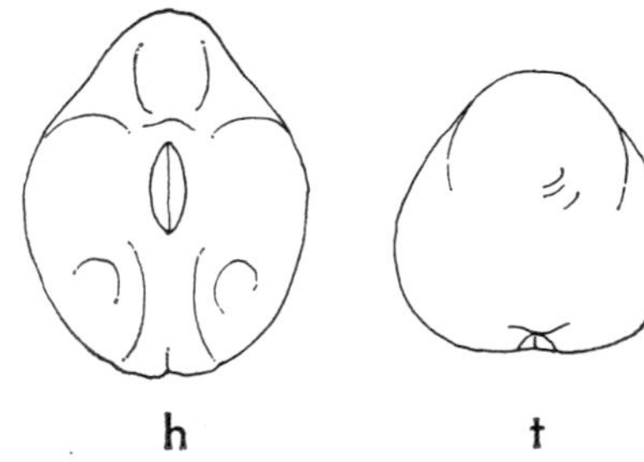

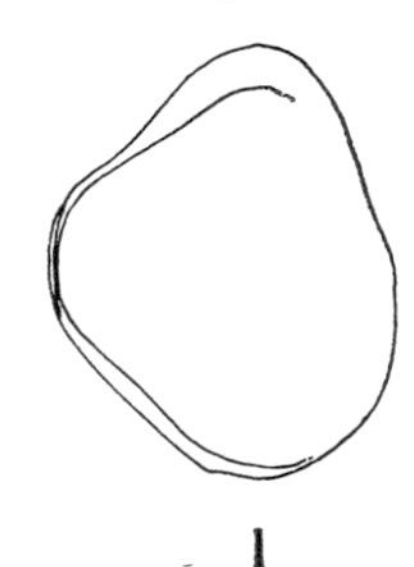

h t l

Fig 72. Bitter Vetch *Vicia ervilia* Wild. Seed in (h) hilum, (t) 'top', (l) lateral view ×5.

around the hilum and elsewhere on the seed (see fig. 72 and pl. 30). They are contained in small pods seldom exceeding an inch in length and strongly constricted: each pod contains three to four seeds (Hector 1936, 655).

Measurements of examples of Vicia ervilia *from archaeological contexts*

Site	*Diameter*
	mm
Tell Qurtass[1]	(2·93–4·03) 3·48
Beycesultan[2]	(2·00–3·26) 2·86
Toszeg[3]	3·00–3·9
Cerveny Hrádek[3]	2·8–3·1
Yassatepe I[4]	2·1–2·7
Karanovo VI[4]	2·9

1. Helbaek 1960E, 80.
2. Idem 1961, 81.
3. Tempír 1964, 91
4. Renfrew, unpublished.

CONDITIONS REQUIRED FOR CULTIVATING *Vicia ervilia*

In general, vetches flourish best in cool temperate conditions, but in mild climates they may be grown as a winter crop. They grow best on fertile loam soils, and they are only moderately sensitive to soil acidity (Martin and Leonard 1967, 710).

Grass Pea: *Lathyrus sativus* L.

Seeds of the grass pea occur quite widely on archaeological sites: Helbaek reports its occurrence at Jarmo in the early seventh millennium B.C. (Helbaek 1965B, 32). It is also found on early neolithic sites in Europe, for example in the Aggtelek Cave, Hungary (Neuweiler 1905, 116), and in Switzerland (Helbaek 1956, 292); it occurs at Troy II (ibid., 292), at Lachish in Palestine (idem. 1958B), and in Italy in the late iron age of pre-urban Rome (idem 1956, 292). It occurs also in Iraq at Tell Bazmosian in the Ur III period (2100–1800 B.C.) (idem 1965B, 32).

The pods of the grass pea are 2·5–4 cm long, broad, flat and glabrous (Hector 1936,

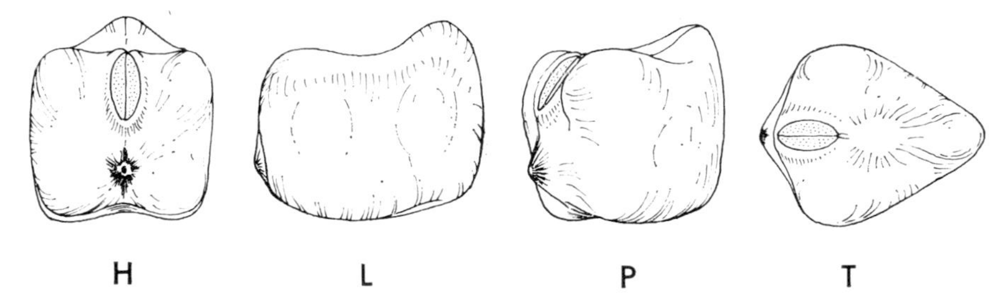

Fig. 73 Grass Pea *Lathyrus sativus* L. Seed in (H) hilum, (L) lateral, (T) top view; also in (P) perspective view showing characteristic appearance when lying on one of the two main lateral facets, with prominent sloping hilum and chalaza beneath it, × 5.

657). They contain two to five seeds which vary greatly in size and shape. The seeds in the middle of the pod are truncated at both ends: those at the ends have one flattened end – towards the middle of the pod – the other end being rounded or slightly pointed in profile (see fig. 73 and pl. 33). All the seeds are approximately triangular in cross-section. The intermediate seeds have a greater breadth than length: the terminal seeds are often longer than they are broad: for example:

Intermediate seeds		*Terminal seeds*	
Length	*Breadth*	*Length*	*Breadth*
mm	mm	mm	mm
2·66	3·16	2·50	2·58
4·00	4·66	5·25	4·41
4·66	5·09	5·92	4·16

(see Helbaek 1965B, 33)

The small oval hilum (0·99 × 0·57 mm in the examples from pre-urban Rome), is placed at one corner of the thick end (Helbaek 1956, 292).

The following measurements for carbonized *Lathyrus sativus* seeds are recorded:

Site	*Length*	*Breadth*	*Thickness*
	mm	mm	mm
Pre-urban Rome	4·54 (3·66–5·31)	3·99 (2·75–4·58)	3·66 (3·11–4·39)
Lachish	5·5	4·0	—
Salamis, Cyprus	4·7 (4·0–5·5)	3·3 (3·0–3·8)	3·9 (3·5–4·9)

At present the grass pea is cultivated in a broad area from south-west Europe to Central Asia; and from Abyssinia to Central Europe, both for human food and animal fodder (Helbaek 1965B, 33). In early deposits it often occurs as a stray admixture, or cultivated in another leguminous crop: in Switzerland mixed with *Pisum sativum* var. *arvense*: in Rome with *Vicia faba* (ibid.). In the Bazmosian find it occurs as a crop in its own right (ibid.). Presumably it was utilized in the same way as the other pulses – in soups and as vegetables supplementing the protein supply in the diet. The seeds contain an alkaloid, and if not carefully boiled before eating lead to a disease known as lathyrism, characterized by a paralysis of the lower limbs (Cobley 1963, 151–3).

Chickpea, Gram *Cicer arietinum* L.

The nearest wild species of the genus is *Cicer judaicum* but Žukovskij suggests that *Cicer arietinum* once existed in the wild state: it is found as escapes in Mesopotamia and Palestine. The form *Cicer arietinum* is diploid: $2n = 14$ (Žukovskij 1950 (1962), 26).

Chickpeas have only occasionally been found on prehistoric sites in Europe and the Near East. They have been found in pre-pottery neolithic B levels, and, more abundantly, in the early bronze age deposits at Jericho (Hopf, in Ucko and Dimbleby 1969, 355 f.). They occur in the bronze age and iron age levels at Lachish in Palestine (Helbaek 1958B), Late Assyrian Nimrud (idem 1966D, 615) and in the Isin Larsan and Horian levels (2000–1500 B.C.) at Tell Bazmosian in Iraqi-Kurdistan (idem 1965B, 28 f.). They have also been found on the sixth–fifth-century B.C. funeral pyres at Salamis, Cyprus (Renfrew, in 1971).

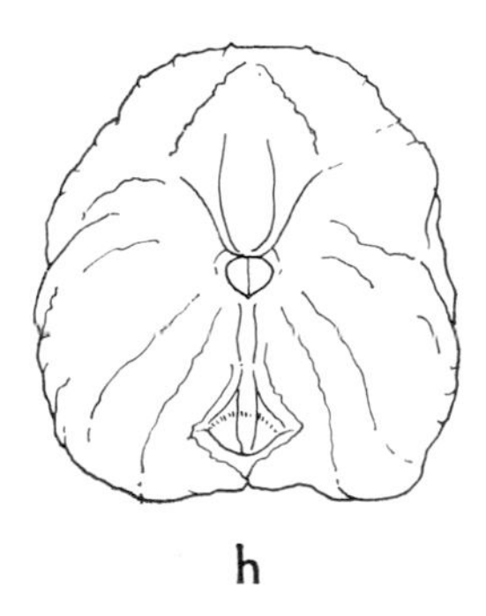

Fig. 74 Chickpea *Cicer arietinum* L. Seed in (h) hilum, (l) lateral view ×4.

Measurements of ancient carbonized chickpeas

Site	*Diameter*
	mm
Lachish	3·5–4·5
Bazmosian	3·66–4·60
Assyrian Nimrud	3·67–5·44
Salamis	3·0–4·5

The seeds of *Cicer arietinum* are contained in a rectangular, swollen pod 1·9 cm long and 1·3 cm wide, with an obliquely situated beak (Hector 1936, 780). Each pod contains one or two squat, angular seeds pointed at one end, and said to resemble a ram's head. The blunt, oblique point formed by the germ is often broken off in carbonized material (see fig. 74 and pl. 32) (Helbaek 1965B, 32).

The chickpea is best adapted to warm, semi-arid conditions (Martin and Leonard 1967, 680). It is very drought-resistant and requires a cool, dry climate and light well-aerated soils. It cannot tolerate heavy rains, and grows best on heavy clay soils with a rough seed bed (Purseglove 1968, 247).

The seeds may be eaten roasted and salted, or ground and made into a porridge or soup. The young plant is also used as a pot-herb. In India a medicinal vinegar is made from the plant (Helbaek 1965B, 33). The chickpea was roasted and ground, to add to soups by the Romans (André 1963, 38), and no doubt was similarly utilized in prehistoric times.

Chapter 15

Flax

GENETIC AND ARCHAEOLOGICAL EVIDENCE FOR THE ORIGINS OF CULTIVATED FLAX

The origins of cultivated flax have been much discussed. Heer suggested in 1865 that *Linum bienne* (then known as *Linum angustifolium*) might be the progenitor of cultivated flax. Neuweiler (1905) put forward a case for *Linum austriachum*, since this species is found distributed in the Alpine region and has seeds which are morphologically similar to *Linum usitatissimum.* Helbaek in his comprehensive discussion of the problem in 1959 supports the case for *Linum bienne* as the ancestral form for the following reasons. It occurs wild in the area in which we know from archaeological evidence that agriculture first became established; the seeds of the ancient cultivated plants are all within the same dimensional order in ecologically comparable zones from south-west Iran to south Britain; there is morphological similarity between the seeds of this species and the palaeoethnobotanical material; both *Linum usitatissimum* and *Linum bienne* have the same number of chromosomes – 30, and *Linum bienne* is the only wild species to produce fertile hybrids with *Linum usitatissimum* (Helbaek 1960, 115–16).

The earliest finds of *Linum usitatissimum* are from Tepe Sabz in south-west Iran, and Tell es-Sawwan on the Middle Tigris in Iraq (idem 1965A), and subsequently it occurs in the Halafian levels at Arpachiya and at Tell Brak before 4500 B.C. It is found at Ur, Hama, Khafajah and Nimrud in the Near East in later periods (idem 1960B, 193).

In Europe it occurs on neolithic sites in Switzerland, Germany and north Italy, for example in the Danubian I settlement at Heilbronn on the River Neckar (Bertsch and Bertsch 1949, 208), at Egozwil in Switzerland (Helbaek 1960A, 115), at Riedschachen and Moordorf Taubried on the Federsee (Bertsch and Bertsch 1949, 208), and at a number of sites on the Bodensee: Sipplingen, Nussdorf, Unteruhldingen, Bodman and Steckborn (ibid.), for example. Flax also reached Holland in neolithic times (two impressions were found at Drouwen – Helbaek 1960B, 193) and across the Channel it occurs in Britain at Windmill Hill in our earliest neolithic culture (idem 1952B, 205).

Slightly later flax reached northern Italy (Lagozza, Parma, Emilia (Bertsch and Bertsch 1949, 209)), and also spread to Spain (El Argar) (ibid.).

It does not appear in Poland, eastern Germany or Scandinavia until the first millennium B.C., and Helbaek (1960A, 116) argues, on account of the larger size of the seeds in this region, that flax may have been cultivated early in Russia and that this large-seed race had spread into Eastern Europe from Russia during the Hallstatt early iron age.

CRITERIA FOR THE PALAEOETHNOBOTANICAL IDENTIFICATION OF *Linum usitatissimum* L.

The mature fruit of flax is a rounded capsule about 6 mm in diameter and surrounded by the persistent sepals. It contains ten locules, each of which contains a single seed (Hector 1936, 801–2). The locules are arranged in pairs, the five chambers being almost completely divided in half by false septa growing in from the midrib of each carpel and almost meeting the placenta between the ovules (see fig. 75). In most cultivated varieties the capsule is almost completely indehiscent, and the seeds are released only by threshing or irregular shattering (Gill and Vear 1958, 142). The seed is oval, 4–6 mm long by 2–3 mm broad, when fresh. It is lenticular in form and somewhat beaked at one end (see fig. 76 and pl. 33) (Hector 1936, 802).

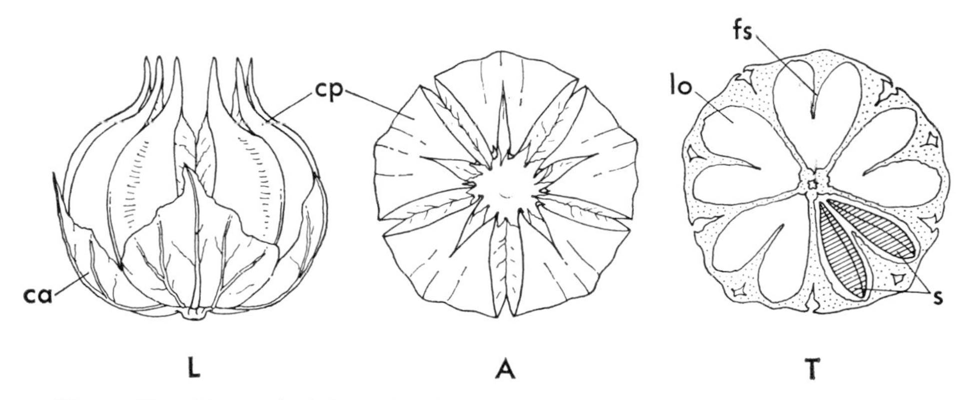

Fig. 75 Flax *Linum usitatissimum* L. Ripe capsule in (L) lateral, (A) axial view and (T) diagrammatic transverse section ×5. (ca) calyx, (cp) carpel, (fs) false septum, (lo) two-chambered loculus, (s) seed.

The seed consists of an outer seed coat or testa, consisting of five layers, four of which may be removed when the seed is ground. The epidermis consists of a single layer of rectangular cells with their long axes at right angles to the surface. Viewed from the surface of the seed they appear to be polygonal in outline with a finely granular cuticle (ibid.). Below this are two layers of round cells, then a single layer of longitudinally elongated fibres with thick pitted walls. Below the fibre layer and running at right angles to it are a number of layers of thin-walled elongated cells. The fifth distinctive layer of the testa is composed of square or polygonal cells which are rectangular

Measurements of ancient flax seeds

		Length	
Neolithic period		mm	
Tell-es-Sawwan	C	4·0	(Helbaek, 1965A, 47)
Arpachiyah	I	3·84	(„ 1960B, 93)
Tell Brak	I	4·03	(„ „ „)
Ur	I	4·39	(„ „ „)
Hama	I	4·76	(„ „ „)
Heilbronn	C	3·8	(Bertsch and Bertsch 1949, 203).
Egolzwil	(Wet)	3·84	(Helbaek 1960B, 193)
Drouwen	I	3·84	(„ „ „)
Windmill Hill	I	4·03	(„ 1952B, 203)
Lagozza	(Wet)	3·90	(„ 1960B, 193)[1]
2000–1000 B.C.			
Handley Down	I	3·66	(Helbaek 1952B, 205)
Westwood, Edinburgh	I	3·29	(„ „ „)
Winterbourne Stoke	I	3·48	(„ „ „)
Agfarell, Dublin	I	3·5	(„ „ „)
El Argar, Spain		4·0	(„ 1959 „)
Khafajah B	(Fossil)	4·76	(„ 1960B, 193)[2]
1000–500 B.C.			
Nimrud	C	5·3	(Helbaek 1966D, 616)
Biskupin	?	3·9	(„ 1959)
Bornholm	I	4·39	(„ „)
500 B.C.–A.D. *1000*			
Meare West	I	4·10	(Helbaek 1952B, 205)
Alrum, Jutland	C	3·84	(„ 1959)
„ „	I	4·03	(„ „)
Østerbølle, Jutland	C	4·40	(„ 1939, 222)
„ „	I	4·58	(„ 1959)
Tollund Man	(Wet)	3·98	(„ 1950, 335)
Vallhagar	I	4·03	(„ 1955B, 674)

Notes: C = Carbonized seeds. I = seed impressions.

Measurements given are maximum, not average, in each case.

1 These seeds from Lagozza came from inside a pot, below some that had been carbonized. They were preserved by the damp conditions, but have now dried out. The actual measurement of the largest seeds is 3·34 mm, but Helbaek following Neuweiler calculates that the original size was one-sixth greater than the present length, viz. 3·90 mm.

2. The seeds from Khafajah B are preserved as a 'chalk replica' or fossil embedded in the clay floor of an old Babylonian house.

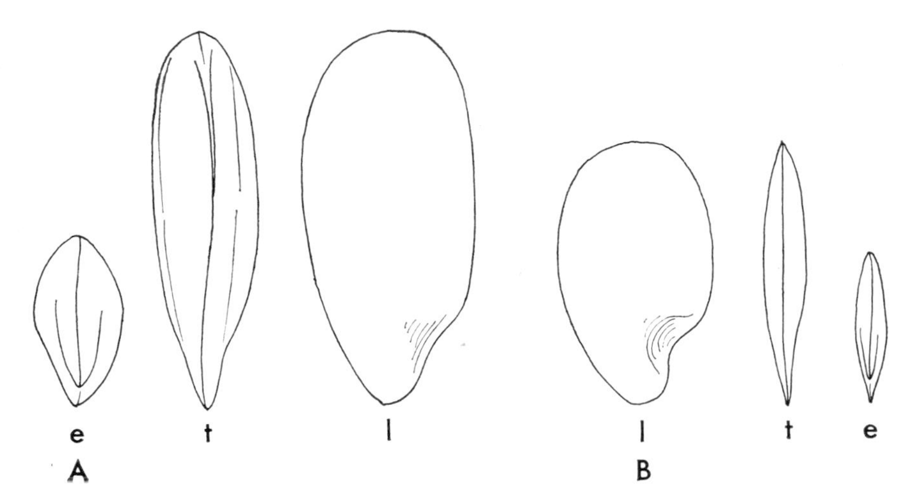

Fig. 76 Flax *Linum usitatissimum* L. (A) seed in (e) end, (t) top, (l) lateral view compared with seed of *L. bienne* L. in similar views (B) (e, t, l), × 10.

in cross-section. The walls are finely pitted and the cavity is filled with pigment (ibid.). Below the testa lies the endosperm consisting of up to six layers of variable thick-walled cells (ibid., 803).

The question of the size of the flax seeds (linseed) is open to various interpretations. In modern agricultural terms the flax grown for its linen fibres has small seeds (1,000 seeds weighing 4 gm – Hunter 1951, 131; illustration in Gill and Vear 1958, 145, fig. 38), whereas that cultivated for its oily, protein-rich linseed has much larger seeds (1,000 seeds weighing 10 gm – Hunter 1951, 131). The greater size of flax cultivated for linseed is thus the result of deliberate selection for the large-seeded forms.

However, another construction can, and has been, put on the appearance of large-seeded forms of flax in certain environments: that is that they reflect artificial irrigation in dry regions. Helbaek claims (1960B, 193) that those seeds ranging in size from 4·39 mm at Ur to 5·3 mm at Assyrian Nimrud demonstrate 'the signal increase in output of linseed oil – and presumably of flax fibre also – which was brought about by the change from the rainwatered uplands to the irrigated plain'.

A few capsules survive more or less entire from prehistoric and early historic contexts, and their measurements may be of interest:

Site		*Maximum diameter*	
		mm	
Lagozza	C	5·49 (5·13–5·86)	23 capsules (Helbaek 1959)
Østerbølle	C	5·50	1 capsule (,, 1938, 223)
Alrum	C	4·16 (4·03–5·12)	5 capsules (,, 1959)

CONDITIONS REQUIRED FOR THE CULTIVATION OF *Linum usitatissimum*

Climate

Flax is a plant of rapid growth requiring ample supplies of moisture during its early stages of development (Hunter 1951, 134). Seed flax is generally grown where the average annual precipitation ranges from 45 to 75 cm, but it may also be cultivated under irrigation in drier climates (Martin and Leonard 1967, 748). Drought and high temperatures of 90°F and over, during the flowering and ripening stages, reduces the yield, size and oil content of the seed, as well as the quality of the oil (ibid.).

Flax may be damaged or killed by temperatures of 18–26°F at the seedling stage, or by a light frost, say 30°F, during the flowering period, but in between these stages the plants may survive temperatures of 15°F or even lower.

Soil

Flax is best suited to fertile, deep, well-drained loams, especially silt loams, clay loams and silty clays. Light soils are unsuited to seed flax, particularly in areas of deficient rainfall. Since the root system is so shallow, the plants rely on the water supplied in the top two feet of soil (Martin and Leonard 1967, 749). It does best when sown after a cleaning crop, as flax is unable to compete satisfactorily with weeds (Hunter 1951, 134).

USES OF FLAX TO PREHISTORIC MAN

Linum usitatissimum is cultivated today for two qualities: the oily linseeds containing 35–40% oil and about 20% protein (Gill and Vear 1958, 143), used for human and animal food, and for its stem fibres from which linen is manufactured. Both these important qualities were appreciated by prehistoric man. We, however, are concerned at present with the seeds used for food.

It is interesting to note in passing that linseed contains cyanogenetic glycoside and if the seeds are allowed to soak in cold water enzymes may break down the glycoside, resulting in the production of poisonous hydrogen cyanide (prussic acid). It is thus important that linseed gruel should be prepared with boiling water (ibid.).

Linum usitatissimum L. was found in substantial quantities in the gruel consumed by both Tollund and Grauballe men in Jutland (Helbaek 1950, 335; 1958, 85), and a piece of linseed cake was recovered from Robenhausen (Heer 1866, 353 and pl. LXXXVIII, 76). Otherwise flax seeds do not appear to have been found in prehistoric food residues.

The Romans used flax to make porridge: they pounded the seed in a mortar and mixed it with barley grains and coriander seed, using these quantities for 9 kg of barley: 1·4 kg of flax seed to 0·2 kg of coriander and two tablespoons of salt. All these ingredients had been previously roasted (Pliny, *Natural History* XVIII, xiv). Columella regarded flax as being among the most agreeable of food plants (André 1961, 40); and Pliny remarks on the extremely sweet taste of the rustic porridge made from linseed in northern Italy (*Natural History* XIX, iii).

Chapter 16

Cultivated and Wild Fruits

Grape Vine: *Vitis vinifera* L.

GENETIC AND ARCHAEOLOGICAL EVIDENCE FOR THE ORIGINS OF *Vitis vinifera* L.

The origin of the cultivated vine is not certain, but it seems clear that it has some relationship with *Vitis silvestris* Gmel., the wild vine found widely throughout Eurasia (see distribution map, fig. 77). In *Vitis silvestris* the plants are monosexual – the female flowers only, ripening into grapes. In *Vitis vinifera* the flowers are hermaphrodite. Occasionally cultivated vines escape into the wild and revert back to *Vitis silvestris* in all characteristics with the exception of remaining hermaphrodite: the seeds of these escaped plants are much closer to those of the cultivated form (see below) (cf. Žukovskij, trans. Hudson 1962, 43–4; Logothetis 1962, 33 f.).

Vitis vinifera is diploid ($2n = 38$) (Žukovskij, trans. Hudson 1962, 43). Žukovskij (ibid.) gives an account of Baranov's theory that its progenitor was a bushy form inhabiting open sunny habitats. With the onset of more humid conditions in the Eocene, the forests became more extensive and the vine adopted a creeping habit – searching for sunny positions. The terminal inflorescences became sterile – those which evolved into tendrils survived. Deliberate selection for type of flower led to the development of modern vines with hermaphrodite flowers. Since originally adapted to sunny habitats it was quite tolerant of its removal from the forests by man (ibid.).

Vitis silvestris grows in northern Greece chiefly on *Cercis siliquaestrum*, *Laurus nobilis*, *Arbutus unedo* and *Arbutus andrachnae*, *Olea oleaster*, *Platanus orientalis*, *Quercus coccifera*, *Pistacia terebinthus*, *Clematis vitalba*, *Ulmus campestris*, *Cornus mas* and *Cornus sanguinea*, *Corylus avellana*, *Rubus* sp., *Fagus silvatica*, *Pyrus amygdaliformis*, *Rosa* sp., *Castanea sativa* (Logothetis 1962, 20; distribution map, fig. 78). From the distribution of *Vitis* in Pleistocene deposits in Europe Troels-Smith has suggested that it requires a temperature of at least 16–17°C in the warmest summer month (Turner 1968, 334). *Vitis silvestris* grows at heights up to 400 m above sea level – on occasion being found as high as 800 m above sea level (Savulesai 1958, VI, 300 f.). *Vitis vinifera* has a more restricted distribution than *Vitis silvestris* (ibid.).

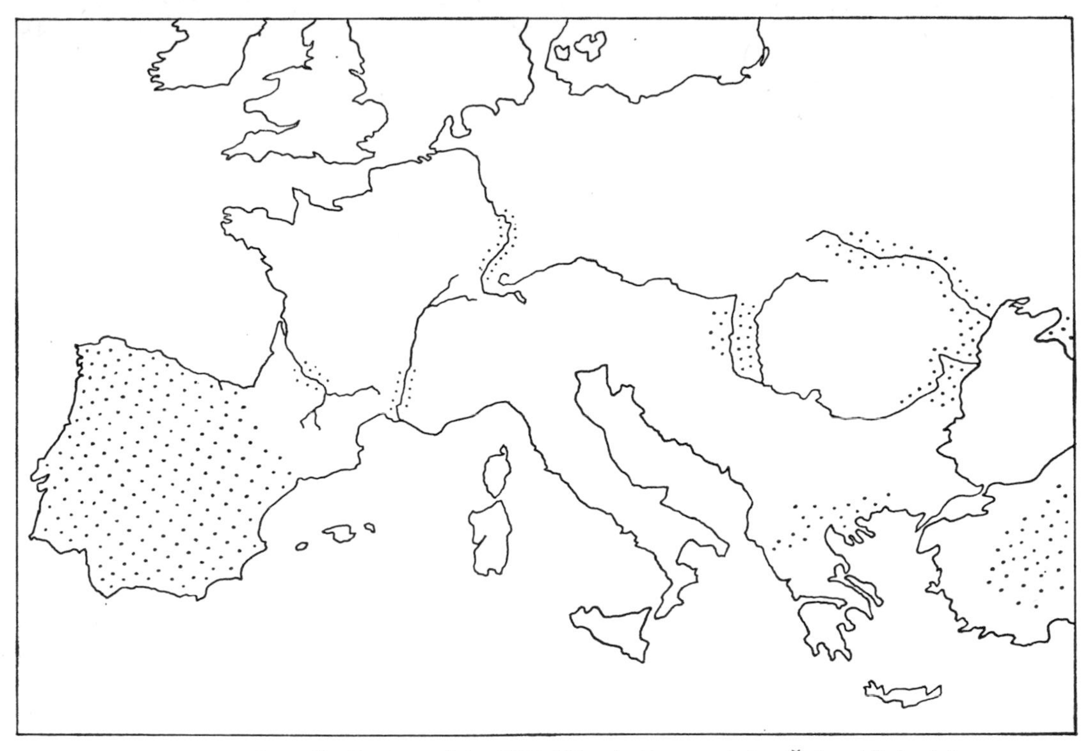

Fig. 77 Modern distribution of the Wild Vine in Europe (after Žukovskij (1962).

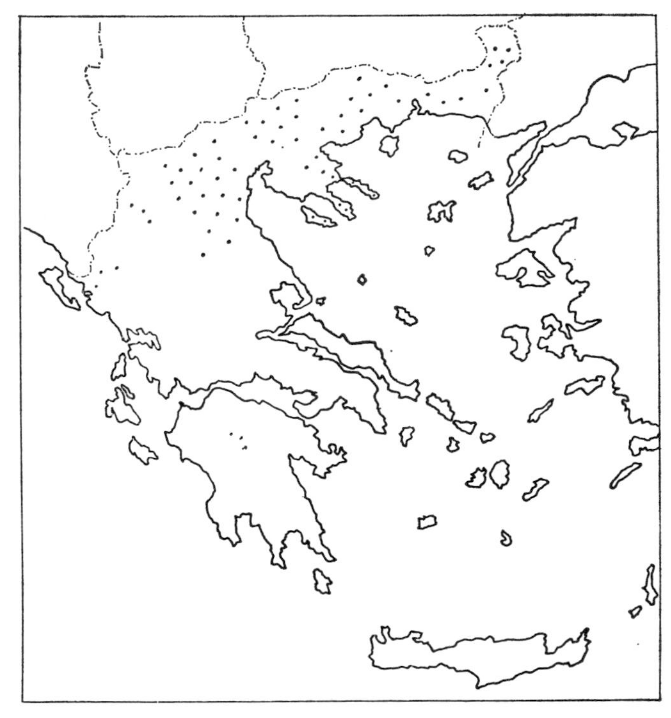

Fig. 78 Wild Vine *Vitis silvestris* Gmel. Distribution map of modern occurrence in Greece (after Logothetis 1962).

Seeds of both *Vitis silvestris* (*Vitis vinifera* ssp. *silvestris*) and *Vitis vinifera* have occurred on prehistoric sites in Europe. Seeds of *Vitis silvestris* were found on many of the neolithic and bronze age lake-side villages in Switzerland, north Italy and Yugoslavia: for example, at St Blaise, Neuenburgeresee, Casale, Polada, Puegnano, Cazzago, Peschiera, Lago di Finon, Isola Virginia on the Varesesee, Bor by Pacengo, Castione near Parma, St Ambrogio, Cogozzo, Fontinellata, Ripac (Bosnia), and Donja Dolina (Bosnia, cf. Stummer 1911, 290–1). Wild grape pips occurred also at Wangen, Bodensee (Heer 1966); Steckborn, Untersee (Neuweiler 1905); Ledro (Bertsch and Bertsch 1949, 128), at sites on the Lac de Bourget, Savoy (ibid.); and also at Valeggio am Mincio (Villaret-von Rochow 1958, 107). Apart from these finds *Vitis silvestris* pips have been recovered from sites further north in Europe at Bovere in the Scheldt valley, Belgium (Stummer 1911, 290); in the neolithic site of Rahnsdorf/Niederbarnim in Mark Brandenburg (Schiemann 1953, 318 f.), and at Inselquelle, Stuttgart (ibid., 320). In the excavations at Sitagroi, northern Greece, a series of grape pips have been recovered dating between 4500 and 2000 B.C. which show the utilization of wild grapes in the lower levels, and the beginnings of cultivation of the vine well before 2000 B.C. (Renfrew, unpublished).

Pips of cultivated grapes have also been found in archaeological deposits. The earliest evidence of cultivated grapes comes from el Omari, Egypt (Helbaek 1966E, 22), and Hama in Syria (idem 1948, 207) in the fourth millennium B.C. They also occur at Lachish and at Jericho in the early bronze age (idem 1962A, 181); at Troy II and Beycesultan in Anatolia during the bronze age (idem 1961, 80–1); and at Apliki and Kalopsidha in Cyprus (idem 1962A and 1966E). Further east cultivated grape pips occur uncarbonized at Nimrud (idem 1966D, 616) and carbonized at Hasanlu, Iran, about 1000 B.C. (idem 1966E, 22).

During the bronze age viticulture was well established in the Aegean. For example, in the Peloponnese grape pips have been found in Early and Middle Helladic levels at Lerna (Hopf 1964, 4) and in Late Helladic deposits at Mycenae and Tiryns. Grape pips, stalks and skins have been found in Early Minoan contexts at Myrtos, Crete (Renfrew, unpublished) and in Middle Minoan contexts at Phaistos (Levi 1956, 255 n. 29). It appears to have been introduced into Italy by the Etruscans (cf. finds from Campo di Servirola and pre-urban Rome; Helbaek 1956, 293).

CRITERIA FOR THE PALAEOETHNOBOTANICAL IDENTIFICATION OF PIPS OF *Vitis silvestris* AND *Vitis vinifera*

The pips of these two species of *Vitis* are pyriform in shape with a short stalk. On the dorsal face they bear an oval-circular chalazal scar or 'shield', while on the ventral side two narrow, deep furrows flank a central longitudinal 'bridge' (see figs. 79–81, pl. 34).

The problem of distinguishing the wild pips from those of the cultivated form has been the subject of discussion among palaeoethnobotanists for the past half century. Stummer (1911) pointed out that those of *Vitis silvestris* were small, short and broad, whereas pips of cultivated grapes had longer stalks and were narrower in relation to

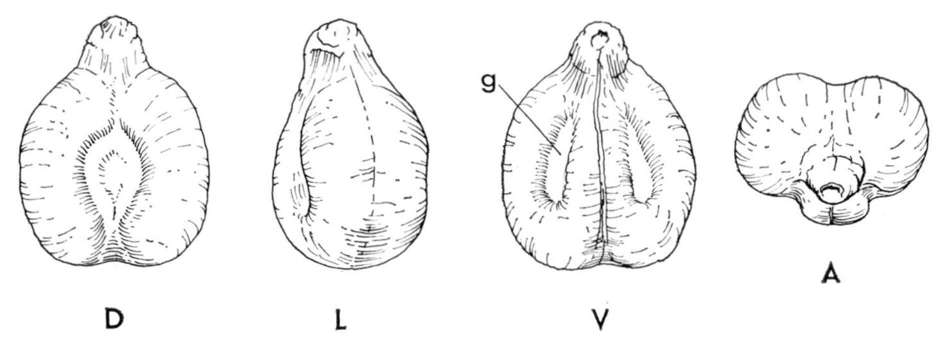

Fig. 79 Wild Vine *Vitis silvestris* L. Seed ('pip') in (D) dorsal or 'outer', (L) lateral, (V) ventral or 'inner', (A) axial view, $\times 5\frac{1}{2}$.

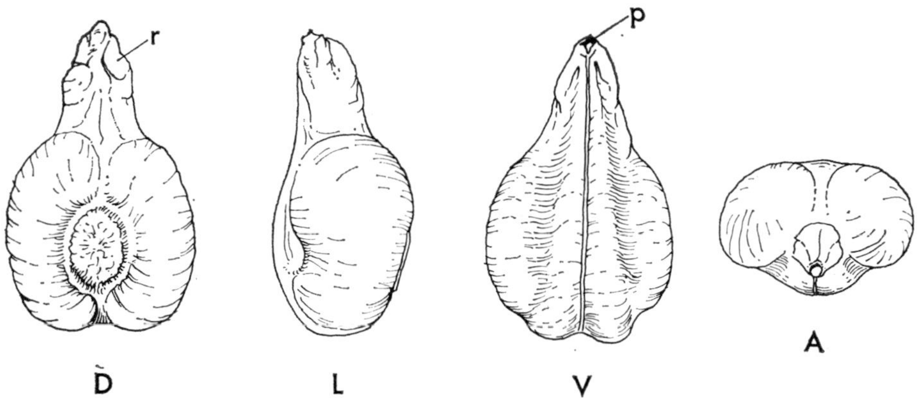

Fig. 80 Cultivated Vine *Vitis vinifera* L. Seed in (D) dorsel, (L) lateral, (V) inner, (A) axial view, $\times 5\frac{1}{2}$. (p) point of attachment of vascular tissue. (r) remnants of fleshy mesocarp adhering to woody endocarp. Compare with Fig. 79.

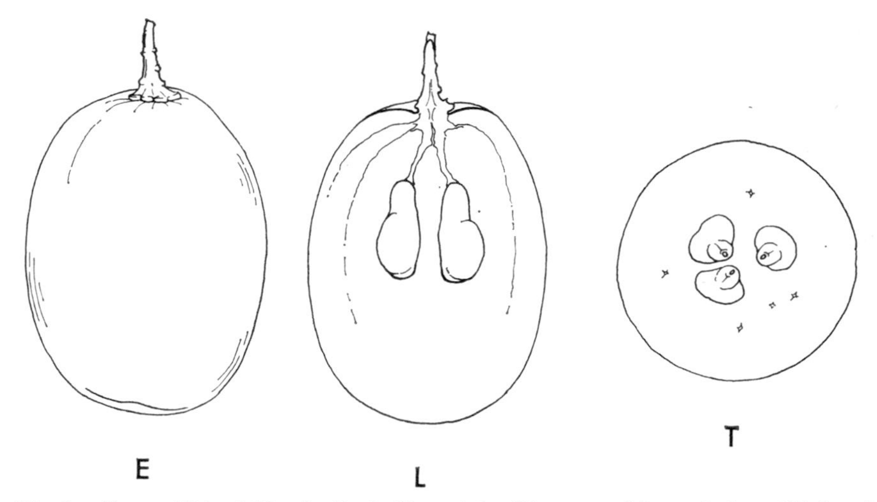

Fig. 81 Grape *Vitis vinifera* L. Fruit (drupe) in (E) external lateral view, (L) longitudinal section, (T) transverse section, $\times 1\frac{1}{2}$, showing seeds (pips) embedded in fleshy tissue of mesocarp.

their length. On some of the wild pips warts occur on the narrow stalk portion (ibid., 290, Abb. 9). He was able to distinguish them best on their length/breadth indices: *Vitis silvestris* has an index of 54–82 with the peak around 64–65. The length/breadth index of *Vitis vinifera* ranges from 44 to 75 with the peak around 55 (ibid., 286).

Thus when the pips are recovered in quantity they can be roughly assigned to the species on the basis of this ratio. When only a few pips occur the problem becomes more complex unless they have indices between 44 and 54 (*Vitis vinifera*) or 75 and 82 (*Vitis silvestris*).

Schiemann (1953, 320) gives a further series of measurements obtained from wild and cultivated grape pips. The averages of fresh seeds of *Vitis silvestris* measured from thirteen locations in the Rhine and Danube valleys are as follows: length 5·3–6·4 mm, breadth 3·5–4·2 mm, with length/breadth indices 0·64–0·83. Fresh cultivated grape

Measurements for grape pips from prehistoric sites

	Length	*Breadth*	*L/B index*
	mm	mm	
Neolithic			
Inselquelle Stuttgart[1]	6·2	3·6	0·58
Basel[1]	4·0–6·0	3·0–5·0	0·70
Bronze age			
Tell el-Duweir, Palestine[1]	4·76–5·67	2·93–4·39	0·60–0·67
St Ambrogio[1]	3·9	3·2	0·82*
Tiryns[1]	4·1	3·0	0·73
Fimonsee[1]	4·8	3·8	0·79*
Troy II[1]	5·0	4·0	0·80*
Castione[1]	5·6	4·5	0·80*
Parma[1]	5·6	4·5	0·80*
Plauen[1]	4·6	2·7	0·59
Kalopsidha, Cyprus[2]	5·33–6·66	4·0–4·5	0·75–0·67
Apliki, Cyprus[3]	3·51	—	—
Beycesultan, Anatolia[4]	5·18	3·40	0·67
Valeggio, N. Italy[7]	—	—	6·4–9·2*
Iron age			
Heilbronn[1]	6·4	2·7	0·42
Brixen, Austria[1]	5·0	3·5	0·70
Schwäbisch Hall[1]	7·0	4·9	0·70
Lachish[6]	4·58–4·94	3·66–3·84	0·80–0·77*
Pre-urban Rome[5]	4·82	3·33	0·68
Salamis, Cyprus[8]	5·1 (4·2–6·0)	3·9 (3·0–4·9)	5·4–9·4

1. Schiemann, E. 1953.
2. Helbaek, H. 1966E, 122.
3. idem 1962A, 180.
4. idem 1961, 80.
5. idem 1956, 288.
6. idem 1958B.
7. Villaret-von Rochow 1958, 108.
8. Renfrew, in press.

* In L/B index column indicates *Vitis silvestris* on criteria indicated above.

pips from eleven locations in the same regions measured: Length 5·3–8·0 mm, breadth 3·2–4·8 mm, with the length/breadth index 0·54–0·70.

At Sitagroi, eastern Macedonia, Greece, the occupation of the site extends from 4700 to 1900 B.C. (in radiocarbon years). The prehistoric sequence at this site has been divided into five phases, the earliest being phase I. Grape pips were found in all phases of occupation at this site, and show a gradual change from definitely wild forms in the first three phases to probably domesticated forms in phases IV and V. The following dimensions were obtained for these grape pips:

Length		
Phases I and II	(8 pips)	= 4·5 (4·0–4·9)
Phase III	(11 pips)	= 4·5 (4·0–5·0)
Phase IV	(11 pips)	= 5·3 (4.2–6·0)
Phase V	(2 pips)	= 5·0 (4·5–5·5)
Length of stalk		
Phases I and II	(7 pips)	= 0·6 (0·3–1·0)
Phase III	(11 pips)	= 0·7 (0·5–1·0)
Phase IV	(11 pips)	= 1·0 (0·5–1·5)
Phase V	(2 pips)	= 0·8 (0·5–1·0)
Diameter of chalaza		
Phases I and II	(5 pips)	= 1·2 (1·0–1·5)
Phase III	(9 pips)	= 1·2 (1·0–1·5)
Phase IV	(11 pips)	= 1·8 (1·0–2·0)
Phase V	(1 pip)	= 1·7.
Length/breadth ratio		
Phases I and II		= 81·8 (92·4–65·2)
Phase III		= 80·6 (90·0–66·0)
Phase IV		= 70·5 (92·8–50·0)
Phase V		= 75·2 (77·7–72·7)

It should be noted that in phases IV and V there are also included pips which are clearly still wild, in addition to those which may be regarded as domesticated.

CONDITIONS REQUIRED FOR THE CULTIVATION OF THE VINE

Climate

It requires average temperatures of at least 16–17°C in the warmest summer month (Turner 1968, 334). The plants are extremely sensitive, requiring moisture, but not in excessive amounts at the wrong time, and heat for ripening the fruit which can, however, be ruined by the dry summer winds. 'Vines require a greater degree of tendance and control of the environment than any other Mediterranean crop.' (White 1970, 229).

Soils

It prefers soils which are heavy and tend to retain moisture (Logothetis 1962, 34), but the best wines come from vineyards with a high proportion of stones or gravel in the soil: Virgil says 'bury in the ground thirsty stones or rough shells: for the water will glide between them and invigorate the plants' (White 1970, 229 f.).

POSSIBLE USES OF GRAPES TO PREHISTORIC MAN

Both wild and cultivated grapes may be eaten as fresh fruits. The cultivated grapes are larger and more succulent; and they may also be dried in the sun to form currants or sweet raisins which can be stored for consumption out of season.

That wild grapes were used for food in neolithic and early bronze age times is demonstrated by their occurrence in settlements in north Greece and the lake-side villages of Switzerland and north Italy. In some places they occur together with the pips and stones of other succulent wild fruits – possibly the residues of wine-making or brewing.

The cultivation of the vine led to the development of wine-making. The find of empty grape skins together with pips and stalks at Myrtos, Crete (Early Minoan), may possibly represent the residues from wine-making (pl. 34b). This process involves the fermentation of the sugars in the grape juice (the must) after it has been expressed from the grapes, and has been allowed to settle. The wine yeast *Saccharomyces ellipsoideus,* usually present naturally in the 'bloom' on the grape skin, is used. Red and white wines are distinguished by the presence or absence of anthocyanin pigments in the grape skins, and whether the skins are present during the early stages of fermentation. Sweet wines may contain up to 18% sugar, whereas dry wines may have less than 2%. In sweet wines the fermentation may be stopped early with the addition of alcohol: in dry wines it continues until most of the sugar is used up (Baker 1965, 121f.). The wine is subsequently 'aged' in wooden casks which involves slow chemical changes which improve the flavour and bouquet.

The Greeks and Romans improved their wines in various ways. Homer describes in *The Iliad,* Book XI, how Hecamede prepared a pottage with Pramnian wine for Nestor – consisting of goat's milk cheese grated into the wine and white barley grains sprinkled on top, the drink first being flavoured with onion (trans. E. V. Rieu 1950, 214). The Romans improved the flavour of their wines sometimes by straining them through a linen cloth perfumed with myrtle oil, or filled with barley meal (André 1961, 166). Sometimes the colour of the wine was improved by the addition of aloes or saffron (ibid.).

Olive: *Olea europaea* L.

GENETIC AND ARCHAEOLOGICAL EVIDENCE FOR THE ORIGINS OF *Olea europaea* L.

Olea europaea ($2n = 46$) (Turrill 1952, 439) probably derives from the wild *Olea chrysophylla* Lam. found widely over Asia and Africa to the Canary Islands. It is highly

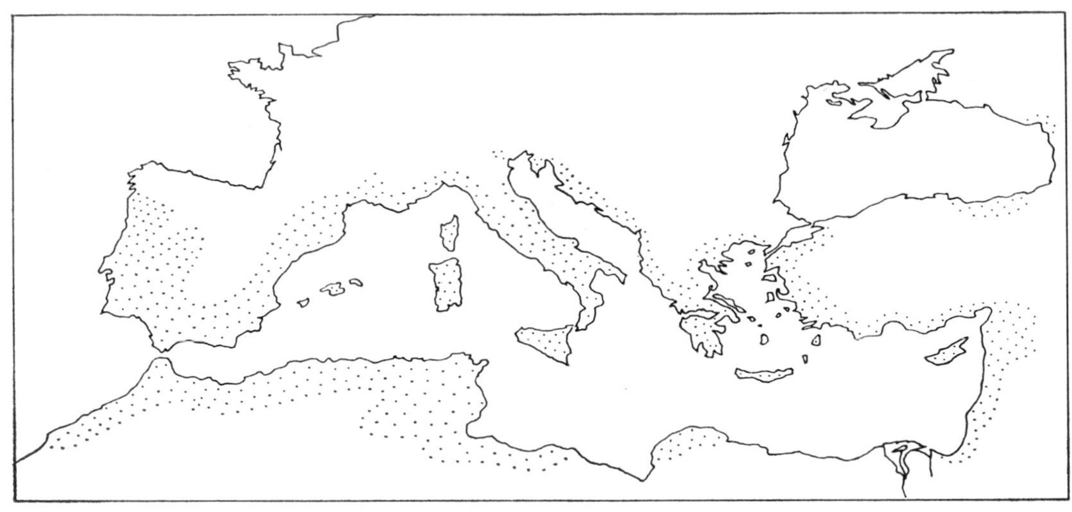

Fig. 82 Distribution of the Olive in the Mediterranean basin. After Polunin (1965).

polymorphic and Žukovskij (trans. Hudson 1962, 21) described nine botanical varieties. *Olea oleaster* or *Olea europaea* var. *oleaster* was widely held to be the progenitor of the cultivated olive. This variety shows considerable variations parallel to those of the cultivated forms. It seems clear that it represents escapes from cultivation – either from trees which have ceased to be cultivated, or from their suckers or seeds (Turrill 1952, 441).

We have no clear evidence of the stages of evolution under cultivation for *Olea europaea*, from *Olea chrysophylla*, except the find of a morphologically intermediate form *Olea laperrini* growing in the Hoggar Mountains of the central Sahara (ibid., 439). Olive cultivation is now concentrated around the Mediterranean (see fig. 82).

Palaeoethnobotanical finds of olive stones have not been very numerous so far. Helbaek reports olive stones from the third millennium B.C. contexts at Lachish, Palestine, and Tell Sukas, Syria (Helbaek 1958B; 1962B, 186). They occur in Early Minoan Crete, at Myrtos, and have been found in Cyprus in bronze age contexts at Apliki (idem 1962A, 182) and in late Assyrian Nimrud (idem 1966D, 616). Olives occur also on the funeral pyres at Salamis, Cyprus: pl. 35 shows characteristic stones split in two by carbonization. Olives occur in bronze age Greece, for example at Mycenae, Tiryns (Vickery 1936, 29) and Iolkos (Renfrew 1966, 33), and in north Italy during the same period: they are recorded from Bor near Pacengo (Neuweiler 1905, 114), Mincio (ibid., 115), at Mentone (ibid., 115), and at Peschiera (ibid., 103). Olive stones have also been found in bronze age Spain at El Garcel (Childe 1956, 267), but Helbaek (1962A, 183) suggests that these small stones may come from wild fruits.

CRITERIA FOR PALAEOETHNOBOTANICAL IDENTIFICATION OF OLIVE STONES

The structure of the olive fruit and characteristics of the stone are shown in fig. 83. I have not had the opportunity to obtain measurements from stones of *Olea chrysophylla*

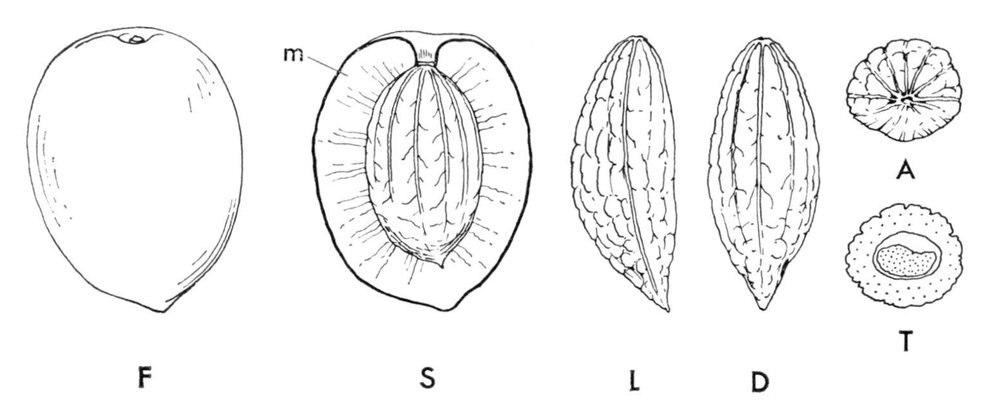

Fig. 83 Cultivated olive *Olea europaea.* (F) entire fruit, (S) fruit in section showing hard 'stone' contained within fleshy mesocarp (m), ×1½. Spindle-shaped stone is shown in (L) lateral, (D) dorsal, (A) axial view and in (T) transverse section with kernel typically not filling the space within the hard endocarp, ×2.

Lam. Buschan (1895) gave measurements of wild olive stones (possibly *Olea oleaster*) from Turkey: length 8·0–10·4 mm, 4·8–6·4 mm thick. Modern cultivated olive stones from Patras, Greece, in my comparative collection, measure 17·0–23·0 mm, average 19·1 mm, in length by 7·0–8·6 mm, average 7·6 mm, thick. All the above measurements are for uncarbonized specimens. A few measurements are available for ancient olive stones:

		Length	*Thickness*
		mm	mm
El Garcel	Bronze age	7·2–8·3	4·0–5·6
Myrtos	Early bronze age	12·4	5·8
Lachish	Bronze age	9·0–13·2 (av. 11·2)	5·0–6·2 (av. 5·7)
Apliki	Bronze age	13·2	8·0
Lachish	Iron age	8·2–12·6 (av. 10·1)	5·1–6·4 (av. 5·7)
Nimrud	Assyrian	10·7–12·5	5·7–7·5
Iolkos	Late bronze age	11·5	5·0–6·0
Salamis	Iron age	7·1–14·1 (av. 11·1)	4·8–7·1 (av. 6·1)

CONDITIONS REQUIRED FOR CULTIVATION OF THE OLIVE

Climate

The olive is best adapted to the Mediterranean climate with warm, wet winters and hot, dry summers; indeed its distribution has often been taken as indicating the exact area of this type of climate (Polunin and Huxley 1965, 3). It will not tolerate an average temperature below 37·4°F (3°C) for the coldest winter month (ibid.). Olives will grow up to 550 m above sea level, some varieties even up to 600 m (Allbaugh 1953, 271).

Soil

It thrives best on calcareous, schistose sandy or even rocky soils, in well-drained situations (Turrill 1952, 438). It has an extensive root system and cultivated trees must be spaced over 10 m apart (ibid.).

USES OF THE OLIVE IN PREHISTORIC TIMES

The fruit of the olive contains a high proportion of oil – a rare commodity in the vegetable aspect of diet – and they are greatly valued for this property (Tibbles 1912, 649). The olives may be eaten fresh or pickled in brine.

The oil is extracted by pressing it from the fruit and stone; olive presses are found fairly frequently in Minoan Crete, for example at Palaikastro (Vickery 1936, 58). To extract their oil the olives are first drenched in hot water, and this is allowed to settle in large vats so that the oil rises to the surface of the water – the water is subsequently drawn off through a spout at the bottom. The fruits and stones are then pressed. Olive trees in Crete yielded an average of 2·5 kg of oil per tree before the war (Allbaugh 1963, 269).

Olives were certainly eaten and olive oil appears to have been expressed from them in prehistoric times in Europe. We can only speculate that the oil was used in cooking: it was also almost certainly used as fuel in lamps. André (1961, 185) describes how the Romans used olive oil with every dish they served and it was also used in cakes.

Fig: *Ficus carica* L.

GENETIC AND ARCHAEOLOGICAL EVIDENCE FOR THE ORIGINS OF *Ficus carica* L.

Figs have been found fossilized in Tertiary and Quaternary deposits in France and Italy. These fossil figs are small and resemble modern cultivated figs in shape (Condit 1947, 9). At the present day *Ficus carica* is found wild in Transcaucasia, Central Asia, Crimea, Anatolia, Iran, Baluchistan and north-west India as well as in the Mediterranean region (Žukovskij, trans. Hudson 1962, 58), for example on the south-west coast of France (de Candolle 1884, 296).

The origin of cultivated *Ficus carica* is obscure: it is thought to have derived from various species of *Ficus* through natural hybridization. *Ficus persica*, *Ficus johannis* and *Ficus virgata* occur wild with *Ficus carica* in south-west Persia, Mesopotamia and Arabia, and natural hybridization occurs between all these species. However, their fruits are often inedible (Condit, 1947, 24). In Central Asia wild figs are collected and planted outside houses – but it is suggested that these are cultivated figs which had first gone wild (Žukovskij, trans. Hudson 1962, 59).

Figs are rarely found in palaeoethnobotanical material – when they survive carbonized, however, they are often remarkably well preserved. They are recorded from neolithic Jericho (Hopf, in Ucko and Dimbleby 1969, 355 f.), and from Gezer, Palestine about 5000 B.C. (Goor 1965, 124), and they occur on several sites in Greece from the late neolithic: Rachmani (see pl. 36), Sesklo and Dhimini in Thessaly: Olynthos in Macedonia and from Early Helladic levels at Lerna (Hopf 1964, 4) for example.

CRITERIA FOR THE PALAEOETHNOBOTANICAL IDENTIFICATION OF FIGS

Morphologically the fig is a shortened fleshy branch (Condit 1947, 23), forming a hollow receptacle protecting the ovaries on the inner surface (see fig. 84). Botanically,

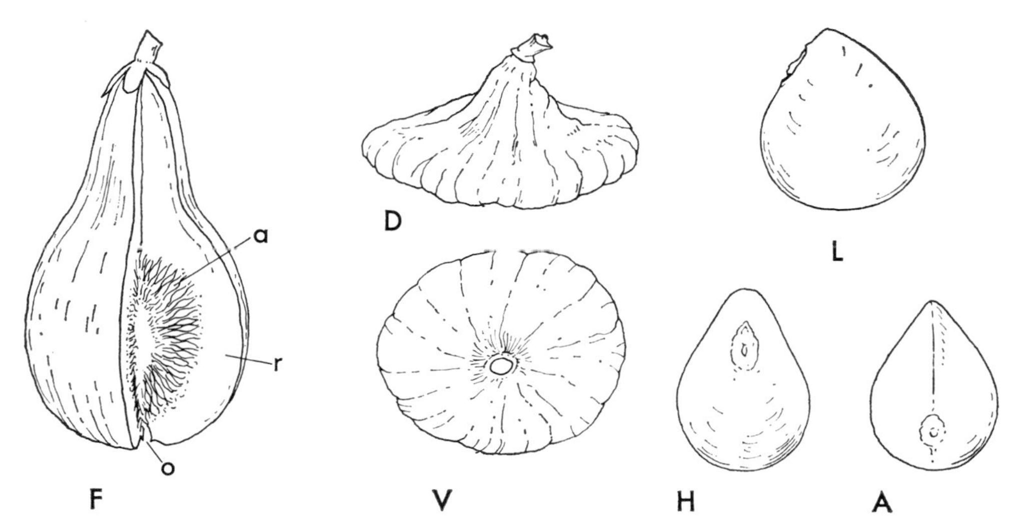

Fig. 84 Fig *Ficus carica* L. (F) fresh fruit ('syconium') in part longitudinal section, $\times \frac{1}{2}$. (a) achenes borne on inner surface of swollen, fleshy, deeply concave receptacle (r), in base of syconium a hole ('orifice' or 'ostiole') (o) gives access to the interior. (D) dried fruit with characteristic short blunt stalk, (V) in ventral view with ostiole, $\times \frac{3}{4}$. Pip 'achene' in (H) hilum, (L) lateral, (A) axial view $\times$ 14. (Note poorly defined hilar area and well-marked 'dorsal' ridge.)

the fruits of the figs are the single-seeded ovaries which line the inside of the receptacle. One unique feature of the fig is the *orifice* or *ostiole* which connects the cavity of the receptacle with the exterior. The whole fruit structure is known as a *syconium* (ibid., 36 f.). The seeds of figs are usually flattened and slightly pointed at the hilum: the surface of the seed is pitted (ibid., 40). A single fig may contain 1,600 or more seeds.

CONDITIONS REQUIRED FOR CULTIVATION OF FIGS

Climate

Figs require an annual rainfall of 86–122 cm, and this should not fall in August or September when it may cause serious damage to the fruit (Condit 1947, 86).

Soil

In Greece figs grow on stony ground, often near water, or on well-watered land (Geographical Handbook II, 72). The fig thrives best on well-drained hill slopes with a thin soil cover. The spreading root system and small leaf area make it ideal for semi-arid conditions where it can draw on moisture from a wide area without losing too much by evaporation from the leaves (White 1970, 228).

POSSIBLE USES OF FIGS TO PREHISTORIC MAN

The fig can be eaten fresh, straight from the tree, but they are chiefly sun-dried. It is their ease of drying and storing which make figs such an attractive fruit.

When the fig is dried the syrup which fills the interior of the syconium evaporates and condenses – some of the sugar crystallizes on the surface and acts as a preservative. Dried figs contain over 50% sugar and thus form an important source of sweetening for societies without either sugar-cane or sugar-beet (Tibbles 1912, 665 f.). In Central Europe roast figs are used to flavour coffee (Larousse Gastronomique, 1961, 411), and may well have had a similar use, not of course in coffee, in prehistoric times. A dish of apparently stewed figs was found in the Second Dynasty funeral meal at Saqqarah, Egypt.

Homer (*Iliad*, Book V, 116) describes how 'the busy fig-juice thickens milk and curdles the white liquid as one stirs'. This property may also have been utilized in prehistoric times.

Figs have also been used for fodder – especially for pigs – in Classical Greece, Majorca and south Spain (Condit 1947, 154).

Apple: *Pyrus malus* L. (*Malus silvestris*, *Malus acerba* and *Malus paradisiaca*)

GENETIC AND ARCHAEOLOGICAL EVIDENCE FOR THE ORIGINS OF CULTIVATED APPLES

The wild apple is found throughout Europe at the present day (Žukovskij, trans. Hudson 1962, 28), and there is no reason to suppose that this distribution did not also exist in prehistoric times. Finds of wild apples are common in temperate Europe from neolithic times onwards: Neuweiler (1905, 110–18) mentions finds from thirty-two localities in Switzerland, Italy and east France. The earliest finds include those from neolithic levels at Vallon des Vaux, Egozwil 4, Zürich–Utoquai–Fäberstrasse (Villaret-von Rochow 1969, 203), Heilbronn, Wangen, Bodman and Ruhestetten (Bertsch and Bertsch 1949, 93–4). The maximum diameter of these wild apples does not exceed 3 cm. Oswald Heer (1866, 340, 341) also reported 'cultivated' apples from the late neolithic lake-side settlement at Bodman, and an elongated form from Robenhausen (early bronze age). These carbonized apples have maximum dimensions in excess of 3 cm (see pl. 1). The 'cultivation' of these apples has been questioned by Helbaek (1952C, 112), who found a wide range in the size of wild apples growing on Bornholm suggesting that Heer's larger apples could equally well be wild and this conclusion has been reinforced by Madame Villaret-von Rochow's more recent researches in Switzerland (Villaret-von Rochow 1969, 201 f.). The apples from neolithic Alvastra, Sweden, were also of two sizes (Wulfe 1910, 65–70).

CRITERIA FOR THE PALAEOETHNOBOTANICAL IDENTIFICATION OF *Pyrus malus* L.

Apples are roughly spherical, with a depression at each end of the fruit: the depression around the stem is wider than that at the calyx (the reverse of the situation in pears). Soft hairs cover the calyx and the epidermis around it (some authors consider the presence of hairs on the stalk, receptacle and around the calyx as an indication that these trees represent cultivated apples which have gone wild) (Clapham, Tutin and Warburg 1964, 186). In cross-section the fruit shows ten bundles running through the fruit flesh, longitudinally forming a loop beginning at the stem and ending at the calyx (see figs. 85 and 86). From these, secondary bundles branch off towards the surface.

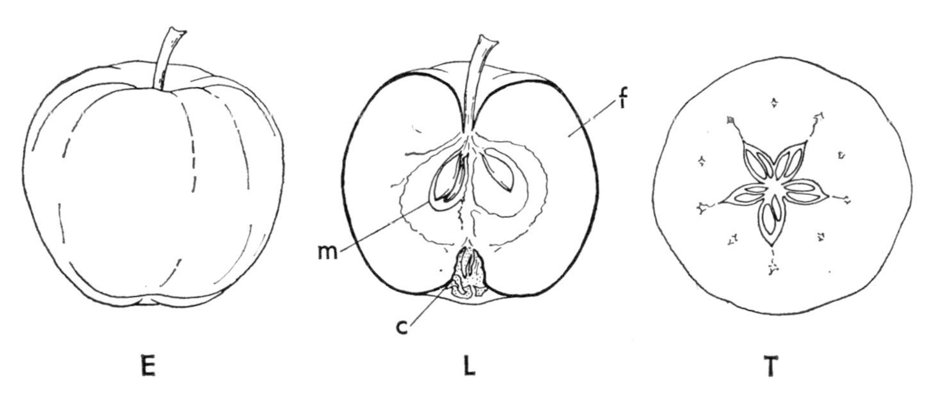

Fig. 85 Apple *Pyrus malus* L. Fruit (pome) in (E) lateral view, (L) longitudinal section, (T) transverse section ×½. (c) 'calyx tube' with sepals and fragments of styles and stamens, (m) stiff membranous lining to loculus of carpel containing a pair of seeds, (f) fleshy tissue of 'pome' developed from lower part of receptacle of flower. (T) shows symmetrical disposition of five carpels.

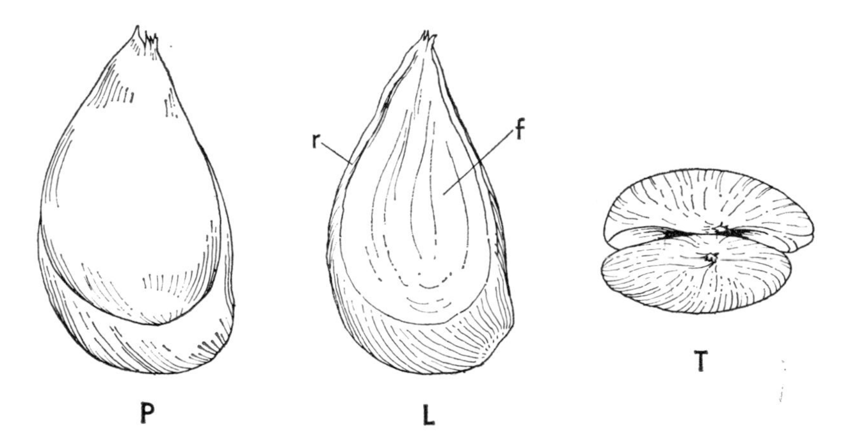

Fig. 86 Apple *Pyrus malus* L. (P) lateral view of pair of seeds ('pips') from one loculus. (L) larger pip with smaller removed from same aspect showing (f) large facet and (r) prominent ridges formed on seed coat by mutual pressure. (T) pair of pips as in (P) viewed from points of attachment. All ×5.

The core consists of a parchment-like endocarp enclosing the five locules: each containing one or two seeds. The endocarp often splits along the central sutures and a mass of hairs may appear in these cracks. The seeds are obovate, flattened, with a pointed base and rounded apex (cf. Winton 1935, Vol. II, 560 f.). The carbonized apples from Nørre Sandegaarde, Bornholm, ranged in length from 15·0 to 23·5 mm, and in diameter from 15·0 to 26·5 mm, thus falling into Heer's group of smaller wild apples (Helbaek 1952C, 111).

Apart from the finds of whole, or halved apples, apple pips occur quite commonly both carbonized and as impressions in pottery. Impressions of pips have been found in neolithic pottery from Windmill Hill, England, measuring:

Length:	7·95	7·32	6·95	6·41	5·49 mm
Breadth:	3·66	3·29	4·03	3·66	2·93 mm

(Helbaek 1952B, 198)

They have also been found at Troldebjerg, Blandebjerg and Lindø in Denmark (ibid.); also from Sweden and Holland (ibid., 200). No measurements are available for these examples. Carbonized apple pips were found in a late La Tène deposit at Nussdorf, Vienna: four pips measured:

Length:	8·2	8·1	8·6	4·8 mm
Breadth:	5·1	4·4	5·0	3·6 mm

(Werneck 1961, 112)

Uncarbonized pips from neolithic Ehrenstein average: length 6·9, breadth 4·0, thickness 2·4 mm (Hopf 1968, 60).

CONDITIONS REQUIRED FOR GROWTH OF *Pyrus malus* L.

Wild apples are found throughout temperate Europe (except in the extreme north), in Anatolia, the southern Caucasus and parts of Persia. They are also said to grow in the mountains of north-west India (de Candolle 1884, 233–4). They grow in woods and hedges and on scrub land (Clapham, Tutin and Warburg 1964, 186).

USES OF *Pyrus malus* TO PREHISTORIC MAN

At the Swiss lake-side village of Wangen Mr Löhle found several areas of 'brownish material an inch thick and several feet wide chiefly composed of apple cores' (Keller, trans. Lee 1866, 68). This material which also contained 'the undigested rinds of apples' may well represent the residue or 'marc' of apples used in making a fermented drink. Unripe fruits of wild apples are used to make verjuice – especially in France – which when fermented and sweetened makes a pleasant drink (Boulger, undated, 31).

In the sixteenth century crab apples were roasted around Christmas time and were served in bowls of hot ale (ibid.).

Apart from this specialized use in prehistoric times, crab apples were frequently cut in half and dried to provide nourishment during the winter months (Helbaek 1952C, 111). Crab apples can also be used to make a sweet jelly which is especially well flavoured if the apples have been slightly frosted (Hill 1941, 39). A soft drink ('Crab Apple Tea') can also be made by slicing the apples and simmering them in water, sweetening it with honey (ibid., 40).

Pear: *Pyrus communis* L.

GENETIC AND ARCHAEOLOGICAL EVIDENCE FOR THE ORIGINS OF CULTIVATED PEARS

A great number of species of wild pear are reported by Žukovskij from the Caucasus, Central Asia and the Far East. Among them *Pyrus communis* is found in great expanses in parts of European Russia and in Central Europe and he suggests that pear cultivars may all have originated by domestication of the best forms of *Pyrus communis* (Žukovskij,

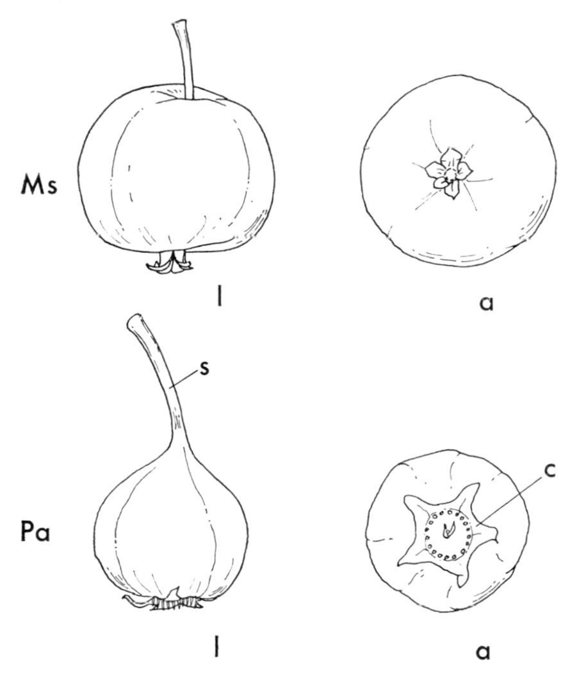

Fig. 87 Fruits of Crab Apple *Malus sylvestris* Miller and Wild Pear *Pyrus amygdaliformis* Vill. compared in (l) lateral and (a) axial views, to same scale ×1.

trans. Hudson 1962, 30). Another form of wild pear common in Southern Europe and the Mediterranean area is *Pyrus amygdaliformis* Vill. (fig. 87 and pl. 4c) which was utilized in late neolithic times at Sitagroi and Dikili Tash, East Macedonia, Greece for example (Renfrew unpublished), but does not appear to have been domesticated. Theophrastus (III, ii, 1) observed that this wild pear produced more fruit than its cultivated counterpart, although it may not all ripen.

The fruits of *Pyrus communis* were occasionally recovered from the Swiss village sites: Heer (1866, 340) reports them from Robenhausen and Wangen (pl. 1). Bertsch and Bertsch (1949, 105) give another from Ruhestetten, and also describe an interesting find of an apparently cultivated pear from Bodman on the Bodensee (early bronze age). In all cases the pears appear to have been sliced in half or in quarters to be dried for storage, in a similar manner to crab apples.

CRITERIA FOR PALAEOETHNOBOTANICAL IDENTIFICATION OF *Pyrus communis* AND *Pyrus amygdaliformis* Vill.

Pears differ from apples in usually having a more elongated form, lacking the depression around the stalk, and having only a slight indentation around the calyx which

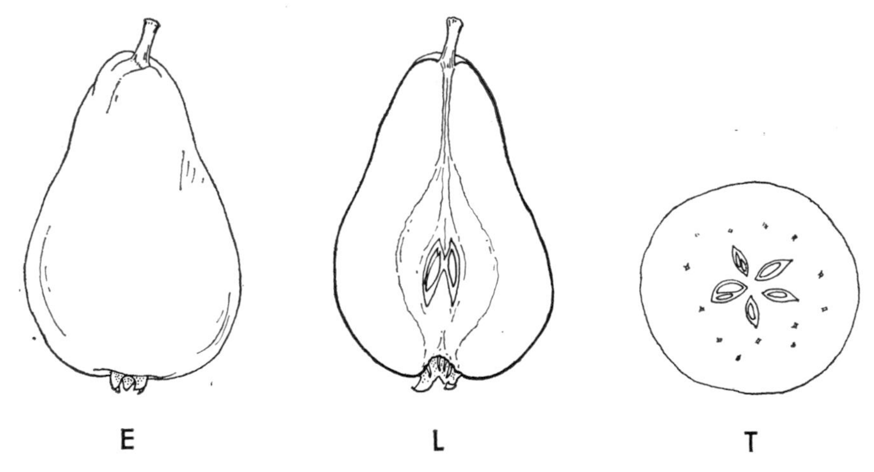

Fig. 88 Pear *Pyrus communis* L. Fruit in (E) lateral view, (L) longitudinal section and (T) transverse section, $\times\frac{1}{2}$.

usually protrudes from the base of the fruit (fig. 88). The flesh has characteristic pockets of hard 'stone cells' (Clapham, Tutin and Warburg 1964, 185).

The fruits of wild *Pyrus communis* from the Alpine villages measured:

	Length	*Breadth*
	mm	mm
Ruhestetten (neolithic)	23	27
Wangen (neolithic)	28	19
St Blaise	23	17·5
Baradello, Como	25	16

(Bertsch and Bertsch 1949, 104–5)

Pips of this wild species from neolithic Ehrenstein measured: length 4·5–6·5 mm, breadth 2–3 mm, thickness 1·3–2·2 mm. They are thus longer and more slender than

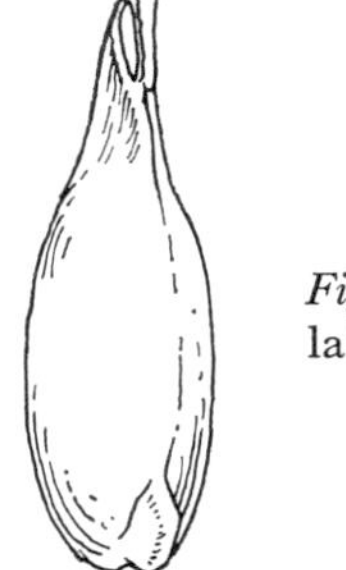

Fig. 89 Pear *Pyrus communis* L. 'Pip' in (L) lateral, (V) ventral view, ×5.

those of *Pyrus malus* (fig. 89) having a length/breadth index of 2·0 compared with 1·7 for crab apples (Hopf 1968, 60). The point of attachment lies obliquely to the pointed end of the seed, whereas in apples it lies at the extremity.

The cultivated form of *Pyrus communis* found at Bodman (early bronze age) is considerably larger and more elongated than the wild fruits: it measured: length 42 mm, thickness 21 mm.

The fruits of *Pyrus amygdaliformis* are even smaller than those of the wild *Pyrus communis* (see fig. 87). Fresh fruits picked in the plain of Drama, northern Greece, in 1968, measured:

Length:	14·0	16·0	15·0 mm
Diameter:	20·0	22·0	21·0 mm

CONDITIONS FOR GROWTH OF *Pyrus communis* AND *Pyrus amygdaliformis*

Both species are found widely dispersed in their environments (*Pyrus communis* in north temperate Europe: *Pyrus amygdaliformis* in the Balkan peninsula), growing in woods and hedges, and also as isolated trees (Clapham, Tutin and Warburg 1964, 186).

USES OF WILD PEARS TO PREHISTORIC MAN

The fruits of *Pyrus communis* were certainly cut in half and dried in prehistoric times to store away for winter use. There is no evidence for their use in preparing an alcoholic beverage although this seems likely, since crab apples appear to have been used for that purpose.

The fruits of *Pyrus amygdaliformis* are usually found whole, and were possibly eaten as fresh fruit.

Cornelian Cherry: *Cornus mas* L.

Cornus mas grows wild over a large area in south Russia from the Caucasus to the Ukraine and Moldavia. It also occurs in Southern, Central and Western Europe (Žukovskij, trans. Hudson 1962, 41 f.).

Its fruits were enjoyed as food in the neolithic and early bronze age in the north Italian lake-side villages: Neuweiler (1905, 114–16) lists finds of cornelian cherry stones from sixteen sites in this region, including Lagozza, Casale, Peschiera, Castione and St Ambrogio. They have also been found at Steckborn (Switzerland) and from Laibach Moor, Lengyel, and other sites further east (ibid.; Werneck 1961, 116). It has been

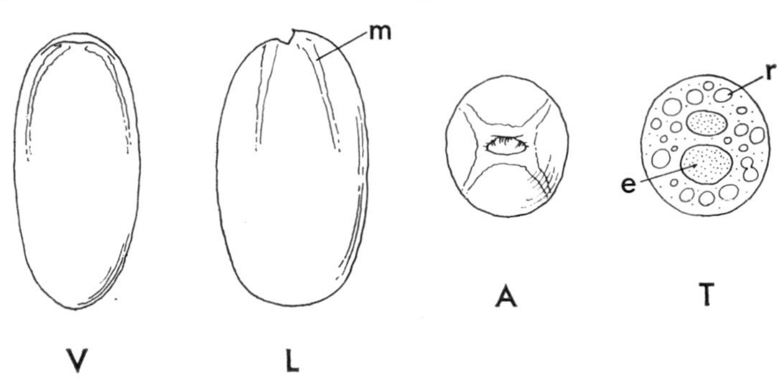

Fig. 90 Cornelian Cherry *Cornus mas*. Stone in (V) ventral, (L) lateral, (A) axial (hilum) view showing smooth surface with four pale linear markings (m) radiating from hilum. (T) transverse section, showing two loculi, containing endosperm (e), thick wall of stony endocarp with numerous spherical resinous cavities (r). All ×3.

found in the neolithic deposits at Nea Nikomedeia (van Zeist and Bottema 1971) and Sitagroi, northern Greece (see pl. 39), and was present at Starčevo, Kakanj, Obre I in neolithic Yugoslavia (Renfrew in press).

The stones of the Cornelian cherry survive, often carbonized, in prehistoric deposits. They are ovoid in outline with a blunt apex and rounded base. Four or more lines radiate from the base towards the apex, being most clearly visible on the lower part of the stone. The stones are hollow and contain two loculi. Modern (fresh) stones from fruits collected in 1968 in north Greece (see fig. 90), measured:

Length:	13·5	11·9	11·8	12·9	14·2	12·2	12·9 mm
Breadth:	7·2	5·8	5·9	6·4	7·0	7·1	6·0 mm

A carbonized stone from Burgschleinitz, late bronge age (Austria), measured: length 10·4 mm, breadth 3·9 mm (Werneck 1961, 109).

Internally the stones are divided into two locules containing one or two cylindrical seeds, each with a bulky endosperm, and flat, thin cotyledons (Winton 1936, Vol. II, 813).

The unripe fruit is nowadays pickled in brine and eaten like olives, having a similar bitter taste (Tibbles 1912, 661). The ripe fruit is sweet and may be eaten fresh or preserved – in Russia in the form of pancakes prepared from sun-dried fruit (Žukovskij, trans. Hudson 1962, 41 f.); in France the fruits are preserved in honey or sugar, or used to make a jelly (*Larousse Gastronomique* 1965, 307). In Sitagroi, north Greece, the fruits are used to make a type of cherry brandy.

Their uses as fresh and dried fruit, and as a constituent of wine, were known to the Romans (André 1961, 83, 176), and were presumably known in prehistoric times also.

Mazzard or Gean Cherry: *Prunus avium*

The cultivated forms of the sweet cherries probably all originate from the mazzard. The wild mazzard has small dark red fruits which may be sweet or bitter but not acid in flavour. It grows in well-drained woodlands on both acid and basic soils, and in districts with heavy rainfall. It has a wider range of distribution than the cultivated forms (Masefield, Wallis, Harrison and Nicholson 1969, 64).

The earliest finds are of three round stones from a neolithic hearth at Kempen on the lower Rhine, which measured 7 mm in diameter. Cherry stones 8·5 mm long were found at Inselquelle near Stuttgart in a late neolithic context. Two forms of cherry stones were recovered from the early bronze age levels of Robenhausen: a round one 7·5–8·0 mm in diameter and an oval one with a maximum length of 8·0–10·0 mm and a maximum breadth of 6·0–7·5 mm. Other late neolithic finds are known from Arbon and Steckborn on the Bodensee, Schussentals near Ravensberg, Mondsee in Austria, Hauersee near Offingen, Schweizersbild near Schaffhausen, Wauwilersee, Moosseedorf, Greing, Sutz and St Blaise. Cherry stones have also been found in bronze age deposits

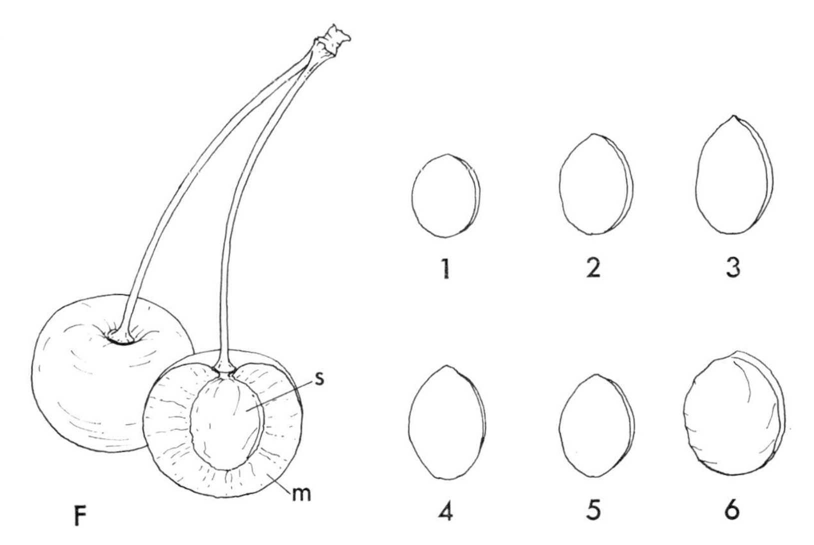

Fig. 91 Cherry *Prunus avium* L. (F) a pair of fruits (single-seeded drupes), one in section exposing (s) 'stone' (hard stony endocarp), (m) fleshy mesocarp, × 1. 1–5 outline drawings of prehistoric cherry stones after Bertsch and Bertsch (1949) compared with a modern cherry stone, 6. × 1½.

1. from mesolithic hearth at Kempen on the lower Rhine
2. from late neolithic deposits at Inselquelle near Stuttgart
3. from Celtic hearth at Schwabisch-Hall
4. from Roman well at Aalen
5. from Roman harbour at Xanten on the lower Rhine
6. modern cherry stone.

at Cortaillod, Yverdon and Montellier and from Le Bourget in Savoy (Bertsch and Bertsch 1949, 112 f., and see fig. 91).

Cultivated cherries occur first in Roman times; they are known from the following sites:

Basel	length of stones	9–10 mm
Aalen	,, ,, ,,	8–10 mm
Xanten	,, ,, ,,	9 mm

(Bertsch and Bertsch 1949, 113)

The chief difference between the stones of the wild and cultivated mazzard cherries lies in the length of their stones: the wild mazzard cherry stones range in length from 6·5 to 9·25 mm with an average of 8 mm. Cultivated cherries have stones ranging in length from 8·5 to over 13 mm (see fig. 92). The length/breadth ratio of the wild mazzard cherry stones ranges between 1·10 and 1·30.

Bullace: *Prunus institia* (*Prunus domestica* L. ssp. *institia*)

The fruits of this slender deciduous tree, the wild plum, are similar to those of the sloe, *Prunus spinosa* (see below), but are larger in size. The fruits are sharp flavoured and are often left on the tree until the early winter frosts have mellowed their flavour. The tree grows wild, and is seldom planted (see Masefield, Wallis, Harrison and Nicholson 1969, 66).

Stones of *Prunus institia* have been found on various prehistoric sites in Switzerland and Germany: three sizes of stones are reported from Sipplingen on the Bodensee:

	Length	*Breadth*	*Thickness*
	mm	mm	mm
Sipplingen	8–9	5	4·5 (neolithic)
,,	11	7	5 (,,)
,,	13	8·5	6 (,,)
Ravensberg	7–8	4	3 (,,)
,,	6	4·5	3 (,,)
Robenhausen	12	10	5 (early bronze age)
Schwabisch-Hall	12–17	—	— (celtic ?cultivated)

(After Bertsch and Bertsch 1949, 108 f; see also fig. 93 and pl. 38)

Blackthorn or Sloe: *Prunus spinosa* L.

The fruits of the sloe are small and round and the black skins are covered with a blue waxy bloom. They are extremely acid-tasting. They measure from 10 to 15 mm in diameter, and are chiefly used today for making sloe wine and sloe gin. The fruits can,

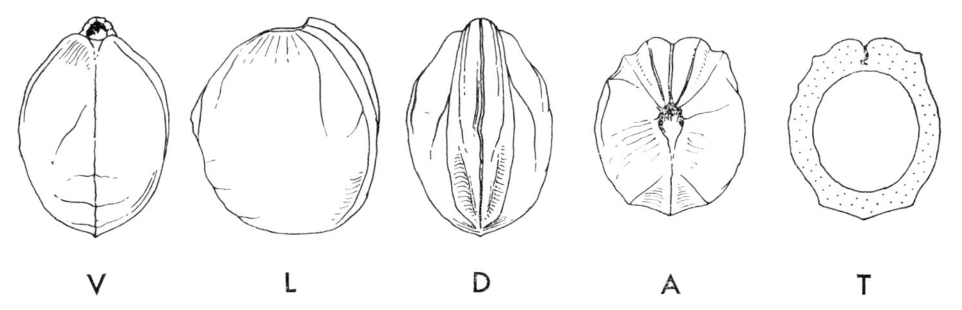

Fig. 92 Cultivated Sweet Cherry *Prunus avium*. 'Stone' in (V) ventral, (L) lateral, (D) dorsal, (A) axial (hilum) views and (T) transverse section. All $\times 2\frac{1}{2}$.

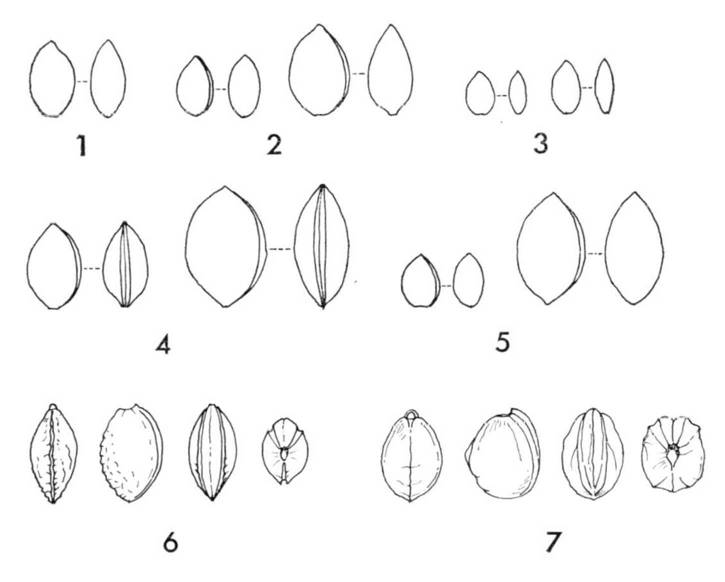

Fig. 93 Bullace (Wild Plum) *Prunus institia* L. 1–5 outline drawings of prehistoric plum stones reproduced from Bertsch and Bertsch (1949).
1. from Stone Age lake-dwellings at Wangen
2. from Stone Age lake-dwellings at Sipplingen
3. from Stone Age deposits at Ravensburg
4. from Celtic furnace at Schwabisch-Hall
5. from Roman well at Aalen
6. carbonized stone from Salamis, Cyprus
7. modern stone of cultivated *P. avium* for comparison.
All $\times 1$.

however, be slowly stewed to extract the flavour of the kernels, and the juice added to cooked elderberries, or whortleberries, or used alone to make an excellent kissel. A tasty jelly can also be made from a mixture of sloes and crab apples. (See Masefield, Wallis, Harrison and Nicholson 1969, 66; Hill 1939, 42.)

The stones of *Prunus spinosa* differ from those of *Prunus institia* in being much more rounded in outline, and having a very rough surface; this latter feature together with the deep furrow along one margin and the prominent dorsal ridge distinguishes them from the stones of *Prunus avium* (see fig. 94 and compare with figs. 91–3).

The modern cultivated plum *Prunus domestica*, the bullace *Prunus institia* and the greengage *Prunus italica* are all thought to have derived from a hybrid between *Prunus cerasifera* Ehrh. the cherry plum, and *Prunus spinosa* L. (Bretaudeau, Barton and Le Faou 1966, 95).

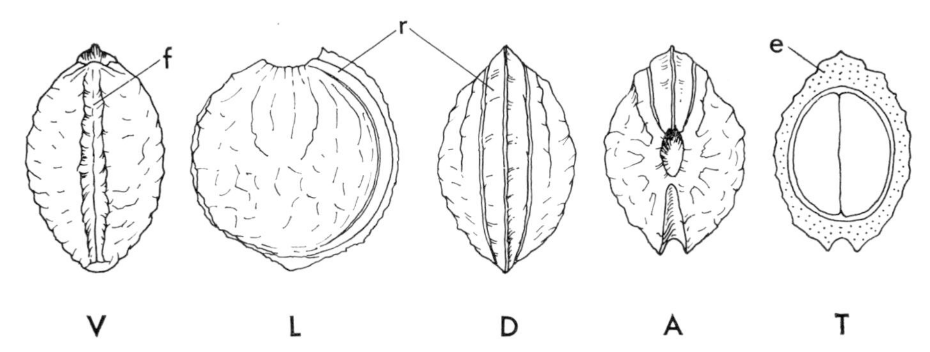

Fig. 94 Blackthorn or Sloe *Prunus spinosa* L. 'Stone' in (V) 'ventral', (L) lateral, (D) 'dorsal', (A) axial (hilum) views and (T) transverse section. All ×3. (f) sharply cut furrow, (r) prominent dorsal ridge, (e) cut surface of hard, stony, rough-surfaced endocarp surrounding seed.

Sloe stones have been recovered from a number of prehistoric sites in Europe: they are known in Britain from the neolithic – at Whitehawk Camp, Sussex, and Bryn Celli Ddu, Anglesey (Godwin 1956, 100), and perhaps the most notable of later prehistoric finds in Britain was almost a wheelbarrow load of sloe stones recovered from Mound V of the iron age village at Glastonbury (ibid.). Elsewhere sloe stones are known chiefly from the Alpine region of Switzerland: Heer reports finds of stones from Wangen, Robenhausen, Moosseedorf, Chamounix and Brügger von Thurnwalden. The following measurements have been obtained:

	Length	*Breadth*
	mm	mm
Robenhausen	7·5–9	—
Wangen	10	8·5
Chamounix	8·5	—
Brügger von Thurwalden	8	—

Neuweiler (1905, 82 f.) adds another ten sites from Switzerland with sloe stones, and also mentions finds from several of the lake-side settlements and terramare of north Italy, including Parma, Bor, Mincio, Casale, Isola Virginia and Fimonsee.

Raspberry: *Rubus idaeus* L.

The fruits of the genus *Rubus* are berries composed of numerous single-seeded drupelets which are closely set together on a small conical receptacle or core.

The fruits of *Rubus idaeus* are borne on canes springing from adventitious buds on the shallow, wide-spreading root system. The canes come to fruition at the end of the first summer, or during the second summer, and then die back. The pips or seeds of the drupelets range in size from 1·8 to 3·0 mm in length, with an average of 2·24: in breadth they range from 1·0 to 1·9 mm, averaging 1·54 (Neuweiler 1905, 79). The seeds are kidney-shaped in outline, with a prominent, raised, reticulate surface to the hard endocarp (see fig. 95).

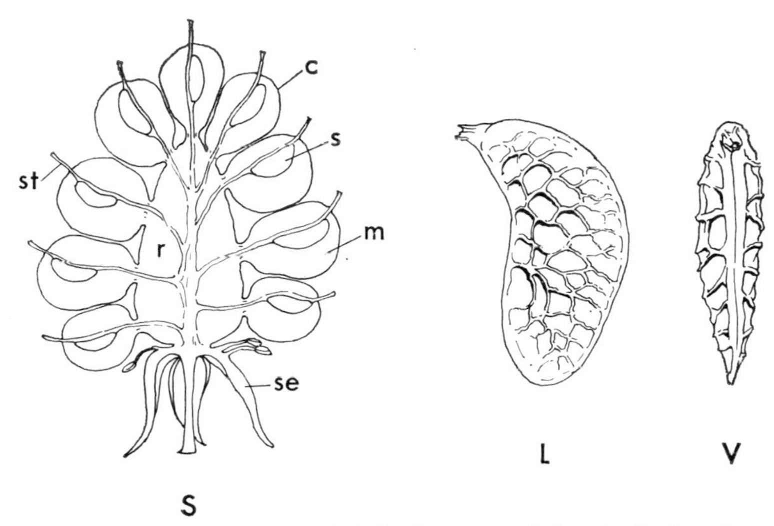

Fig. 95 Raspberry *Rubus idaeus* L. (S) fruit in diagrammatic longitudinal section ×4. (r) receptacle, (c) carpels developed into small drupelets with (m) succulent mesocarp each containing (s) a single hard 'pip', (se) sepals, (st) styles. Pip in (L) lateral and (V) ventral view ×14 showing prominent raised reticulate surface of hard endocarp.

Seeds of *Rubus idaeus* were found in large numbers at the mesolithic site of Muldbjerg, Denmark (Troels-Smith 1959, 593), and no doubt wild raspberries formed an important item in the summer diet from early post-glacial times in Europe. Raspberry seeds were commonly found in the later prehistoric settlements in the Alpine region. Neuweiler (1905, 80) reports them from Steckborn, Wangen, Lützelstetten, Neiderwil, Zug, Wauwil, Oberkirch-Sempachersee, Baldegersee, Burgäschi, Inkwil, Moosseedorf, St Blaise, Lattrigen, Schussenriedt, Mörigen, Greing, Peschiera, Fimonsee, Mercurago and Laibach Moor, Olmutz and Ratibor. Raspberry seeds from the neolithic settlement at Ehrenstein measured 2·27–2·37 mm in length (Hopf 1968, 61).

Blackberry: *Rubus fruticosus* L.

The fruits of the blackberry are remarkably similar to those of the raspberry: but the seeds of the drupelets are more or less triangular in outline and larger in size (see fig. 96).

Neuweiler (1905, 79) gives the following measurements for fresh blackberry seeds: length 2·6–3·7, average 2·95 mm; breadth 1·6–2·5, average 1·19 mm. A seed from neolithic Ehrenstein measured 2·41 mm in length (Hopf 1968, 61). The surface of the seed is coarsely reticulated – to an even greater degree than that of *Rubus idaeus*.

Seeds of the blackberry have been recovered from a large number of sites, especially in the Alpine region of Europe where they are often found associated with remains of other wild fruits. Neuweiler reports them from the following sites on the Swiss lakes: Steckborn, Lutzelstetten, Bodman, Arbon, Robenhausen, Meilen, Zug, Wauwil, Baldeggersee, Bürgaschi, Moosseedorf, St Blaise, Bevaix, Lattrigen, Wangen, Möringen,

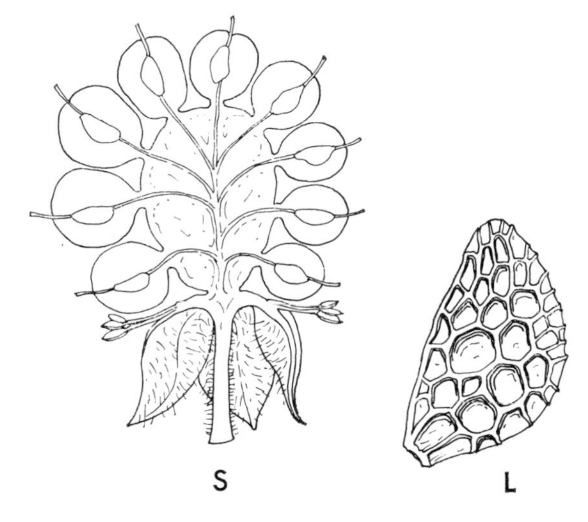

Fig. 96 Blackberry *Rubus fruticosus* L. (S) fruit in diagrammatic longitudinal section ×3 showing similar structure to that of *R. idaeus*, a cluster of small 'drupelets' attached to a central swollen receptacle. (L) 'pip' in lateral view ×14, irregularly triangular with marked coarsely reticulate surface to endocarp.

Greing and Castione. Uncarbonized seeds of *Rubus fruticosus* were found at Vallegio am Mincio (Villaret-von Rochow 1958, 106 f.). Blackberry seeds have also been found in Britain in prehistoric contexts, most notably in the stomach region of a bronze age burial at Walton-on-the-Naze, Essex (Hazzeldine Warren 1911), and in the iron age village at Glastonbury (Reid 1917).

Seeds of the closely related dewberry – *Rubus caesius* L. – have been reported from several prehistoric contexts in Europe; for example from neolithic Ehrenstein (Hopf 1968, 61), bronze age Möringen (Neuweiler 1905, 80) and iron age Glastonbury (Reid 1917, 627). The seeds are similar in shape and have a reticulated surface like those of *Rubus idaeus*, but are smaller in size.

Wild Strawberry: *Fragaria vesca* L.

The succulent fruit of the wild strawberry consists of a swollen, juicy receptacle on the surface of which small achenes or true fruits are embedded (see fig. 97). The plant is a perennial herb, with a leafy crown from which prostrate stems or runners radiate – these bear small leaf clusters which take root and will grow into new plants.

The achenes are roughly oval in shape with a faint trace of reticulation on the seed surface especially noticeable towards the point of attachment of the seed to the receptacle.

Oswald Heer (1865, 29) reports finding wild stawberry seeds at Robenhausen which measured 1·5 mm in length. Neuweiler (1905, 78) reports finds from eighteen localities in Switzerland, Austria and north Italy. Three hundred uncarbonized seeds were found

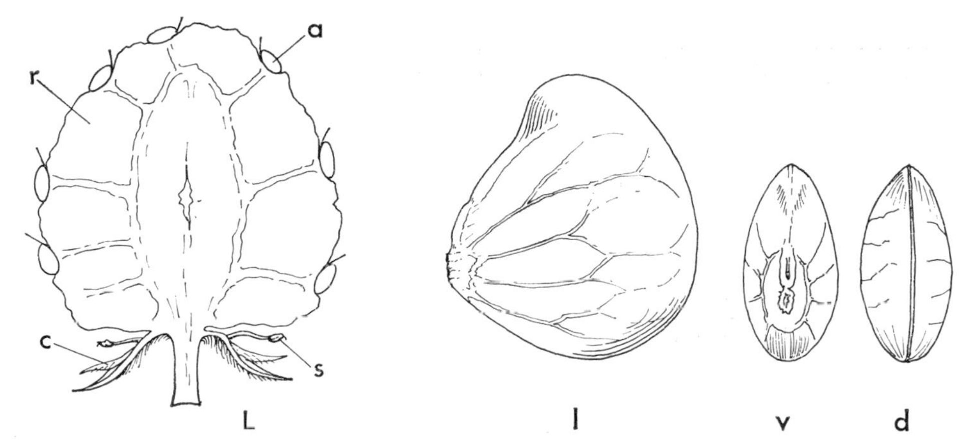

Fig. 97 Wild Strawberry *Fragaria vesca* L. 'False fruit' in (L) longitudinal section ×4. (a) achenes or true fruits, (c) calyx and epicalyx, (r) succulent tissue of swollen receptacle, (s) stamens. Achene in (l) lateral view ×32, (v) ventral, (d) dorsal view ×22.

at Ehrenstein measuring 1·22 × 0·9 × 0·51 mm (Hopf 1968, 62). Bertsch and Bertsch (1949, 155) also mention the occurrence of wild strawberry seeds from the Danubian I settlements at Öhringen and Böckingen, and from a number of late neolithic sites on the Federsee and Bodensee. Wild strawberry seeds were also recovered from the mesolithic settlement at Muldbjerg, Denmark.

The fruits discussed above are those which have been found most frequently on prehistoric sites in Europe and the Near East, but there are a number of others which deserve mention in passing.

Hackberry, Nettle Tree: *Celtis australis* L.

Fruits have been found on a number of Near Eastern sites. The stones of this fruit were found associated with the remains of Pekin man in the Middle Pleistocene deposits at Choukoutien – perhaps the earliest record of plant utilization by man (Brothwell and Brothwell 1969, 130). More recent documentation for the utilization of this fruit comes from Çatal Hüyük, Anatolia (Helbaek 1964A, 123), and Hacilar (idem 1970, 230). The stones of this cherry-like fruit have a strongly sculptured design. The stones are usually divided into equal quarters on the surface by two ridges meeting at right angles. The pedicel scar is oblique and crested and the Hacilar stones measured 4·67–7·17 mm in length and 4·50–6·50 mm in breadth (Helbaek 1970, 230). The fruits are 9–12 mm long, globular in form and borne on long stalks. The tree is common today in most of Mediterranean Europe, except Greece and Turkey (Polunin 1969, 56).

Hawthorn: *Crataegus* sp.

Fruit stones of *Crataegus* sp. are occasionally found in prehistoric contexts: Helbaek reports them from Lachish, Palestine and Nørre Sandegaard, Bornholm (Helbaek 1958B, 312; 1954, 253). The fruit stone is roughly hemispherical: the domed dorsal side has faint longitudinal grooves, the ventral side is flattened with a medial suture for two-thirds of its length terminating in a small pit. The Lachish stone measured: length 7·7 mm, breadth 7·5 mm, thickness 5·3 mm, with a seed cavity of 4·9 × 3·1 mm, circular in cross-section. Fruits of *Crataegus oxyacantha* L. have been used dried to add to flour in some parts of Europe, and the seeds used as a substitute for coffee. The young leaves have also been utilized as substitutes for tea and tobacco. The succulent fruits of *Crataegus azarolus* L. are often eaten fresh or preserved and are sold in markets in Europe and the Near East (Uphof 1959, 112).

Rose hip: *Rosa canina* L.

Rose hips belonging to *Rosa canina* L. have been found on a number of sites in the Alpine region. Heer (1866, 29) reports then from Moosseedorf and Robenhausen, and Neuweiler (1905, 78) added finds from Steckborn, Wangen, Baldeggersee, Burgaschi, St Blaise, Lattrigen and Mondsee. In Britain they are known from the early bronze age burial at Walton-on-the-Naze, Essex, and from the iron age village at Glastonbury, Somerset (Godwin 1956, 119). The true seeds or achenes are enclosed in an urn-shaped receptacle 50–60 mm long. The internal cavity is hair-lined and contains 20 to 25 achenes. They are 7–8 mm long, roughly triangular in outline and covered with hairs. Rose hips are full of vitamins and are used to make syrup, jellies, preserves and sauces (Masefield, Wallis, Harrison and Nicholson 1969, 62).

Elderberry: *Sambucus nigra* L.

Uncarbonized seeds of elder have been found on a number of prehistoric sites chiefly in the Alpine region. Neuweiler (1905, 107) reports them from Bodman, Steckborn, Wangen, Neiderwil, Robenhausen, Zug, Wauwil, Bürgaschi-see, Moosseedorf, St Blaise, Castione and Parma. More recently they have been found at Vallegio am Mincio (Villaret-von Rochow 1958, 111), and at Ehrenstein near Ulm, Germany (Hopf 1968, 64). The seeds from Vallegio measured: length 3·8 mm, breadth 1·8 mm. Fresh seeds measure: length 4·0–4·5 mm, breadth 2 mm (Neuweiler 1905, 107). They have been found fairly frequently on later prehistoric sites in Britain, for example at the Glastonbury and Meare villages in Somerset. The fruits are borne in large flat-topped clusters with five primary branches. The juicy black fruits are 6–8 mm in diameter and contain a number of seeds (see fig. 98). Elderberries can be preserved in sugar, but are chiefly used for making wine (Uphof 1959, 324).

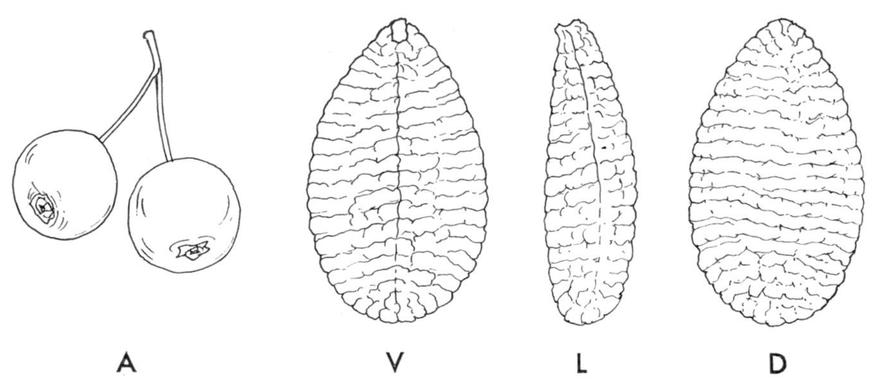

Fig. 98 Elderberry *Sambucus nigra.* (A) fruit ×2. Seed in (V) ventral, (L) lateral, (D) dorsal view ×14 showing coarsely ridged surface.

Water Chestnut (Caltrops): *Trapa natans* L.

The strange-shaped fruits of this water plant were used for food in the Alpine region in prehistoric times: they were recovered from Robenhausen, Moosseedorf, Lützelstettin, Burgäschi-see, Isolino Varesesee and Laibach Moor (Neuweiler 1905, 99). The hard fruits are 1–2 in across and have two opposite pairs of hornlike projections. They can be eaten raw or roasted like sweet chestnuts or boiled. They have a floury texture and a mild flavour (Masefield, Wallis, Harrison and Nicholson 1969, 32).

Bilberry: *Vaccinium myrtillus* L.

Fruits of the bilberry and also of its close relative the cowberry (*Vaccinium vitis idaea* L.) have been recovered from a few sites in temperate Europe. They grow on low shrubs – up to 60 cm – on moorlands and heaths. The fruits of the bilberry are globose, about

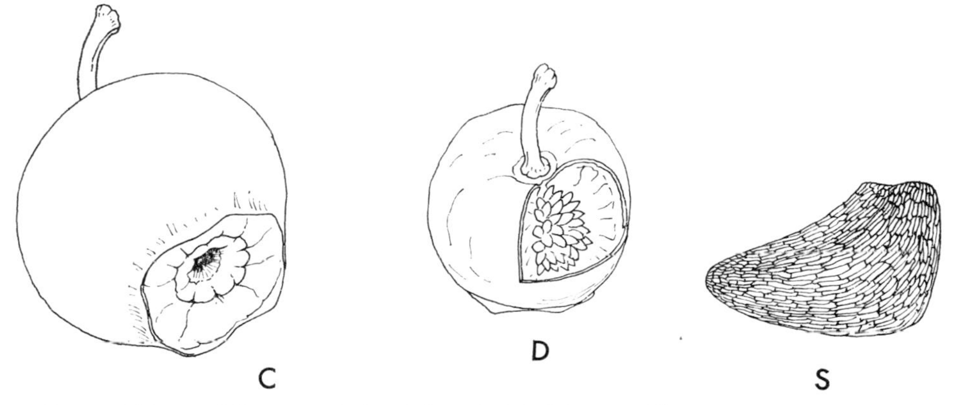

Fig. 99 Bilberry (Blaeberry, Whortleberry) *Vaccinium myrtillus* L. (C) fruit (berry) showing calyx end with five shallow persistent lobes and deep axial stylar depression, ×5. (D) fruit with one loculus in part section showing seeds borne in axial placentation, ×4. (S) seed showing characteristic shape and minute linear reticulation of surface, ×30.

7 mm in diameter, with numerous seeds borne in axial placentation (see fig. 99). The fruits of the cowberry are slightly larger. Seeds of *Vaccinium myrtillus* and a fragment of a fruit were found at Robenhausen (Neuweiler 1905, 102): seeds of *Vaccinium vitis idaea* were recovered from Moosseedorf. Bilberries are rather acid when raw, but can be made into palatable tarts and jam.

Strawberry Tree: *Arbutus unedo* L.

One wild fruit was found at Lerna, Greece, in the late neolithic deposit (Hopf 1964, 4). This tree is characteristic of maquis vegetation in the Mediterranean (Polunin and Huxley 1965, 139). It is an erect shrub or tree growing up to 10 m high. The fruits are red, globose with a more or less warted strawberry-like berry 15 to 20 cm long (Clapham, Tutin and Warburg 1964, 270). They have five cells each containing four to five angular seeds (Ward 1908, 149). The fruits are edible and in Corsica and Italy are used to produce an alcoholic distillation (Polunin and Huxley 1965, 139). In Roman times the fruits were picked in the forests during the winter (André 1961, 83). Theophrastus (III, xvi, 4) describes the edible fruit of arbutus, called *memaikylon*, and observes that it takes a year to ripen so that both flowers and fruit may be found on the tree together.

Date palm: *Phoenix dactylifera*

This tall palm reaches up to 24 m in height. The male and female trees are separate and it is necessary to have one male tree to the acre (or to 50–100 females). Date palms are propagated by offsets or suckers. They begin to bear fruit about five years after planting and reach full bearing by fifteen years. They continue to bear fruit for about eighty years. The average annual yield of a single tree is about 45 kg of fruit. The fruits are borne in bunches, each of which may consist of about forty strands, with 25–35 dates per strand. At the present time the fruits are divided into three categories: soft dates which are sold pressed into masses, semi-dry dates which are boxed with the fruits still attached to the strand, and dry dates which are hard and can be kept for long periods, and may be ground into flour. Their chief nutritional value is the high sugar content of the fruits – 60% in soft dates and 70% in dry dates; they are also rich in vitamins A, B_1, B_2 and in nicotinic acid. In the Old World they are grown chiefly in Saudi Arabia, Iran, Egypt and Algeria. They have been found in ancient plant remains from middle bronze age Jericho (Hopf 1969), and at late Assyrian Nimrud (Helbaek 1966D, 613 f.).

Pomegranate: *Punica granatum* L.

The fruits of this spiny shrub, or small tree, are large – up to 9 cm across – with a thick, leathery skin and a persistent crown-like calyx. Internally the fruit is divided into several cells each containing several seeds embedded in the juicy pink flesh. The fruits are usually eaten fresh, and the juice may be used in cool drinks, conserves or syrups. Remains of pomegranates have been found at Nimrud (Helbaek 1966D, 613 f.), and at Jericho in the middle bronze age levels (Hopf 1969). Remains of a carbonized pomegranate from Sassanian contexts at Quimis, Iran, measured 44 mm in diameter, and the loose pips associated with it were 6·4–8·1 mm long, and 2·9–3·9 mm in diameter (Renfrew, unpublished).

Cucumber: *Cucumis sativus* L.

This trailing or climbing plant bears long cylindrical fruit, with a thick, rough and sometimes prickly skin. The flat, white seeds are embedded in the centre of the juicy white flesh. Seeds of cucumber have been found at Nimrud (Helbaek 1966D, 613 f.). Fresh seeds are 8–10 mm long, white and smooth (Whitaker and Davis 1962, 41).

Caper: *Capparis spinosa* L.

Seeds of this shrub have been found in quantity at Tell es-Sawwan, Iraq, dating to c. 5800 B.C. (Helbaek 1965A, 46) and a few are known from Ali Kosh and Tepe Sabz. They were found at Hacilar (idem 1970, 30), and also at Beidha, Jordan: Hama, Syria, and Nimrud. The caper is a small shrub growing in semi-deserts and mountain regions in Mediterranean Europe and all over the Near East. It has a fruit like a small fig which bursts when ripe, exposing the seeds embedded in red jelly-like flesh. The fruits are collected and eaten in autumn. The flower buds are also cooked and pickled to produce the 'capers' commonly used in cooking and flavouring. The seeds of the caper are small and round with a curved, protruding radicle; they measure 2·5–3·0 mm in diameter.

Prosopis: *Prosopis stephaniana*

Fruits and seeds of prosopis have been recovered in quantity from the earliest neolithic sites in the Deh Luran plain in Iranian Khuzistan – from Ali Kosh and Tepe Sabz. They have also been recovered from the fertile plain of Mesopotamia at Tell es-Sawwan (Helbaek 1965A, 46) and from Late Assyrian Nimrud (idem 1966D, 617). The seeds are contained in thick, blunt pods, each having from one to five seeds inside. The seeds have a distinctive horsehoe-shaped incision in the flat side of the seed coat. The medium-sized prosopis bushes have very deep root systems which enable them to survive in arid regions where most other plants cannot survive the dry season.

Chapter 17

Nuts

Acorns: *Quercus* sp.

Acorns have been found in many archaeological deposits in the Near East (e.g. Çatal Hüyük), and in Europe they occur in the earliest neolithic sites in Greece (Renfrew 1966, 21). Neuweiler (1905, 57 f.) lists over thirty sites in Switzerland, southern Germany and northern Italy yielding acorns; they do appear to have been used as human food. Theophrastus (III, viii, 3) in describing different species of oak says: 'All these bear fruit: but the fruits of Valonia oak [*Quercus aegilops* Boiss.] are the sweetest . . . , second to these of *hemeris* [gall oak – *Quercus infectoria*], third those of the broad-leaved oak [scrub oak – *Quercus lanuginosa* Thuill.], fourth sea-bark oak [*Quercus pseudo-robur*] and last aegilops [Turkey oak – *Quercus cerris*], whose fruits are very bitter.' In Macedonia he reports that *Quercus robur* L. has the sweetest acorns (III, viii, 7); he also describes the acorns of *Quercus ilex* (holm oak) found in Arcadia, as being sweeter than those of the kermes oak (*Quercus coccifera* L.), but more bitter than those of *Quercus robur* (III, xvi, 3). Hesiod says that honest men should not starve since: 'The earth bears them victuals in plenty, and on the mountains the oak bears acorns upon the top and bees in the midst' (*Works and Days*, 233–4). Acorns have been used in times of famine to provide flour for making bread; Du Bellay, Bishop of Mans in 1546, complained to Francis I of the misery of the peasants who had nothing to eat but bread made from acorns (Soyer 1853, 24). The Bahktyari tribe of south Persia still make acorn bread today (using acorns of *Quercus persica*, the manna oak). Roasted acorns have also formed a substitute for coffee (Tibbles, 1912, 683 f.), and fresh acorns are used for malt in making beer in some places (Howes 1948, 172). The problem in using them for food is to extract the bitter tannin: this can be done by boiling or roasting them (Helbaek 1964, 123), or by burying them in the ground (Tibbles 1912, 684). In North America acorns of over a dozen species have been used for food: the Klickitat Indians of the Pacific coast prepare them by removing the husks, grinding the cotyledons to flour and then placing the meal in hollows in the sand lined with pine needles, hot water is then poured over it, thus removing the bitter substances. It is then washed, kneaded and baked into flat cakes (Howes 1948, 174). In prehistoric times they were presumably often roasted, since the cotyledons usually survive in carbonized form. At Çatal Hüyük Helbaek found them closely associated with a hearth (Helbaek 1964, 123). At the late bronze age site of Raskopanitza, near Plovdiv, Bulgaria, grains of einkorn and six-row barley were found mixed with acorns between the upper and lower stones of a saddle quern: presumably they were being ground together into meal in preparation for making porridge or bread (Renfrew, unpublished).

It is possible that some acorns may have been collected for making dyes or for tanning leather: this cannot be discounted in interpreting finds of large quantities of acorns. Where they are mixed with other seeds used for food as at Raskopanitza it is likely that they played a part in human diet. It is unlikely that they would have been

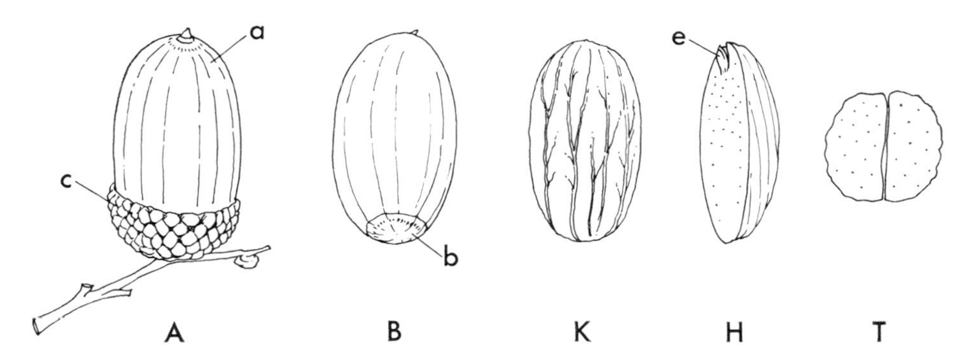

Fig. 100 Oak *Quercus robur* var. *pedunculata*. (A) acorn (a) borne in sub-hemispherical scaly cupule (c), (B) acorn showing pale area at base (b), (K) kernel showing thin brown membranous skin with pronounced branching veins, (H) half of kernel (one cotyledon) showing embryo at apex (e), and longitudinally striated outer surface, (T) kernel in transverse section showing two cotyledons. All ×1.

collected to feed to pigs, since it would be more economical of labour to send the pigs into the woods to forage for themselves under the supervision of a swineherd.

It is quite evident that many species of *Quercus* may yield edible acorns: but it should be remembered that their palatability can vary greatly even between trees of the same species in the same locality (Howes 1948, 172). Well-developed trees may yield between 700 and 1,000 litres of acorns each year (ibid., 173).

In archaeological deposits usually only the inner cotyledons of the acorn survive: the cup and outer covering having been removed (see fig. 100 and pl. 40). This makes identification of species rather difficult, since the form of the cupule and the shape of its scales are often as diagnostic as the shape of the acorn it contains. The cupules of *Quercus ilex* L. are long, narrow, spreading and hairy: those of *Quercus cerris* L. are moss-like (Ward 1908, 109–10), whilst those of *Quercus coccifera* have prickly scales, for example. The acorns of *Quercus robur* are narrow and 20–40 mm long (ibid., 109): those of *Quercus aegilops* Boiss. are very large (Polunin and Huxley 1965, 55). Usually it is not possible to give a specific identification of an acorn cotyledon.

Walnut: *Juglans regia* L.

Walnuts have occasionally been found in archaeological deposits: they were found at Mečkur, Bulgaria, in Gumelnitsa culture levels (Arnaudov 1948/1949 and 1936; Gaul 1948), and at Sadowetz, north-west Bulgaria, in the dark age deposits (Arnaudov 1937/1938). Neuweiler (1905, 58), reports finds from the iron age settlement at Fontinellato. They are also known from Wangen, Untersee; Bleiche near Arbon on Bodensee, and Haltnau on Lake Constance (Bertsch and Bertsch 1949, 120–1).

Walnuts are native to Central Asia, Iran, the Caucasus, Anatolia, the Balkan peninsula and Southern Europe (Žukovskij, trans. Hudson 1962, 48). They require an

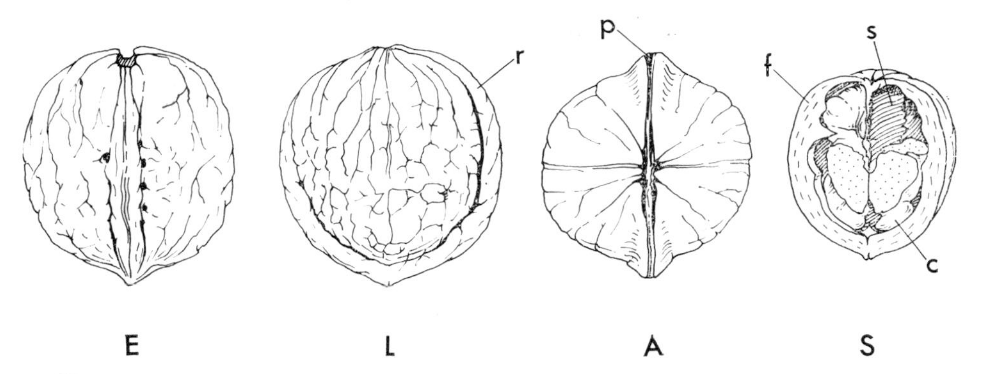

Fig. 101 Walnut *Juglans regia* L. Nut in (E) 'end', (L) lateral, (A) axial (hilum) view ×1 showing irregular furrowed reticulate surface of hard shell (endocarp) and prominent symmetrical ridges (r) bounding fracture plane (p) along midribs of carpels. (S) perspective view, ×¾ of half of nut showing (f) surface of fracture plane, (s) septum exposed by removal of part of (c) corrugated cotyledon.

annual rainfall of 20–30 in, are resistant to extreme heat and cold, but are most susceptible to late spring frosts (Howes 1948, 120).

The fruit of *Juglans regia* is a large indehiscent drupe with the stone 'walnut' incompletely two- to four-celled, containing a cotyledon with a strongly convoluted surface (Clapham, Tutin and Warburg 1964, 251). The shell, endocarp, is grooved and pitted irregularly and splits into two halves down the midrib of the carpel (see fig. 101) (Ward 1908, 123). When ripe, in favourable conditions, the husk cracks open while still on the tree, releasing the nuts which fall to the ground (Howes, op. cit., 121).

The kernels are eaten fresh – or dried: the percentage of oil increases with keeping, thus making them less digestible. Walnut oil is also extracted (Howes 1948, 122). The fleshy husk of the walnut, the shuck, is used in France for making a liqueur (*Larousse Gastronomique*, 1004).

Almond: *Prunus dulcis* (Miller) D. A. Webb (*Prunus amygdalus* Batsch.)

A few dozen shells of possibly wild almonds *Amygdalus orientale* were found in level VI at Çatal Hüyük, and they also occurred at Hacilar: almonds have been reported from various prehistoric sites in south-east Europe, e.g. Dimini and Sesklo, Greece (Wace and Thompson 1912, 73 and 85), and were found at Apliki (bronze age) and Salamis (iron age) in Cyprus.

Prunus dulcis grows wild from Central Asia and Persia to Anatolia (Žukovskij, trans. Hudson 1962, 34). It is now cultivated in the Mediterranean region and sometimes escapes to the wild. It is tolerant of most climates but is extremely susceptible to early spring frosts. Deep, well-drained, rich soils suit it best: it does not thrive in alkaline

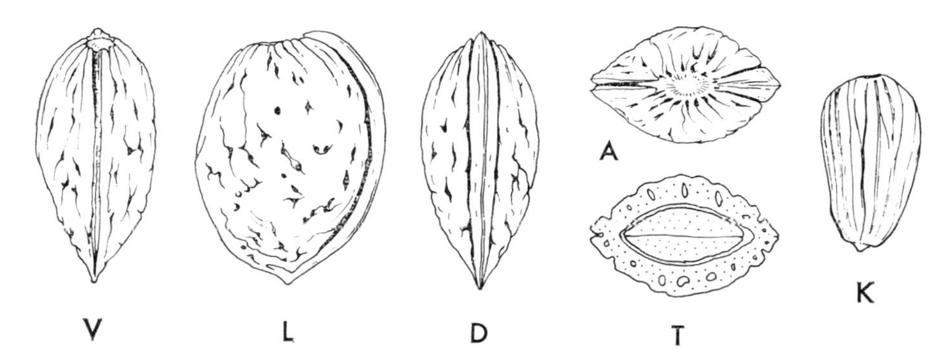

Fig. 102 Almond *Prunus amygdalus* Batsch. 'Stone' in (V) ventral, (L) lateral, (D) dorsal, (A) axial (hilum) views and (T) transverse section. Kernel in (K) lateral view. All ×1.

conditions. Almonds grow most successfully in areas with a rainfall of 16–40 in per annum (Howes 1948, 106–7).

The fruit of the almond is a drupe, ovoid in shape and slightly compressed. The tough, dry, velvety pericarp may split as the fruit matures. The fruit is about 40 mm long when complete. The endocarp is fibrous, deeply pitted and woody, and contains a small oval kernel; occasionally two kernels may develop within the same shell (fig. 102 and pl. 40).

The kernels are eaten fresh, or used ground into flour for culinary purposes – in almond paste, biscuits, bread and marzipan, for example (Tibbles 1912, 685). The fat averages about 50% of their total composition and is extracted as a bland oil with the same nutritive value as olive oil (ibid., 686). The nuts of wild almonds were sold in the bazaar in Dizful, south-west Iran (1969), and doubtless elsewhere in the Near East.

Pistachio: *Pistacia atlantica* Desf.

The small nutlets of *Pistacia atlantica* have been found in many of the early farming settlements in the Near East: they occur at Beidha (Helbaek 1966A, 63), Jarmo, Sarab, Çatal Hüyük (idem 1964A, 123). They also occur in the earliest farming sites in Thessaly – at Ghediki, Sesklo (see pl. 41; Renfrew 1966, 24–5).

Pistacia atlantica occurs now in Palestine, Syria, Cyprus, Greece and along the North

African coast to the Canary Islands (Zohary 1951, 35). It grows at altitudes ranging from −30 below to +1,400 m above sea level on calcareous as well as on basaltic soils (idem 1940, 160).

The nutlets are quite small: the following dimensions are available for ancient carbonized nutlets:

		Length	*Breadth*	*Thickness*
		mm	mm	mm
Sesklo	Neolithic	4·06 (av.) 4·6 (max.) 3·5 (min.)	3·4 (av.) 4·2 (max.) 2·7 (min.)	
Hacilar	,,	5·00	4·67	3·68
Lachish	Early bronze age	5·9 3·9	4·3 3·5	

They have a thickened flat base with a conspicuous duct in or near the centre – fig. 103 (Helbaek 1958, 311). These small nutlets are still collected and consumed as a delicacy all over the Near East and they are sold on a large scale in the bazaars (idem

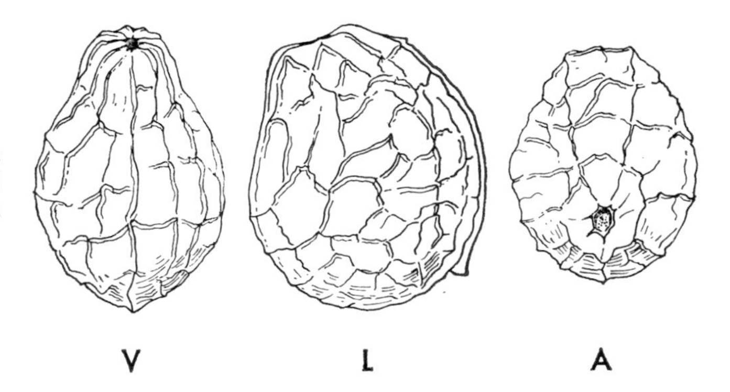

Fig. 103 Pistacia atlantica Desf. Nutlet in (V) ventral, (L) lateral, (A) axial (hilum) view ×6 showing prominent raised surface reticulation.

1964, 123). They may be used in baking or confectionery, or eaten roasted. In some places the young fruits are pickled. Turpentine, resin and gum are made from the sap of the tree and the bedouin use the root bark for tanning (idem 1970, 231).

Pistachio: *Pistacia vera* L.

A single nut of *Pistacia vera* was found in the late neolithic levels at Sesklo (Plate 42), otherwise it has not occurred in palaeoethnobotanical samples.

This species of pistachio is found wild in Central Asia, Turkmenia, Uzbekistan and Kazakstan (Žukovsky, trans. Hudson 1962, 49). It is now widely cultivated in the Mediterranean region (Polunin and Huxley 1965, 119). These are among the most drought-resistant fruit trees known, and some of them have lived for over 400 years (Žukovskij, trans. Hudson 1962, 49). In eastern Persia 15-year-old trees yield about 12 lb of nuts per annum; and 20-year-old trees between 20 and 60 lb per annum (Howes 1948, 104).

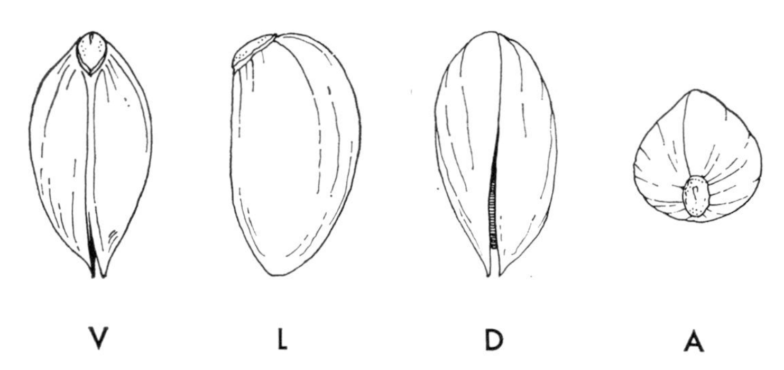

Fig. 104 Pistachio *Pistacia vera* L. Nut in (V) ventral, (L) lateral, (D) dorsal, (A) axial (hilum) view, $\times 1\frac{1}{2}$.

The nuts from the wild trees are smaller than the cultivated forms and have a different flavour: they are highly prized as winter food by the nomads of North Persia and Afghanistan (ibid., 100).

The fruits resemble olives, each containing a single nut. They are harvested when the outer green husk begins to shrivel and turn yellow (ibid., 14). Normally the thin, horny shell splits open at one end when the nut is ripe (ibid., 101). The nuts vary in size up to 2·5 cm long and 1·3 cm in diameter (ibid.). They usually have a scar of attachment at the base (see fig. 104). The kernel consists of two greenish cotyledons covered by a reddish, papery seed coat. The kernel is chiefly used for food: it contains high percentages of both fat and protein (Winton I, 1935, 547 F.).

Hazelnut: *Corylus avellana* L. (fig. 105)

Two forms of hazelnuts were first noted by Heer (1865, 30) in the finds from the Swiss lake-side settlements: *C. avellana* L. var. *ovata* Willd. with round to oval nuts; called *f. silvestris* in the later literature; and *C. avellana* L. f. *oblonga* G. And. with longer nuts. Both forms appear to have been popular and occur together at the following Alpine sites: Wangen, Hornstad, Rauenegg, Bodman, Robenhausen, Wollishofen, Wauwil, Oberkirch-Sempachersee, Baldeggersee, Mauensee, Mörigen, Sütz, Lüscherz, Vinelz, Schönfeld, St Blaise, Bevaix, Concise, Montelier, Greing and Bor (Neuweiler 1905, 53 f.). Hazelnuts have provided a useful source of food since early post-glacial times. They occur on many mesolithic sites – large quantities of hazelnut shells were found in the Maglemosian settlement at Holmegaard, Denmark, for example – and they are frequently found on neolithic and later prehistoric sites too – at Alvastra, Sweden (Helbaek 1955B, 694), at Glastonbury (Reid 1917), and at Ehrenstein, Germany (Hopf 1968, 58).

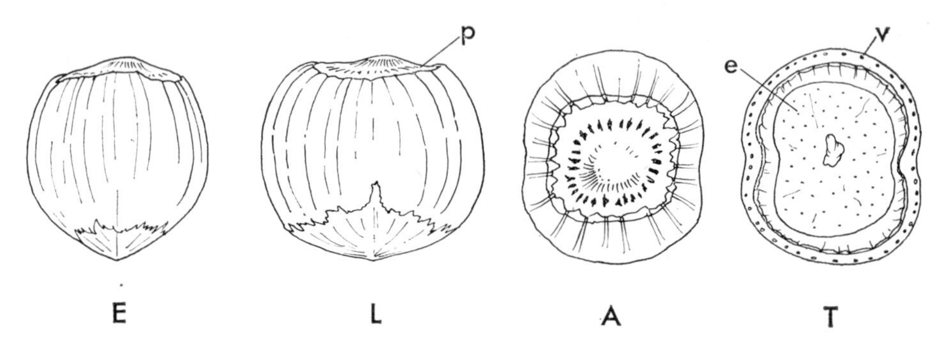

Fig. 105 Hazelnut *Corylus avellana* L. Nut in (E) end, (L) lateral, (A) axial view, (T) transverse section, $\times 1\frac{1}{2}$. (p) pale area of attachment to husk or involucre, (e) cut surface of endosperm of kernel, (v) longitudinal vesicles in woody pericarp.

Beech Mast: *Fagus sylvatica* L. (fig. 106)

Beech mast occasionally occurs in prehistoric food plant remains: it was found at Wangen, Steckborn, Bodman, Robenhausen, Wollishofen, Zug, Wauwil, Moosseedorf, Mörigen, St Blaise, Cortaillod, Greing and Schonfeld in the Swiss lake-side villages,

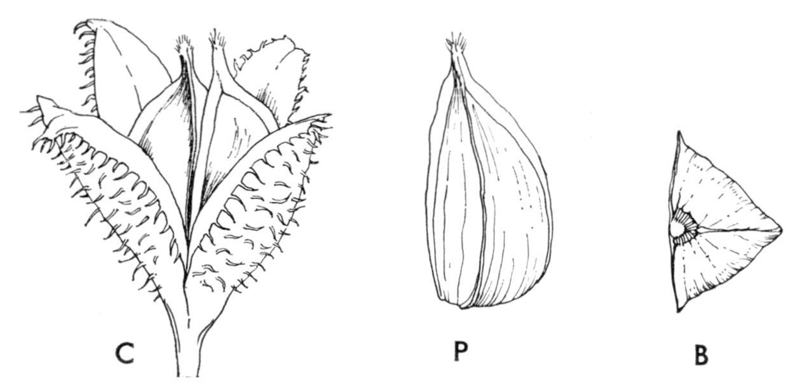

Fig. 106 Beech *Fagus sylvatica* L. (C) cupule, splitting into four valves exposing the two fruits (beech-nuts) within, (P) trigural nut in perspective view, (B) in basal view showing attachment scar. All $\times 2\frac{1}{2}$.

and Hopf (1968, 59) reports it also from the neolithic site at Ehrenstein, Germany. The nuts are contained in a four-lobed, woody cupule; they are triangular in cross-section. Beech mast is used for human food after being baked and salted or roasted (Hill 1941, 39).

Chapter 18

Drug Plants

Opium Poppy: *Papaver somniferum* L.

Finds of capsules and seeds of *Papaver somniferum* L. have been most numerous in the prehistoric lake-side villages of Switzerland. Bertsch and Bertsch (1949, 198) give a list of seventeen finds from neolithic contexts chiefly in Switzerland, but with some in south-west Germany, north Italy and from the Murcielagos Cave in Spain. They also list ten finds from the bronze age in Switzerland. Seeds of *Papaver setigerum* were found at five Bandkeramik settlements in Germany (Oekoven, Aldenhoven, Lamersdorf, Garsdorf and Langweiler) (Knörzer 1971, 34), and *Papaver somniferum* at Langweiler and Wickrath in iron age contexts (idem). Helbaek (1952B, 222) has reported finding *Papaver somniferum* L. at the iron age settlement at Fifield Bavant in southern England, and he refers to Hallstatt period finds at Biskupin, Poland, indicating the expansion of this species into Northern Europe during the iron age.

Capsules of *Papaver somniferum* were found at Murcielagos, Spain, and Robenhausen in Switzerland: the latter measured 1·2 cm in height and 1 cm in diameter (Heer 1965, 32). On the basis of the size of fresh seeds Hartwich has divided *Papaver somniferum* into three varieties:

var. *setigerum*	with seeds	0·66–0·97 mm long
var. *nigrum*	,, ,,	0·88–1·00 mm long
var. *album*	,, ,,	1·17–1·29/1·41 mm long

The finds from the lake-side villages have seeds ranging in length from 0·75 to 1·00 mm (Neuweiler 1905, 73). The Bandkeramik seeds measure 0·75–0·85 (average 0·78 mm) long (Knörzer 1971, 36).

The seeds of *Papaver somniferum* are borne inside a globular capsule (fig. 107). A large capsule may contain over 11,000 seeds, but on average the capsules produce between 6,000 and 7,000 seeds each (Salisbury 1961, 172).

Opium, the congealed latex derived from the juice of the unripe capsules, contains more than thirty alkaloids, of which morphine and codeine are the most widely used for medical purposes. Among the other constituents narceine is hypnotic, whilst thebaine, papaverine and narcotine have an action resembling that of strychnine. Opium is the source of the toxic and extremely habit-forming narcotic, heroin or diamorphine.

Opium is obtained by making fine slits on the outer surface of the capsule – taking care not to pierce the interior (pl. 43a). Pliny (*Natural History* XX, lxxvi, 199) recommends that 'the incision be made beneath the head and calyx, and in no variety is an incision made into the head itself'. The exuded juice is collected next day by scraping it off the cut surface and it is then squeezed into small lumps and left to dry in the shade. The capsules are only capable of yielding the drug for a very short period (9–15 days) after the flowers have fallen and before they ripen and dry.

The seeds of *Papaver somniferum* contain no opium. They are chiefly used in baking

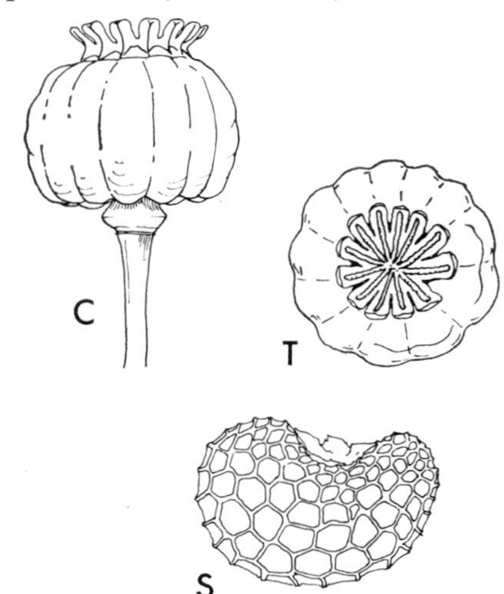

Fig. 107 Opium Poppy *Papaver somniferum* L. Capsule (C) in side view and (T) top view $\times\frac{1}{2}$, (S) seed in side view.

and for sprinkling on the tops of loaves of bread; they have an oil content of up to 56%. The oil obtained from the seeds is similar to olive oil and can be used for cooking and also in lamps – in which it is said to burn for much longer than olive oil (Johnson 1862, 20). The seeds found in the lake-side settlement at Robenhausen, Switzerland, were in the form of a cake and may have been pressed for oil (Heer 1866, 342). Although there is no direct proof that opium was known in prehistoric times it seems likely that its properties were known, and perhaps exploited, by those who collected poppy seeds for food. Certainly the statuette from the late Minoan III site at Gazi, Greece (pl. 43b) strongly suggests the drug was being taken by this time.

In Roman times opium was taken in several ways: the calyx of the white-flowered variety was pounded and added to wine to induce sleep (Pliny, *Natural History* XX, lxxvi, 198), the latex was dried in lozenges which were swallowed (ibid., 199). The seeds were also pounded with milk and made into lozenges which were taken to induce sleep. 'But that liquor is ye best, which is thick & heavy & sleepy in ye smell, bitter in the taste, easily pierced with water, smooth, white, not sharp, neither clotty, nor growing thick in ye straining as wax does, & being set in the sun flowing abroad, & being lighted at a candle, not of a dark flame, & keeping after it is put out, strength in the smell.' (Dioscorides, Book IV, translated Goodyear 1655, edited Gunther 1959, 459.)

Hemp: *Cannabis sativa* L. (fig. 108)

Seeds of hemp have occasionally been recovered from prehistoric contexts in Europe. The earliest find is that from the Bandkeramik site at Eisenberg, Thuringia, Germany, and other neolithic finds come from Thaingen, Switzerland, and Voslau, Lower Austria, and from Frumusica, Roumania (Willerding 1970, 358; Dumitrescu, personal communication).

Male and female flowers are produced on separate plants. Hemp is cultivated for its tough bast fibres, which are used in the manufacture of sacking, sailcloth and other coarse textiles. The seeds contain a fatty oil which is sometimes used as an adulterant or substitute for linseed oil. The upper parts of the pistillate plants are sparsely covered with glandular hairs, which in hot, dry climates secrete a volatile oil and a resin with narcotic properties. In cooler climates the secretion is almost devoid of poisonous

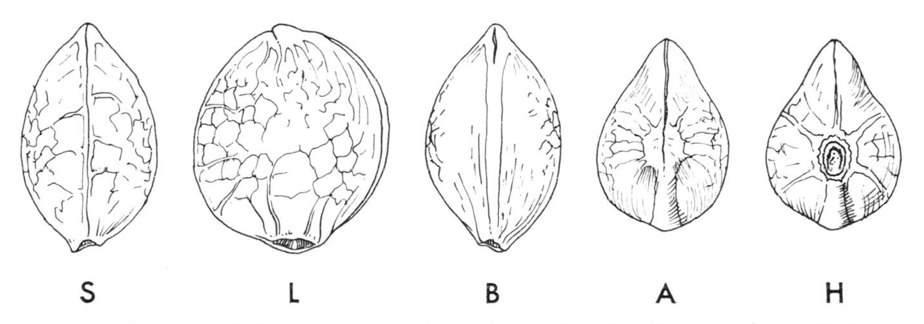

Fig. 108 Hemp seed *Cannabis sativa* L. Fruit (achene) or 'seed' viewed from (S) sharp edge, (L) side, (B) blunt edge, (A) apex, (H) hilum ×6.

properties although the plant still possesses a peculiar stupefying odour (Percival 1900 (1947), 348). Dried flowering tops of the pistillate plants are used medicinally, being sedative, analgesic and narcotic. They are also a source of marijuana or hashish, excessive smoking of which is physically and mentally injurious.

Herodotus (Book IV) describes how the Scythians used hemp seeds to make a vapour bath. They placed a bowl of red-hot stones under a small conical tent of woollen cloth, and crept inside, throwing the hemp seed on the hot stones. It smoked instantly, giving off a vapour 'unsurpassed by any vapour bath one could find in Greece. The Scythians enjoy it so much that they howl with pleasure.' Remains of a conical tent standing over a copper censer together with a pot containing hemp seed were found in the frozen Scythian tomb at Pazyryk (Mound 2), Altai Mts (Artamonov 1965, 100 f.). It is impossible to say how early this practice began, but it is quite possible that it was known to some prehistoric communities.

Chapter 19

Edible Wild Plants

A catalogue of the seeds of wild plants most commonly found on prehistoric sites in Europe and the Near East

This catalogue is by no means exhaustive, but it aims to describe some of the seeds most frequently found in prehistoric grain deposits in Europe and the Near East, which appear to have been used for food. They vary widely in the reasons for their utilization: some are purely the weeds of the cornfield, *Agrostemma githago*, *Sinapis arvensis* and *Stellaria media* for example; or weeds of the flax field – in particular *Camelina sativa* and *Spergula arvensis*; some are deliberately collected flavourings – *Coriandrum sativum*, *Pimpinella anisum* and *Papaver rhoeas* for instance. Some may be seeds of plants whose green parts were possibly eaten, *Rumex acetosa*, *Lapsana communis*; or whose tuberous roots were used for food: for example *Daucus carota*. The possible uses of each species are noted in the catalogue. For ease of reference for non-specialists the catalogue of species which follows is arranged by alphabetical order of the Latin names, and not by botanical families.

Couch Grass, Twitch Grass: *Agropyron repens* (L.) Beauv., *Triticum repens* L.

Spikelets of this species were found carbonized at Mörigen (Neuweiler 1905, 51), and are also reported from the Gumelnitsa levels at Junatcite, Bulgaria, in a sample of hulled six-row barley.

The spikelets of this grass are up to 20 mm long and comprise three to eight florets; they fall entire at maturity. The glumes are lanceolate, 7–12 mm long, with rough keels. The lemmas are also keeled towards the apex; they are 8–13 mm long with five nerves. The grain remains tightly invested in the lemma and palea (Hubbard 1968, 99; see fig. 109). The roots are said to contain a certain amount of nutritive material and can be classed as 'famine food' (Hill 1941, 67).

Corncockle: *Agrostemma githago* L.

Seeds of this species are reported from Robenhausen, Switzerland, and Lengyel, Hungary (Neuweiler 1905, 66), and from Gumelnitsa culture levels at Junatcite and at

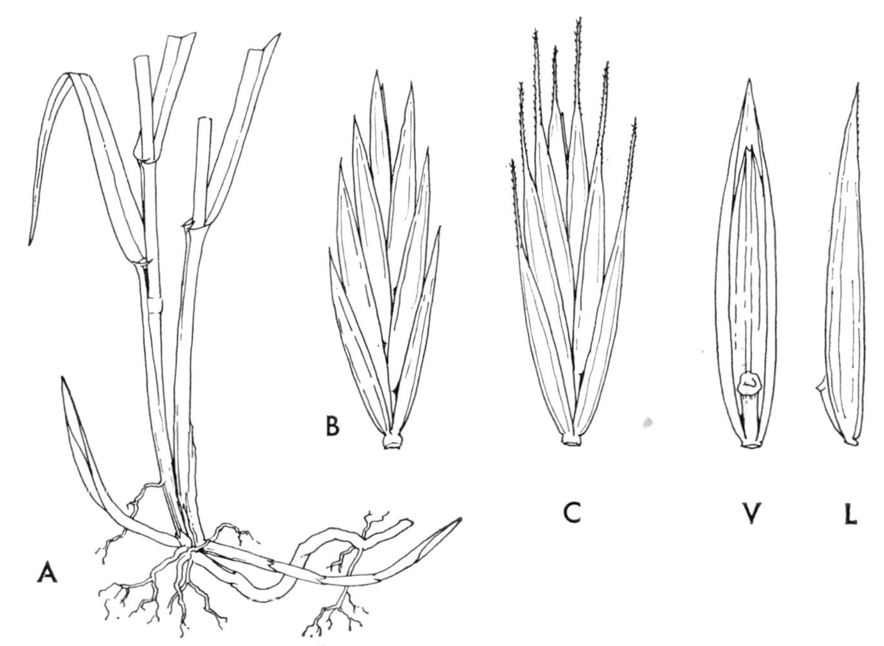

Fig. 109 Couch, Twitch *Agropyron repens* (L.) Beauv. (A) base of plant with rhizomes and young shoots ×½. (B, C) spikelets of awnless and awned forms ×5. Seed in (V) ventral, (L) lateral view ×7.

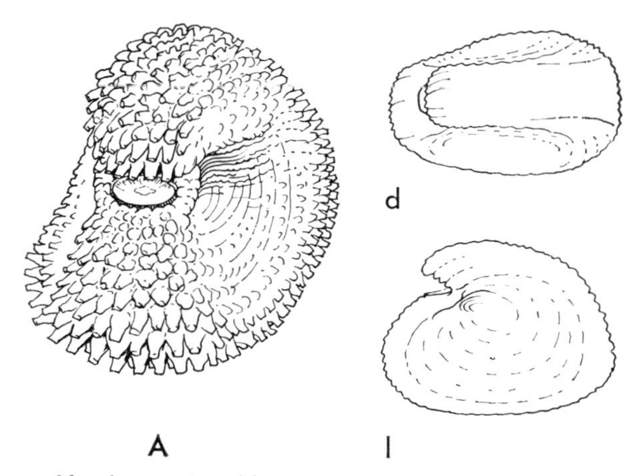

Fig. 110 Corncockle *Agrostemma githago* L. (A) seed ×15 showing surface covered with coarse black projections; (d, l) seed in outline ×10 in dorsal and lateral views. Projections are reduced in size in concave areas on both sides of seeds.

Svetikyrillovo, Bulgaria, where it was found in a deposit of wheat (Gawril 1914, 88).

This is often a troublesome weed in cereal crops: the seeds are poisonous and can damage the physical properties of wheat flour (Polunin and Huxley 1965, 64).

The seeds are contained in a large, ovoid capsule which opens with five almost erect teeth (Gill and Vear 1958, 416). Each capsule contains about forty rough, black seeds

(Percival 1900 (1947), 600). The seeds are about 3–5 mm in diameter and are covered with coarse projections (see fig. 110; Gill and Vear 1958, 416).

A carbonized seed from Roman Isca (Caerleon) Wales measured 3·08 × 2·75 mm, and Helbaek noted that the rough seed coat formed a blunt bulge above the micropyle of the seed (Helbaek 1964B, 162).

Common Orache (Lambs' Quarters): *Atriplex patula* L.

Seeds of common orache were found at Itford Hill, Sussex, and at the Meare lake-side village, dating to the late bronze age and early iron age of Britain respectively (Helbaek 1952, 230). Neuweiler (1905, 65) refers to finds of seeds of this species from Mörigen, Switzerland – late bronze age, and the Aggtelek Cave, Hungary, neolithic.

The pistil forms an utricle enclosed by a pair of triangular bracts which may be fused at the base, the enclosed smooth seeds may be of two types: about 2% of the total output are large (1·6–2·8 mm), almost circular with a small beak-like projection and flattened; the more common smaller seeds (1·0–2·0 mm) are biconvex and circular in form (Muenscher 1955, 204; Salisbury 1961, 162). They are commonly found as weeds in cereal crops (Gill and Vear 1958, 420).

White Mustard: *Brassica alba* (L.) Rabenhorst (*Sinapis alba* Linn.)

A bag full of seeds of *Brassica alba* was found in the Late Helladic site of Marmariani, Thessaly (Tsountas 1908, 360; Netolitzky 1931; Bertsch and Bertsch 1949, 175).

This species is cultivated today to provide salad leaves, and also for the manufacture of condiment mustard. It has a yellow flower about 1·3 cm in diameter. The

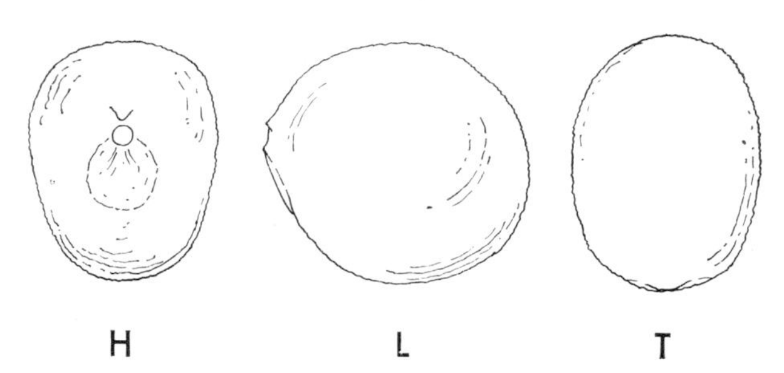

Fig. 111 White Mustard *Brassica alba* (L.) Rabenhorst (= *Sinapis alba* L.) Seed in (H) hilum, (L) lateral, (T) top view ×20.

fruit is a siliqua with strongly nerved valves about 2·5 cm long, covered with stiff hairs and extended to form a sword-like beak. It is found wild throughout Europe and Western Asia (Oldham 1948, 190–1). The seeds are more or less spherical with a thickly sculptured surface (*reticulum*), forming a netting of ridges surrounding small hollows (*interspaces*). The seeds measure between 2 and 3 mm in diameter (Musil 1963, 95–6, and see fig. 111).

Field Brassica (Field Cabbage, Navew): *Brassica campestris* L.

Seeds of this species have been found at Itford Hill, Sussex (late bronze age) (see Helbaek 1952B, 230); at the contemporary site of Nørre Sandegaard, Bornholm (idem

1952C, 110), and were also found in the stomach of Tollund man (idem 1958A, 85). The seed pods are glabrous and have a conical beak about half as long as the valves. The seeds are spherical and about 2 mm in diameter. The surface has a reticulum, and in addition is covered with microscopic pits or stipples (Musil 1963, 95).

Field Brome: *Bromus arvensis* L.

Seeds of this species are reported from the Gumelnitsa levels at Kapitan Dimitrievo III, Bulgaria (Arnaudov 1940/1941). It is also recorded at Lengyel, Hungary (Neuweiler 1905, 51).

The spikelets of this grass are borne in an open panicle – the branches bearing up to eight spikelets each. Each spikelet is 1–2 cm long, 3–4 mm wide and contains four to ten florets. The florets break off at maturity beneath each lemma. The lemmas are 7–9 mm long and have two small teeth at the apex. A straight, rough awn 6–10 mm long arises below the tip of the lemma. The apex of the grain is hairy, but it remains tightly enclosed within the lemma and palea (Hubbard 1968, 77).

Soft Brome, Lop Grass: *Bromus mollis* L.

Seeds of lop grass have been recovered from several prehistoric contexts in Scandinavia: from Vallhagar and Visby in Sweden dating to the iron age (Helbaek 1955B, 681), and from the stomach of Grauballe man, Jutland (idem 1958A, 84). It is also reported

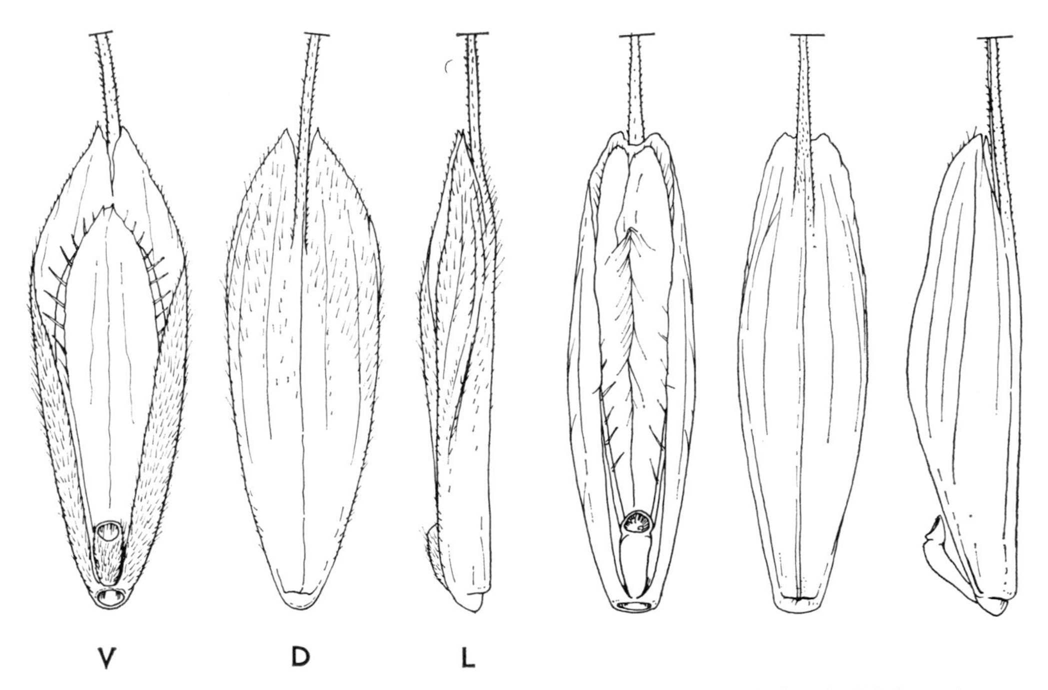

Fig. 112 Soft Brome *Bromus mollis* L. Floret in (V) ventral, (D) dorsal, (L) lateral view showing broad, hairy lemma contrasting with floret of Rye Brome *B. secalinus* L. in similar views showing glabrous lemma and curved rachilla. All ×8.

for the late bronze age at Mörigen in Switzerland (Neuweiler 1905, 51). The spikelets are pubescent and consist of six to twelve florets which break away at maturity from beneath each lemma. The lemmas have five to nine prominent nerves and may split at the apex at maturity. A short, rough awn arises from the back of the lemma just below its apex. The paleas are shorter than the lemmas and have two keels fringed with short stiff hairs (see fig. 112). The caryopsis remains tightly enclosed by the hardened lemma and palea at maturity. The florets are 8–11 mm long and 4·5–5·5 mm wide when fresh (Hubbard 1954, 57). The carbonized 'seeds' from Vallhagar measured 5·4–6·38 (average 6·05) mm long. When carbonized the lemma and palea are liable to break away, exposing the caryopsis, which is thin, tapers towards the base and the small, acutely pointed embryo; the apex is broadly semi-circular in outline.

Chess, Rye Brome: *Bromus secalinus* L.

Seeds of chess were recovered from the early iron age settlement at Fifield Bavant, England (Jessen and Helbaek 1944, 60), and it is reported from the middle bronze age burial at Coity, Bridgend, Wales. In Central Europe it has been found at Lengyel, Hungary; Butmir, northern Yugoslavia, and at several sites in Austria.

The spikelets contain four to eleven florets which break up very slowly at maturity beneath each lemma. The lemmas are glabrous with rounded backs and tightly incurved margins; a fine, straight awn arises from the back of the lemma just below the apex. The paleas are nearly as long as the lemmas and have short, stiff bristles along the keels (see fig. 112). The caryopsis from Fifield Bavant measured: length 5·0 mm, breadth 1·1 mm, thickness 0·9 mm. This is a typical weed of cultivation.

Barren Brome: *Bromus sterilis* L.

Barren brome seeds are very frequently found on prehistoric sites in Britain: Fifield Bavant, Maiden Castle, Winklebury and Itford Hill have all produced seeds of this species (Helbaek 1952A, 230). It is also recorded from Central Europe, for example at Nagyárpás, Hungary, in Vučedol contexts. The seeds are most distinctive, being one of the longest and most slender of the grass caryopses. The caryopsis from Fifield Bavant measured length 9·6 mm, breadth 1·1 mm, thickness 0·7 mm, with a rounded dorsal surface bearing impressions of the nerves of the lemma, and a deep, broad ventral groove (Jesson and Helbaek 1944, 60 f.). The lemmas are seven-nerved, with two small teeth at the apex, and have a fine, rough awn arising from just below the tip. The paleas are shorter than the lemmas with the two keels fringed with short stiff hairs (Hubbard 1954, 43).

Gold of Pleasure: *Camelina sativa* (L.) Crantz

Both the seeds and one or both valves of the silicle are quite frequently found in prehistoric contexts in Europe. They are reported from the Aggtelek Cave in Hungary in neolithic contexts, and occur commonly in iron age deposits in Scandinavia, often mixed with linseed, for example at Østerbølle, Gørding Heath and in the stomachs of the Tollund and Grauballe men in Denmark, and at Vallhagar, Trullhalsar and Rings, Hejnum, Gotland, Sweden (see Helbaek 1938, 1950, 1951, 1955, 1958 for example).

The seeds are fairly large – up to 2 mm long when fresh – oval in outline with a large embryo extending up the whole of one side (see fig. 113); when carbonized the seed shell often bursts, releasing the embryo. A small hilum is often visible at the end of the embryo. The whole seed has a finely pitted epidermis which shows clearly on the

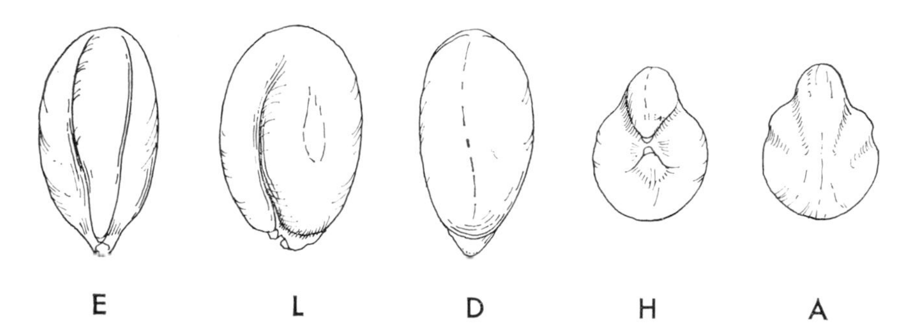

Fig. 113 False Flax, Gold of Pleasure *Camelina sativa* L. Seed in (E) 'embryo', (L) lateral, (D) dorsal, (H) hilum and (A) apical views. Note prominent ridge forming one edge of seed caused by embryo within.

carbonized seeds and also in good grain impressions. The silicle is ovoid, 7–9 mm long, and each valve has a distinct ridge along the margin and a straight, prominent vein running from the base to the apex. The surface of the silicle is irregularly reticulate. The seeds of *Camelina* are rich in edible oil, and it may even have been cultivated for this reason in prehistoric times as it is in some places today. It has recently been found that *Camelina* has a toxic effect on flax, and that even a few plants in the flax field may reduce the yield significantly (Dimbleby 1967, 88).

Shepherd's Purse: *Capsella bursa-pastoris* (L.) Medic.

A pure deposit of these tiny seeds was found at Çatal Hüyük, Anatolia, dating to c. 5800 B.C., and other finds come from iron age Denmark, in particular from the stomachs of the Tollund and Grauballe men. The seeds are up to 1 mm long, more or less oval, with a U-shaped furrow down one side. The surface is minutely granular in appearance. The seeds are contained in a silicula 6–9 mm long, each compartment of which contains ten or more seeds. The seeds are highly oleaginous and may well have been valued for this reason.

Syrian Scabious: *Cephalaria syriaca*

This is a common weed of cereal crops in the Near East today, and its fruits lend a bitter taste to the flour of contaminated grain. Finds of its seeds in ancient deposits are known from late bronze age Beycesultan, late Assyrian Nimrud and from Islamic Quntara (Helbaek 1961, 83). The fruits are roughly grain-shaped with a longitudinal furrow, and often, even when carbonized, bear remains of the long calyx teeth at one end. They measure 4·84–5·68 mm long (carbonized, at Beycesultan) and are thus difficult to remove from cereal grains by sifting.

Fat Hen: *Chenopodium album* L.

The small, shiny black seeds of *Chenopodium album* have often been found on prehistoric sites in Europe: for example Tollund (Helbaek 1950, 334), Vallhagar (idem 1955, 682) and Grauballe (idem 1958A, 84) in Scandinavia; Bertsch and Bertsch (1949, 190) list twelve localities in Switzerland and Germany where these seeds have been found in prehistoric contexts. In some cases large deposits of *Chenopodium album* seeds have been found – for example at Nørre Fjand, Jutland, in a Roman iron age hut, three deposits of grain contained 1,670 c.c. (2,400,000 seeds) of *Chenopodium album.* Inside an early iron age house at Gørding Heath, Jutland, a pot was discovered containing the charred remains of a vegetable soup which, among other seeds, contained 18 c.c. (20% of total number of seeds) of *Chenopodium album* seeds. 7,055 seeds were found in the Bandkeramik settlement at Lamersdorf in the Rhineland (Knörzer 1971B, 93). These large finds of *Chenopodium album* seeds suggest that prehistoric man may have collected them intentionally to supplement his diet (Helbaek 1960C, 17–18).

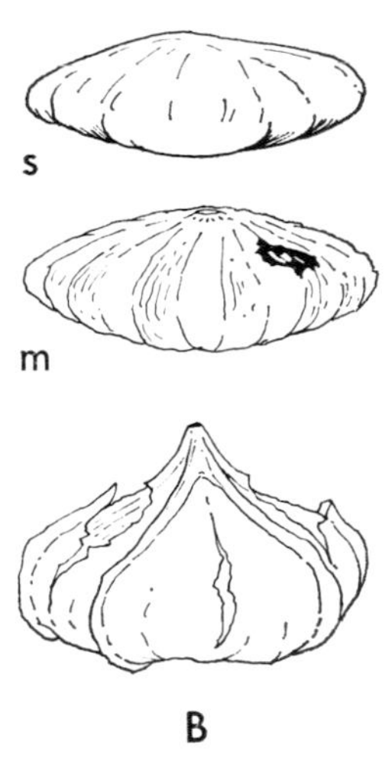

Fig. 114 Fat Hen *Chenopodium album* L. Fruit (achene) loosely surrounded by five-segmented papery perianth, (m) fruit covered with rough, radically striped pericarp with shiny black surface of true seed (s) visible through the break. All ×24.

The seeds (fig. 114) are 1·25–1·85 mm in diameter, the seed coat has a *reticulum* with an elongate or quadrate pattern (Clapham, Tutin and Warburg 1964, 115–16). It occurs commonly as a weed on arable land, especially on nitrogen-rich soils; and is often found as an impurity in clover, barley, oats and rye seed (Gill and Vear 1958, 420). The average plant will produce about 3,000 seeds (Salisbury 1961, 161). The seeds have a high protein and fat content, due to the relatively large embryo (Winton Vol. I, 1932, 325). In some of the Hebrides it is reported to be collected, boiled and eaten as a green vegetable (Pierpoint Johnson 1862, 217).

Coriander: *Coriandrum sativum* L.

Coriander seeds have occasionally been found in ancient deposits. The most notable find was half a litre of coriander in the tomb of Tut-ankh-Amon (fourteenth century B.C., Egypt); seeds were also found in late Assyrian Nimrud (Helbaek 1962A, 183) and from 800 A.D. contexts at Quntara near Khafajah (ibid.). Half a twin fruit was found in the late bronze age site at Apliki, Cyprus. It was carbonized and measured 2·00 mm in diameter and 1·00 mm thick; in form it was a blunt-edged plano-convex disc (ibid.). Coriander seed is also reported from Thera-Therasia (late bronze age, Cyclades). In

Fig. 115 Coriander *Coriandrum sativum* L. Globose fruit showing ridges and rough patterned surface × 10.

Barrows 2 and 5 at Pazyryk seeds of 'cultivated' coriander were found in small leather pouches. Since it does not grow in the Altai region today, these seeds were thought to have been imported (Rudenko 1970, 76 f.). Recently coriander seeds have been found in quantities in the early bronze age levels at Sitagroi (Renfrew, unpublished; and pl. 44).

The fruits of *Coriandrum sativum* L. are ovoid, 3–4 mm in diameter; the primary and secondary ribs are only visible after drying (Polunin 1969, 279; and see fig. 115).

The Romans used coriander seed for flavouring, and it was added to wines, preserves, soups and meat dishes to give flavour in medieval times (Freeman 1943, 6).

Wild Carrot: *Daucus carota* L.

Seeds of this species have been found on neolithic and bronze age village sites in Switzerland (Sipplingen, Lake Constance; Utoquai, Lake Zürich and Robenhausen, for example). One seed was found in iron age Vallhagar, Sweden (Helbaek 1955B, 686).

The seeds are large, 2·5–4·0 mm in length, oval in outline, and flattened on one face. They have four stout, spiny ribs alternating with five slender, hairy ribs on the dorsal surface (see fig. 116); the spines are often rubbed off with handling, and in carbonized material. The ssp. *sativa* (Hoffm.) Arcangeli is the edible, cultivated carrot.

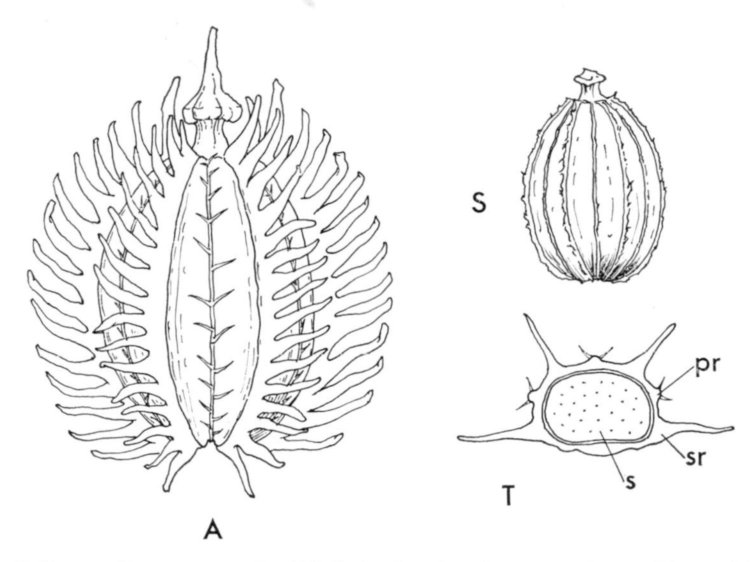

Fig. 116 Carrot *Daucus carota* L. (A) fruit showing four prominent ridges with stiff hooked spines alternating with small ridges bearing short, stiff hairs, ×18. (T) transverse section of fruit ×15, (pr) primary ridge, (sr) secondary ridge, (s) seeds, (S) agricultural seed with spines and most of ridges removed by 'rubbing', ×15.

Common Hemp-nettle: *Galeopsis tetrahit* Mill.

Seeds of this species have been recovered from a number of sites in the Alpine region belonging to the neolithic and bronze ages: Neuweiler (1905, 104, 110–13, 117) reports finds from Steckborn, Robenhausen, Burgäschi, Petersinsel, St Blaise and Schussenried. Seeds have also been found in iron age contexts at Fifield Bavant, England (Helbaek 1952B, 230), in the stomach of Grauballe man (idem 1958A, 85) and 224 seeds were found at Vallhagar, Sweden (idem 1955B, 686 f.).

The seeds are 2·40–3·05 mm long and 1·67–2·05 mm broad, roughly pear-shaped, and where the surface of the seed is revealed it is regularly warted with a conspicuous hilum towards the pointed end (ibid., 687). The seeds are rather similar to those of *Cannabis sativa* L. (*vide supra*).

Hemp-nettle seeds are oily and may have been valued for food in their own right; the young leaves of nettles are sometimes used as a green vegetable and may be made into a purée (Hill 1939, 70). Fibres obtained from the stems are used in cord making (Polunin 1969, 355).

Cleavers: *Galium* sp.

Seeds of *Galium aparine*, *Galium mollugo* L., *Galium spurium* and *Galium tricorne* have all been found in prehistoric contexts in Europe and the Near East. The seeds are almost spherical with a flattened side which is penetrated by a comparatively large aperture leading to the interior. In *Galium aparine* the surface is covered with coarse hooked prickles (see fig. 117A). The fruit splits into two dry one-seeded portions covered with hooked bristles (Gill and Vear 1958, 453). Fresh seeds average 3·5 mm in diameter: carbonized seeds from Vallhagar were 1·97 mm in diameter (Helbaek 1955B, 688). Carbonized seeds from Itford Hill, Sussex, measured 1·90–1·98 mm (idem 1952B, 222). The true seed of *Galium aparine* is spherical with a finely corrugated surface (Parkinson and Smith undated, 74). *Galium mollugo* L. lacks the hooked bristles, but has a reticulate pattern on the testa (see fig. 117c): the seeds are smaller than those of *Galium spurium*

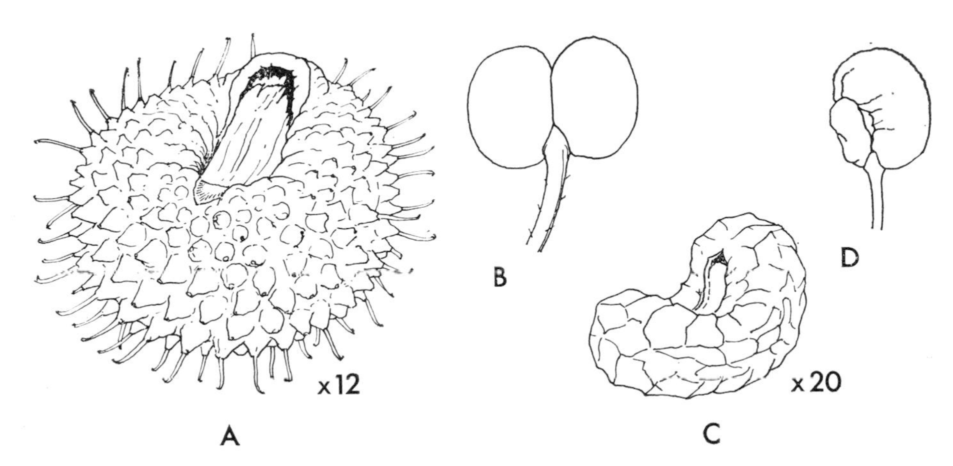

Fig. 117 Goosegrass, Cleavers *Galium aparine* L. (A) half fruit × 12, showing numerous large projections bearing hooked bristles, some rubbed off. (B) fruit in outline × 5. Hedge Bedstraw *G. mollugo* L. (C) half fruit × 20, in similar view showing irregular, coarsely reticulate surface. (D) fruit in outline × 10 showing characteristic under-development of one half.

(Helbaek 1955B, 688). The seeds of *Galium spurium* have a fine reticular pattern which is especially noticeable on the dorsal side opposite the large hole of the scar. The seed is bluntly kidney-shaped in side view. Carbonized seeds from Itford Hill, Sussex, measured: length 1·44–1·52 mm, breadth 0·87–1·03 mm (idem 1952B). Those from Vallhagar: length 1·25–1·85 mm, breadth 1·06–1·56 mm. The carbonized fruits of *Galium tricorne* from Beycesultan measured from 1·82–3·08 mm in greatest diameter (idem 1961, 83); they were distinguished by their amorphous powdery surface, showing only faint traces of the indistinct warts which cover the surface of fresh seeds.

Nipplewort: *Lapsana communis* L.

The long, tapering seeds of this species have been found at the neolithic village at Ehrenstein (Hopf 1968, 65), they occur on many Bandkeramik and Rössen neolithic sites in the Rhineland (Knörzer 1971B, 93), at Schussenriedt and Baden (Neuweiler 1905, 109), and in iron age contexts at Fifield Bavant, England (Helbaek 1952 B,230), in the stomach of Grauballe man (idem 1958A, 85) and at iron age Vallhagar, Sweden (idem 1955B, 689).

The seeds are about 4 mm long when fresh; they are thin, elongated and curved, being pointed at one end and covered with numerous narrow, longitudinal ridges The following measurements have been obtained from the ancient finds:

	Length	*Breadth*
	mm	mm
Fifield Bavant	3·50	0·76
Grauballe man	3·31	1·10
Vallhagar	3·04	—

Dwarf Chickling: *Lathyrus cicera* L.

A number of seeds of this species were found in deposits of early and middle bronze age date at Lerna in the Argolid, Greece (Hopf 1964, 4). Professor Wittmack reported seeds of this species from Bos-öjük measuring 3·7 × 5·0 mm (Neuweiler 1905, 94). One seed of this species was also found at Nimrud (Helbaek 1966D, 620).

The brick-red flowers are borne singly in the axils of the upper leaves on a stalk longer than the leaf-stalk, but shorter than the leaves. The pods are glabrous, 3–4 cm long and 8–10 mm wide (Polunin and Huxley 1965, 108). One coatless seed from Roman Isca, Wales, measured: length 3·67 mm, breadth 4·00 mm, thickness 2·75 mm (Helbaek 1964B, 161). The seeds are angular and bluntly wedge-shaped (see fig. 118). It is sometimes grown as a fodder crop.

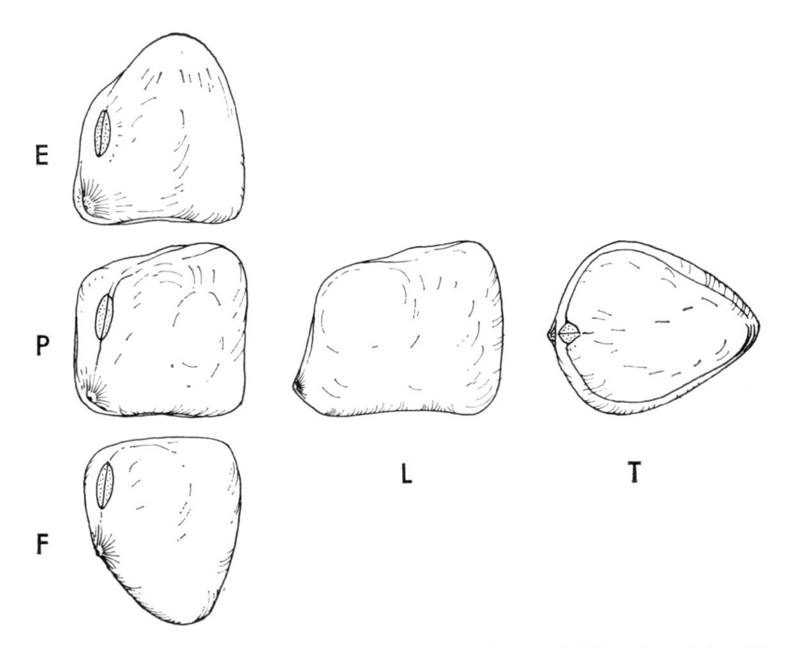

Fig. 118 Dwarf Chickling *Lathyrus cicera* L. Seed from middle of pod in (P) perspective view, (L) lateral and (T) top view; also (E, F) seeds from ends of pod showing characteristic effect on seed shape of position in pod ×6.

Apart from *Lathyrus sativus* (*vide supra*), a number of other species of this family have been found in archaeological contexts: *Lathyrus aphaca* L. (yellow vetchling) is reported from Nimrud (Helbaek 1966D, 620), Roman Isca, Wales (idem 1964, 161). Fresh seeds measure 4 × 3 mm (Musil 1963, 77). *Lathyrus nissola* L. (grass vetchling) was also found at Roman Isca; these seeds have a thick seed coat which is coarsely warted.

Corn Cromwell: *Lithospermum officinale* L.

Seeds of this species were found in a late neolithic burial at Brześć Kujawski (Gimbutas 1956, 118) and over 600 seeds, claimed to have been perforated for use as beads, were recovered from the Gumelnitsa culture burial at Kodjadermen in north-east Bulgaria (Gaul 1948, 133).

This species is a common weed of cereal crops in Eastern Europe – especially in Hungary (Parkinson and Smith, undated, 30). The ovoid nutlets are very hard, smooth or warty, with a truncated base (Clapham, Tutin and Warburg 1964, 301).

Perennial Rye-grass, Tinker-tailor grass: *Lolium perenne* L.

Seeds of this species have been found in iron age contexts at Vallhagar, Sweden (Helbaek 1955B, 682), and it is probably also represented in the stomach of Grauballe man (idem 1958A, 84). Carbonized seeds have also been found in Roman contexts in Britain at Verulamium (idem 1952B, 230) and at Isca (idem 1964, 160). The spikelets are 5–20 mm long; the glumes are always shorter than the spikelet, though their length may vary; the lemmas are lanceolate and may have a slightly pointed blunt or rounded apex, with five prominent veins. The paleas are as long as the lemmas, the two keels being minutely rough. The caryopsis is completely enclosed by the hardened lemma and palea (see Hubbard 1954, 129). The following measurements have been obtained for the ancient finds:

	Length	*Breadth*
	mm	mm
Vallhagar	2·36–2·86	0·95–1·21
Grauballe	2·85–3·53	1·03–1·25
Verulamium	3·04	1·14
Isca	2·65–3·17	1·08

Schrank: *Lolium remotum*

Professor Wittmack of Berlin identified this species in a sample of cereal grains from middle bronze age levels at Orchomenos, Boeotia, Greece (Bulle 1907).

This grass is characteristic of Central Europe: in the past it was a weed of flax crops. It closely resembles *Lolium temulentum* L. but has smaller spikelets (8–11 mm) and smaller, awnless lemmas – 4–5 mm long (Hubbard 1968, 155, and see fig. 119).

Poisonous Rye-grass, Darnel: *Lolium temulentum* L.

Seeds of this species have been found in the Near East – at Tell Bazmosian, Iraq (Helbaek 1965B, 34), Nimrud (idem 1966D, 620), and at Lachish, Palestine (idem 1958); it is also reported from pre-urban Rome (idem 1956, 291).

The spikelets of this species are squat and robust with both the lemma and palea

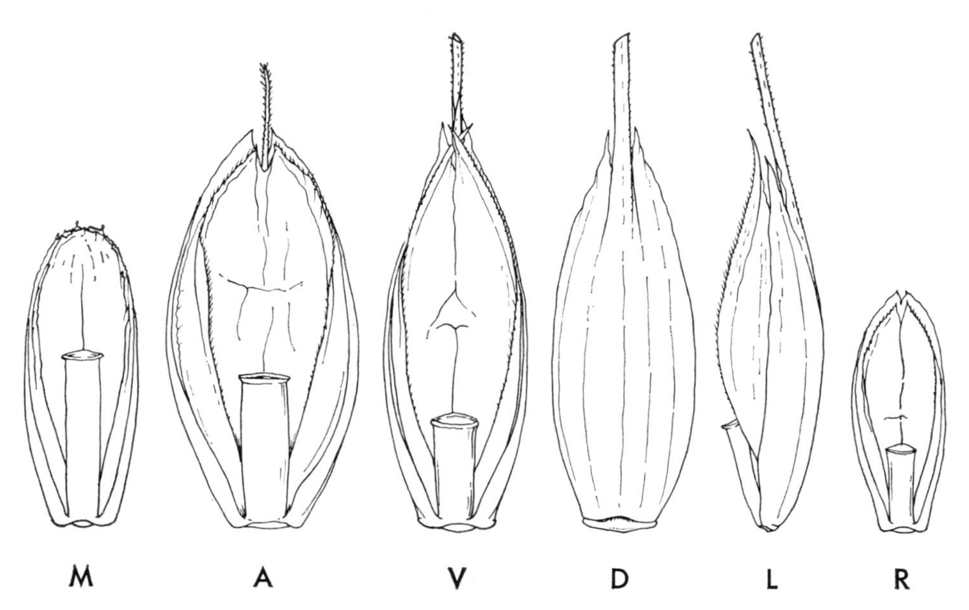

Fig. 119 Darnel *Lolium temulentum* L. Seed in (V) ventral, (D) dorsal, (L) lateral view showing awned form (var. *macrochaeton*) ×7. (A) ventral view of var. *arvense* with awn reduced, (M) smaller-grained form deriving from Mediterranenan area, (R) seed of *L. remotum* Schrank to same scale.

being represented in the carbonized material. It differs from *Lolium perenne* in being awned and having much stouter fruits. A series of florets are illustrated in fig. 119 showing the variation within this species. The following measurements are given for this species:

		Average	*Maximum*	*Minimum*
		mm	mm	mm
Lachish	Length	3·92	4·58	3·48
	Breadth	1·94	2·20	1·65
	Thickness	1·33	1·83	1·10
Rome	Length	4·39	4·58	3·66
	Breadth	1·88	2·20	1·46
	Thickness	1·23	1·46	1·10
Bazmosian	Length	—	4·58	3·84
	Breadth	—	2·01	1·83

Darnel is a particularly obnoxious weed because a fungus is harboured under the seed coat which is toxic to man and animals. When incorporated in the flour of wheat it can cause serious illness to those eating bread made from it. It is a common weed in most of Europe and favours calcareous soils in particular.

Scotch Thistle: *Onopordon acanthum* L.

Seeds of this large biennial thistle were found in the Early and Middle Helladic levels at Lerna (Hopf 1964, 4). The achenes of this species are 4–5 mm long, dark mottled and transversely wrinkled. They are ovoid–oblong in outline, either flattened or 'four-angled' with a pappus of many rows of rough hairs united into a basal ring (Clapham, Tutin and Warburg 1964, 404).

Corn Poppy: *Papaver rhoeas* L.

Seeds of this species have been found occasionally in prehistoric contexts: they occur at Thermi in Lesbos, Aegean, in the early bronze age deposit below House O belonging to Town IV and one or two seeds come from Town III (Lamb 1936, 217).

These seeds should be compared with those of *Papaver somniferum* L. – *vide supra* – which are larger and have been found far more widely in palaeoethnobotanical material in Europe.

This is a common annual on arable land. The fruit is a globose capsule opening by pores beneath the expanded stigmatic disc (Clapham, Tutin and Warburg 1964, 39–40). The seeds are small (about 0·75 mm long) and kidney-shaped. They have a mesh-like pattern over the surface due to concentric rows of minute pits (see fig. 120a; Parkinson and Smith, undated, 58). The seeds are used as a tonic for horses by the

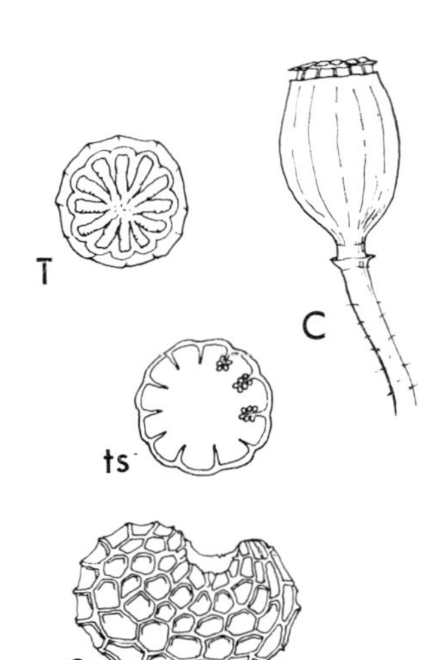

Fig. 120a Field Poppy *Papaver rhoeas* L. (C) capsule in side view; (T) capsule in top view; (ts) transverse section of capsule to show placentation, all ×1; (s) seed ×36.

Arabs and Turks, and infusions from the fruits are good for coughs. A red ink is obtained from the petals (Polunin and Huxley 1965, 74). Compare with *Papaver somniferum* in chapter 18.

Canary Grass: *Phalaris* sp.

Canary grass was one of the main grass species to be found on early sites in south-west Iran; it was most numerous at Ali Kosh in the Bus Mordeh and Mohammad Jaffar phases, and also occurred at Tepe Sabz and Tepe Musiyan (Helbaek 1969, 391). It is an extremely common weed of the Near East today and a few seeds were found in Late Assyrian Nimrud (idem 1966D, 620).

The spikelets of this family of grasses are much flattened, being some 6–10 mm long and up to 6 mm wide. The lemmas are 5–6 mm long, tough and hairy, with five nerves. The palea has two nerves. The caryopsis is retained by the hardened lemma and palea at maturity. *Phalaris paradoxa* L. of the Mediterranean region to which the Iranian finds probably belong is distinguished by having six to seven spikelets in a cluster, the central one of the cluster being fertile, the others sterile, each cluster falling as a unit at maturity (Hubbard 1954, 247).

Anise: *Pimpinella anisum* L.

Aniseed was found in the late bronze age settlement at Thera-Therasia, Greece (Vickery 1936, 26). This seed was highly prized in medieval times for flavouring food: in *The Goodman of Paris* it was sprinkled over meat jellies with bay leaves and cinnamon. In Southern Europe today it is used to flavour bread and cakes (Freeman 1943, 3) – these uses were also known to the Romans (André 1961, 203). The seeds are ovoid, slightly contracted at the top, with a ribbed surface, and a short hard pubescence (*Larousse Gastronomique* 1965, 40; see fig. 120b).

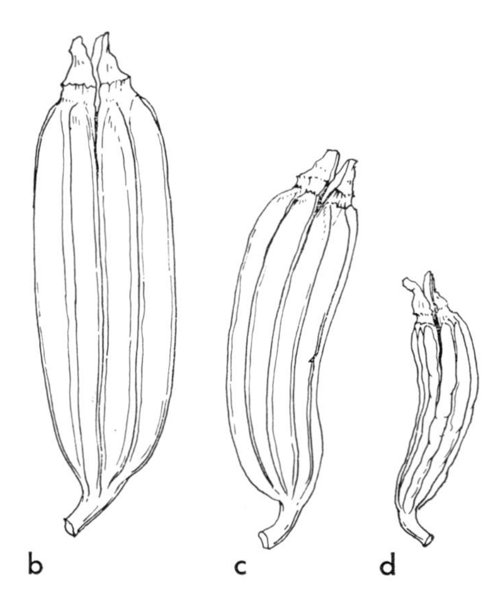

Fig. 120b Aniseed *Pimpinella anisum* L. Range of elongated fruits, (b) large and straight with prominent ridges, (c) smaller, curved, (d) much smaller, wrinkled, ×7.

Ribwort Plantain, Black Plantain: *Plantago lanceolata* L.

Seeds of this species are commonly encountered in archaeological contexts, for example in the stomachs of Tollund and Grauballe men (Helbaek 1950, 337; 1958A, 84 f.); they are also known from Vallhagar, iron age Sweden (idem 1955B, 687), Gørding Heath, Denmark (idem 1951, 65 f.), and from Itford Hill, southern England (idem 1952B, 230). In Central Europe they have been found in neolithic contexts at Aggtelek Cave and Lengyel and at Frehne (Neuweiler 1905, 106).

The seeds are oval, the dorsal side is flatly domed and the ventral side is concave with a conspicuous hilum (see fig. 121); one or two seeds are contained in each capsule. The annual production of seeds per plant is around 2,500 (Salisbury 1961, 221).

Measurements have been obtained for seeds from the following sites:

	Length	*Breadth*
	mm	mm
Tollund	2·17	1·14
Grauballe	2·09–3·31	0·95–1·56
Gørding Heath	2·40	—
Itford Hill	1·75	—
Vallhagar	1·93, 1·97	—

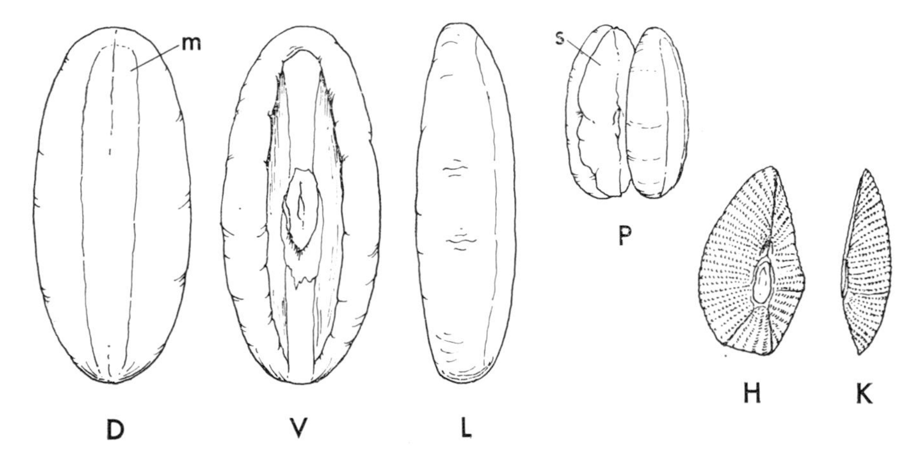

Fig. 121 Ribwort Plantain *Plantago lanceolata* L. Seed in (D) dorsal, (V) ventral, (L) lateral view × 16 (P) attitude of seeds of *P. lanceolata* in capsule. Surface is rough but glossy with pale median streak (m). Broad-leaved Plantain *P. major* L. Seed in (H) hilum, (K) lateral views to same scale. Note row of minute tubercles radiating from the hilum.

Plantago lanceolata requires open ground and cannot exist in naturally wooded country in conditions of sheltered undergrowth. It is a characteristic plant of cleared land, and became very common from neolithic times onwards in temperate Europe. Its mucilaginous leaves are relished by sheep and to some extent by horses and cattle;

the seeds are covered with a coating of mucilage which separates readily when macerated in hot water. The gummy substance thus obtained is used for stiffening certain kinds of muslin and other fabrics (Johnson 1862, 214). The plants are rich in minerals and are sometimes sown for grazing (Gill and Vear 1958, 451).

Broad-leaved Plantain, Waybread: *Plantago major* L.

Seeds of this species were recovered from the stomach of Grauballe man (Helbaek 1958A, 85) and measured 0·84–1·63 × 0·53–0·84 mm. The seeds are variable in shape, angular on one side, with ridges radiating from the scar; the surface is minutely netted (see fig. 121). The capsules are 8–16 seeded, and about 13,000–15,000 seeds are produced per plant annually (Salisbury 1961, 220). The seeds are sometimes found as an impurity in seeds of small legumes (Gill and Vear 1958, 451). It thrives best in grassland where the turf is short, and is common by waysides.

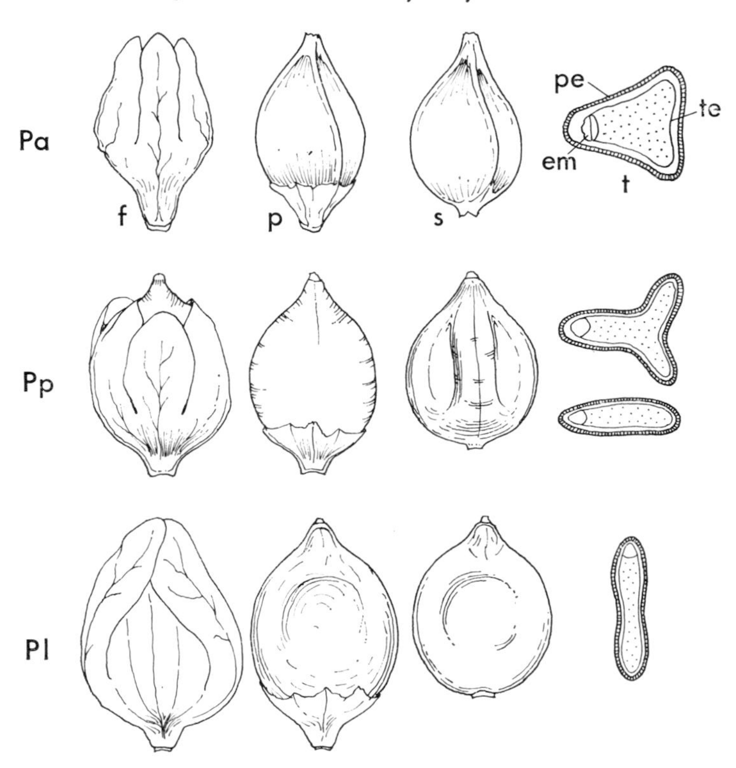

Fig. 122 Polygonum species. *Pa* Knotgrass *Polygonum aviculare* L. (f) fruit (achene) with perianth intact, (p) 'seed' with perianth fragment at base, (s) 'seed' free from perianth, (t) 'seed' in transverse section, (pe) firm, brittle pericarp, (te) thin testa enclosing hard, starchy, semi-transparent endosperm, (em) embryo lying in groove along one angle of endosperm.
Pp Redshank *P. persicaria* L. Fruit, seed in similar views showing the two shapes encountered in this species, trigonous and lenticular.
Pl Pale Polygonum *P. lapathifolium* L. Fruit, seed in similar views showing biconcave, lenticular shape. All to same scale × 12.

Amphibious Bistort: *Polygonum amphibium* L.

A single seed of this species was found at Vallhagar, iron age Sweden; it measured 1·98 × 1·71 mm and was flat, with slightly convex sides, circular/oval in outline with the tip extended into a short point. The seed is usually enveloped in its perianth. This species is usually associated with water, to which its long floating stems are well adapted; in its terrestrial form it may be a troublesome weed on dry ground by river banks (Helbaek 1955B, 682).

Knot Grass, Iron Grass: *Polygonum aviculare*

Carbonized seeds from Beycesultan measured 1·92–2·08 mm long by 1·31–1·68 mm broad (ibid.). Five carbonized seeds from Itford Hill measured 1·98–2·09 mm long: those found in the stomach of Grauballe man measured 2·47 × 1·06 × 0·91 mm (Helbaek 1952B, 230; 1958A, 84). Over 1,900 seeds of *Polygonum aviculare* were found in one deposit in neolithic levels at Sitagroi (see pl. 45).

The seeds of *Polygonum aviculare* frequently occur as impurities in modern seed of cereals, red clover and lucerne (Gill and Vear 1958, 431). The seeds are small achenes 2–4 mm long, triangular in section, with a shiny-punctate surface (Helbaek 1961, 83). Some of them may have one side narrower than the others. Seeds of *Polygonum* are distinguished from those of *Rumex* by the position of the embryo in a groove or channel on one of the angles of the endosperm (Musil 1963, 102; see fig. 122 *Pa*).

Black Bindweed: *Polygonum convolvulus* L.

Seeds of this species have been found in a number of samples from neolithic Bulgaria: in Vesselinovo culture contexts at Yassatepe, Plovdiv (length 2·5, breadth 1·8 mm), Kapitan Dimitrievo III, Gumelnitsa culture (length 1·8, breadth 1·0 mm), and Asmaska Moghila, Gumelnitsa culture levels. It was also found in a number of grain samples from Sitagroi (Renfrew, unpublished). They are also reported from a number of sites in Switzerland and Hungary (Neuweiler 1905, 63) and appear to have been deliberately collected at Ermelo, Holland, early iron age, see van Zeist 1970, 97.

Carbonized Polygonum convolvulus *seeds measure:*

	Length	*Breadth*
	mm	mm
1. Isca, Wales	2·42	1·92
2. Itford Hill, Sussex	2·09–2·58	1·82
3. Gørding Heath	2·89 (av.)	—
4. Grauballe man (uncarbonized)	3·57 (max.)	—
5. Vallhagar	1·90–2·85	1·48–2·01
6. Ermelo	2·2–2·9 (av. 2·49)	1·5–1·9 (av. 1·69)

1. Helbaek, 1964B; 2. idem, 1952B; 3. idem, 1951; 4. idem, 1958B; 5. idem, 1955B; 6. van Zeist 1970, 96.

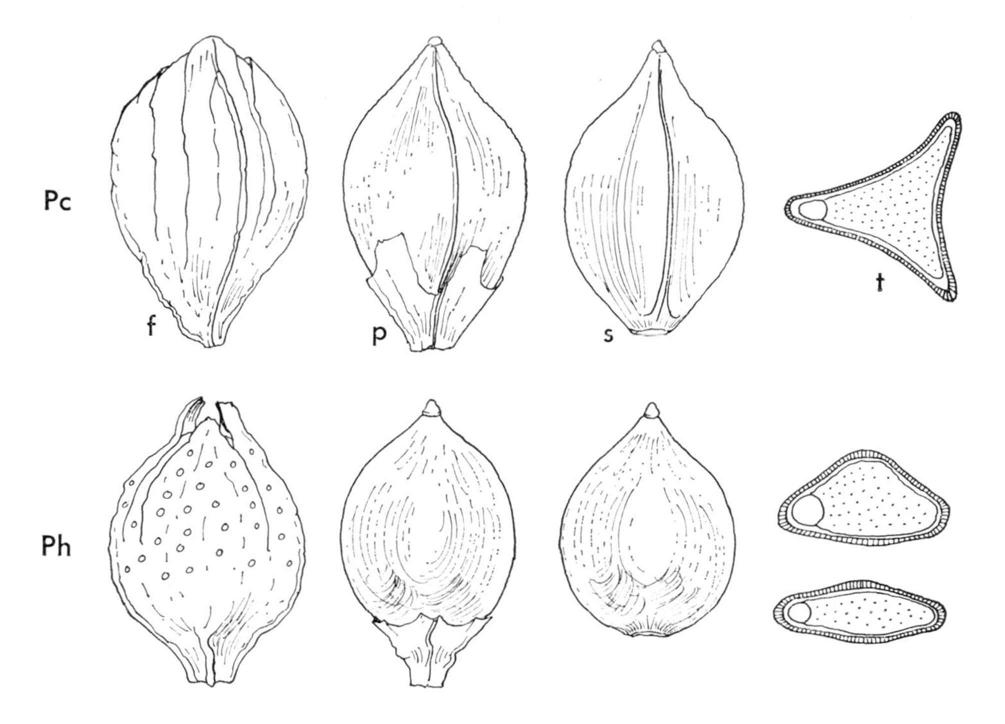

Fig. 123 Polygonum species.
Pc Black Bindweed *Polygonum convolvulus* L. (f) fruit (achene), (p) 'seed' with adherent perianth fragment at base, (s) 'seed' free from perianth, (t) seed in transverse section showing acutely angled trigonous shape.
Ph Water Pepper *P. hydropiper* L. Fruit, seed in similar views to above, showing shape in section which varies from trigonous to plumply lenticular, and also glandular hairs on the perianth.
Both to same scale × 12.

The seeds of this species are similar to those of *Polygonum aviculare* but larger. They have three slightly concave sides and appear triangular in cross-section and elliptical in profile (see fig. 123 *Pc*; Salisbury 1961, 176). Fresh seeds measure about 3·5 mm in length. The surface of the seed is minutely roughened, but is smooth and glossy at the angles (Musil 1963, 104).

The high starch content of the seeds made it attractive as a supplementary food – 7 c.c. of carbonized seeds were found at Gørding Heath (Helbaek 1951): Winton (1932 Vol. I, 318) suggests that a flour comparable in value to buckwheat flour could be made from seeds of *Polygonum convolvulus* L.

Water Pepper: *Polygonum hydropiper* L.

Seeds of this bitter-tasting plant were found at Robenhausen (Neuweiler 1905, 62) and also in a burnt house of iron age date at Sorte Muld (Helbaek 1954, 256). The fruits are about 3 mm long, biconvex or bluntly triangular in section, with a dull surface (see fig. 123 *Ph*).

Pale Persicaria: *Polygonum lapathifolium* L.

Seeds of this species have been found in sizeable deposits in several places in Denmark, strongly suggesting that they were collected deliberately for food in the iron age: 1,000 c.c. of pure *Polygonum lapathifolium* were found at Alrum, Jutland, 1 c.c. of carbonized seeds were recovered from Gørding Heath (Helbaek 1951); 35 complete seeds were found in the stomach of Tollund man (idem 1950, 330 f.), and they were also present in the stomach of Grauballe man (idem 1958A, 84 f.). Besides these Danish finds *Polygonum lapathifolium* has been found in Britain in bronze age contexts at Glenluce Sands, Scotland, and Gorey, Dublin, Ireland, and from early iron age levels at Maiden Castle, Dorset (Helbaek 1952B, 230; Jessen and Helbaek 1944, 26). This species is also reported from Central Europe at Steckborn, Inkwil, Bevaix, Castione, and the Aggtelek Cave, Hungary (Neuweiler 1905, 63). Relatively large amounts of pale persicaria seeds were found at Emmerhout (bronze age) Holland (van Zeist 1970, 166).

The achenes of pale persicaria are usually biconvex in section, with a thick, shiny pericarp (see fig. 122 *Pl*). Between 800 and 1,500 seeds are produced annually per plant (Salisbury 1961, 175 f.). The following measurements have been obtained for the ancient seeds:

	Length
	mm
Gørding Heath	2·04 (2·56–1·52)
Grauballe man	1·75–3·00
Tollund man	2·44 (1·81–2·90)

Pink Persicaria, Redshank: *Polygonum persicaria* L.

Seeds of this species have been found in House A, Østerbølle, Denmark (Helbaek 1938, 217), and from Gørding Heath (idem 1951). It has also been recorded from Steckborn, Bodman, Robenhausen, Burgäschi, St Blaise and Byčiscálahöhle (Neuweiler 1905, 63).

The seeds are usually triangular in section with hollow faces (fig. 122 *Pp*), and about 2·5–3 mm in length when fresh. They occur with cereals and especially as a weed of flax. They thrive best on less calcareous, moist soils (Salisbury 1961, 175).

Common Sorrel: *Rumex acetosa* L.

Seeds of this species were found in the stomach of Tollund man (Helbaek 1950, 333), where the seeds measured 1·75 and 1·94 mm long. The achenes of this species are elliptic in outline with the apex and base being equally pointed (see fig. 124 *Ra*); fresh seeds measure up to 2·5 mm long and 1·5 mm broad (Musil 1963, 103).

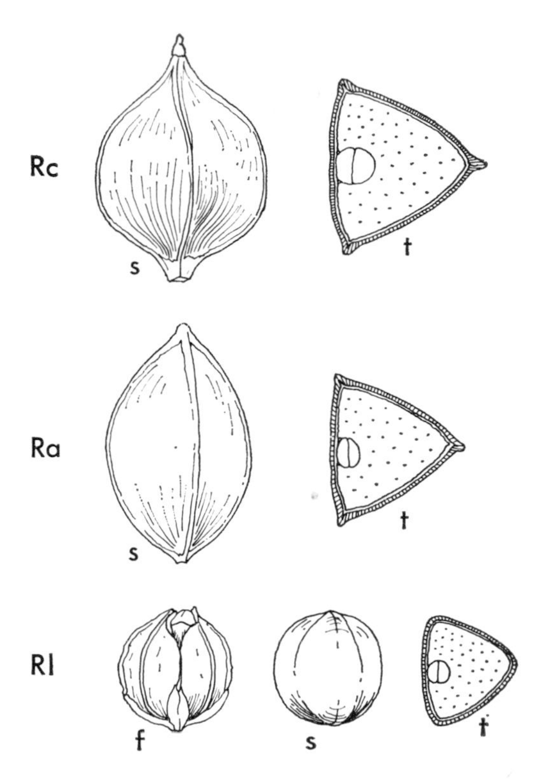

Fig. 124 Rumex species
Rc Curled Dock *Rumex crispus* L.
(s) seed, (t) seed in transverse section showing position of the embryo in centre of one side contrasting with that in *Polygonum* spp.
Ra Common Sorrel *R. acetosa* L. Seed in similar view to above.
Rl Sheep's Sorrel *R. acetosella* L. (s, t) seed as above, (f) fruit (achene) with perianth intact. All to same scale as *Polygonum* spp. (Fig. 122–123) ×12.

Sorrel commonly grows in meadows, especially in damp areas. The sour leaves may be eated in salad, or cooked like spinach (Hill 1938, 21).

Sheep's Sorrel: *Rumex acetosella* L.

Seeds of this species were found in the stomachs of both Tollund and Grauballe man (Helbaek 1950, 333; 1958A, 84) and at Gørding Heath, Denmark, in iron age contexts (idem 1951). The following measurements have been obtained for these finds:

	Length	*Breadth*
	mm	mm
Tollund	0·99–1·29	0·80–0·91
Grauballe	0·91–1·18	0·72–0·87
Gørding Heath	0·95	—

These seeds (achenes) have a rounded triangular section, with ends much less acutely pointed than in the other species of *Rumex* (Helbaek 1950, 333). The perianth often persists round the seed (see fig. 124 *Rl*). It is a weed of arable land and poor sandy soils; it grows most vigorously on lime-deficient land. The taste of the foliage is both acid and bitter and may lead to the development of toxic symptoms when eaten in considerable quantities. Small amounts of sheep's sorrel are known to reduce the milk yield and to affect butter made from the milk (Salisbury 1961, 245).

Curled Dock: *Rumex crispus* L.

Seeds of curled dock were also found in the stomachs of Tollund and Grauballe man (Helbaek 1950, 333; 1958, 84), and they were also identified at Roman Isca (idem 1964, 161), and seeds which may belong to this species, or to *Rumex longifolius*, have been reported from Itford Hill (late bronze age) and from Rivenhall and Wickbourne (Roman) (see idem 1952, 230).

The seeds are trigonous with the style or apex slightly more drawn out than the base (see fig. 124 *Rc*). They occur fairly regularly as impurities in cereal crops and legumes and the plant is scheduled as an 'injurious weed' under seed regulations. A large plant may produce 30,000 seeds in a single season, and since these usually show about 88% germination it is hardly surprising that this is one of the worst agricultural weeds (Salisbury 1961, 179). It is curious that it has not been found more widely in palaeo-ethnobotanical material so far.

Cow Basil, Pink Cockle, Cow Cockle, Cow Herb: *Saponaria vaccaria* L. (*Vaccaria pyramidata* Medicus)

Seeds of this attractive pink-flowered weed have been recovered from Late 'Ubaid/early Uruk levels at Tell Chragh, Shahrazur valley, in the Zagros foothills in Iraq (Helbaek 1960E, 80). In this find the seeds were small and round with a finely warted surface, but had been damaged by puffing during carbonization. The seeds are contained in a capsule which opens by means of four teeth. The seeds are globular, 2 mm in diameter, and covered with minute tubercles; they have a nearly circular depressed hilum (Muenscher 1955, 230 f.).

This is a common weed of cornfields in Central and Southern Europe and in the Near East.

Sea Club-rush: *Scirpus* sp.

Seeds of *Scirpus maritimus* were found in the Bus Mordeh phase at Ali Kosh (Helbaek 1969, 390), and tubers of this species were found at Çatal Hüyük (level 3, sample 4) (idem 1964A, 121 f.). Seeds of *Scirpus tabernaemontani* were found in the latest levels at Nimrud (idem 1966D, 615). These rushes are found frequently by rivers, streams, irrigation canals and open water in the Near East and Europe, especially near the sea or in saline conditions.

The spikelets of *Scirpus maritimus* are 1–2 cm long, oval, and arranged in clusters of two to five. The glumes are a rusty brown colour with a prominent central vein, and are 7 mm long. The apex is bilobed and a central awn arises from the notch (Polunin 1969, 562). In *Scirpus tabernaemontani* the glumes are distinguished by having small swellings on the back.

Annual Knawel, German Knot Grass: *Scleranthus annuus* L.

This species has been identified among the seeds found at Østerbølle House A (Helbaek 1938), Gørding Heath (idem 1951) and in the stomach of Grauballe man (idem 1958A, 84 f.). The seeds are enclosed in a ten-angled perigynous tube which is surmounted by five almost erect sepals. The plants are dominant weeds on markedly acid sandy soils where they may form a complete carpet; they are characteristic on non-calcareous sands. A single plant may produce 2,000 seeds in a season (Salisbury 1961, 248).

Caterpillar: *Scorpiurus subvillosa* L.

Seeds of this species were found in the early bronze age levels at Lachish, Palestine (Helbaek 1958B, 311). Seeds of *Scorpiurus* cf. *sulcata* were found at Nimrud (idem 1966D, 620). The small leguminous seeds of these species are reniform and bluntly pointed at both ends. They are contained in a curved pod, constricted between each seed. Fresh seeds measure 3·30 mm long; the carbonized ones from Lachish were 2·75–2·93 mm long and 2·0 mm wide, and the end of the germ had puffed up into a bubble. It is a typical plant of sandy places in the Mediterranean and Near East.

Herb Sherard, Spurwort, Blue Field Madder: *Sherardia arvensis* L.

One seed of this species was found at Itford Hill, England (late bronze age) (Helbaek 1952B, 230); it is also known from Roman deposits at Baden (Neuweiler 1905, 106). The fruits are about 4 mm long, consisting of two single-seeded halves of the calyx, bearing two or three spiny sepals at the apex. The convex face of the calyx is covered with short bristles; the inner surface has a broad furrow running from the apex to the base. The seed from Itford Hill measured 1·44 mm long and 0·99 mm broad; those from Baden were 1·3–1·7 mm long by 1·0 mm broad.

Charlock, Wild Mustard: (*Brassica arvensis*) *Sinapis arvensis* L.

One round seed with a faint reticulate surface, some 1·32 mm in diameter, was found in a sample of lentils at Beycesultan (Helbaek 1961, 80), and seeds of this species have also been found at Mörigen (Neuweiler 1905, 74). The seeds of this species have a very fine reticulum with minute stipples, and very small interspaces (Musil 1963, 96). The seeds are contained in hairy siliquaes terminating in a long conical beak. The seeds are circular, about 1·6 mm in diameter. Charlock is a very common weed of arable land, being especially common on calcareous soils, clays and heavy land. Before the introduction of chemical herbicides it was one of the worst weeds of arable land, and heavy infestation of spring wheat crops often led to their having to be ploughed in (Salisbury 1961, 186; Gill and Vear 1958, 127 f, 411). This species should be compared with *Brassica alba* and *Brassica campestris* above.

Corn Spurrey: *Spergula arvensis* L.

Seeds of this species are quite frequently found in palaeoethnobotanical material. In several cases in Denmark it has been found in substantial quantities – at Gørding Heath,

Jutland, 2 c.c. of *Spergula* seeds were found with barley grains, *Polygonum* and *Chenopodium* seeds in a small pot, as though it was some kind of vegetable soup. A heap of *Spergula* seeds (some 5,600 c.c.) was found by Knud Jessen in an iron age house at Ginderup, Jutland (Helbaek 1960D, 18). They were also found at Østerbølle, Jutland, in a sample of flax and cameline seeds (idem 1938, 217, 225). Seeds of this species were present in the stomachs of the Tollund and Grauballe men (idem 1950, 334; 1958A, 84); *Spergula arvensis* has also been reported in a sample of flax seed from iron age deposits at Frehne, and at Peschiera, north Italy (Neuweiler 1905, 69). Corn spurrey seeds also occurred in middle bronze age deposits at Emmerhout 215 (van Zeist 1970, 70). The following measurements have been obtained for the ancient seeds:

	Length	*Breadth*
	mm	mm
Gørding Heath	0·90 (1·06–0·80)	—
Grauballe	1·06–1·18	0·87–0·99
Tollund	1·15–1·22	—

Fresh seeds measure 1·8 mm in diameter. They are biconvex and disc-like with a winged rim running around the circumference (Parkinson and Smith, undated, 67). The seeds are contained in a capsule, which splits when ripe into five teeth at the top, exposing the seeds. Each capsule usually contains about fifteen seeds.

Owing to its tolerance of soil acidity the larger-seeded forms have been cultivated as a crop where more remunerative crops cannot survive. It is a common weed of cereal and flax fields. The seeds even in recent times have been ground up for use as human food – hence the Shetland name, meal plant. (See Gill and Vear 1958, 417; Jessen and Helbaek 1944, 61; Salisbury 1961, 246.)

Common Chickweed: *Stellaria media* (L.) Vill.

Seeds of this species occur quite frequently in samples of prehistoric seeds. They were found in the stomachs of Tollund and Grauballe man (Helbaek 1950, 334; 1958A, 84 f.), they were also found at Vallhagar (idem 1955B, 683) and at Gørding Heath (idem 1951), all belonging to the Scandinavian iron age. In Central Europe Neuweiler (1905, 68) reports them from Steckborn, Robenhausen, Baldegersee, Bürgaschi, St Blaise, and Peschiera, north Italy.

This species is especially common on moist, arable land. The seeds are contained in oblong capsules which open at the tip by splitting into six triangular segments, separating when the air is dry, but closing together in damp conditions, so that the seeds only escape under conditions favourable to dispersal. Each capsule contains five to sixteen seeds, about 1·0 mm in diameter, round or kidney-shaped in outline, and covered with rows of warty protuberances. An average plant produces about 240 capsules per annum, though some large plants have produced up to 1,153 capsules.

Excessive consumption of chickweed has led to the death of lambs; it has a tendency to form large indigestible lumps which ferment in the stomach.

Fenugreek: *Trigonella graecum* L.

Seeds of this species have been recovered from Tell Halaf (4000 B.C.), Meadi, Egypt (3000 B.C.) and from iron age levels at Lachish, Palestine (Helbaek 1958B, 311). The seeds from Lachish measured: length 3·84–4·21 mm, breadth 1·83–2·20 mm. They are oblong and angular with a deep groove between the radicle and the cotyledons. Fresh seeds measure 5·5–6 mm long and 3·0–3·5 mm broad. The surface is faintly roughened by minute tubercles or short raised lines. The radicle is about half the length of the cotyledons and the minute hilum lies partly obscured in the deep groove (Musil 1963, 90).

Seeds of fenugreek are highly aromatic and contain about 23% protein, 9% oil, 10% carbohydrate and considerable quantities of fibre and resinous material. They also contain trigonelline and choline; the flavour of the seed is due to the essential oil, coumarin. The plants are grown for fodder in India today. They are used as a constituent of curry powders and as a condiment, and they are also used in local medicine (Cobley 1963, 219 f.).

Cow-herb: *Vaccaria segetalis* L.

Seeds of this decorative plant of the carnation family were encountered in the samples of bitter vetch and barley from Beycesultan, Anatolia. They also occurred at Tell Chragh, Iraqi Kurdistan (Helbaek 1960E), at Quntara, Bagdad, in the early Islamic levels (idem 1961, 80), and late Assyrian Nimrud (idem 1966, 620). The seeds from bronze age Beycesultan were spherical, but had burst on carbonization, and had finely punctate surfaces; they measured 1·52–1·72 mm in diameter. Seeds from neolithic Hacilar were globose, minutely and densely warted, and 1·85 mm in diameter (idem 1970, 234).

Narrow-leafed Vetch: *Vicia angustifolia* L. *Vicia sativa L.* ssp. *angustifolia* (L.) Gaud.

Seeds of this species have been identified in a number of samples from Bulgaria: they were found in the early neolithic levels at Tell Azmak, and also from the eneolithic levels at the same site (Kohl and Quitta 1966, 33–5), and from the eneolithic levels at Karanovo (ibid., 38). Seeds of this species were also recovered from Roman Isca,

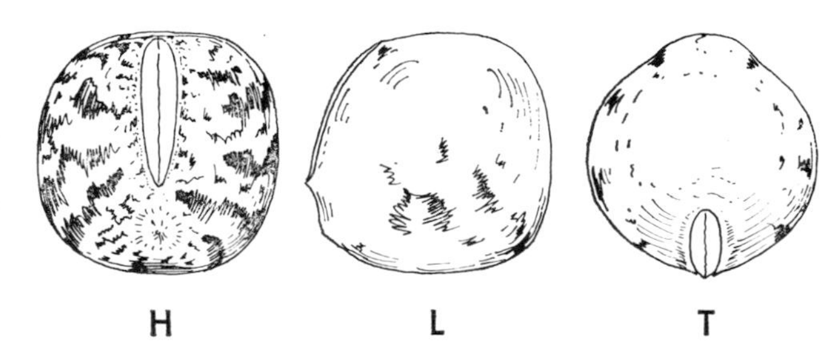

Fig. 125 Narrow-leaved Vetch *Vicia angustifolia* L. Seed in (H) hilum, (L) lateral, (T) top views ×10 showing flattened, spherical shape.

Wales (Helbaek 1964B, 161). The seeds from Isca measured 2·25–2·83 mm diameter and had a wedge-shaped hilum 0·50–1·67 mm long. The fresh seeds measure 2·5 mm in length (Parkinson and Smith, undated, 70; fig. 125). They are more or less globular, or may be cubical and wrinkled. The hilum is about 2·0 mm long, slightly depressed at the margins and raised along the median groove (Musil 1963, 92).

Tufted Vetch: *Vicia cracca* L.

Seeds of this species of vetch have been recovered from a wide range of archaeological contexts: they were found at Meadi, Egypt (3000 B.C.), were present in Central Europe in late neolithic and bronze age contexts at Steckborn, Baden, Lengyel and Hostomits (Neuweiler 1905, 84), and were also identified in iron age deposits at Vallhagar, Sweden (Helbaek 1955B, 685). They were also present at Roman Isca and there the carbonized seeds measured 3·17 mm diameter, with a parallel-sided hilum 2·42 × 0·42 mm (idem 1964B, 161).

Field Pansy: *Viola arvensis* Murr.

The seeds of field pansy were found in the stomach of Tollund man, and because this plant grows low on the ground one must imagine that the seeds were deliberately collected and added to his last meal. The seeds are pear-shaped, with an oblique irregularity at the pointed end. The seeds measured average 1·44 × 0·84 mm (minimum 1·29 × 0·76; maximum 1·60 × 0·99 mm). The seeds are contained in a capsule which opens by means of three valves; the portion of the valves which survives in Tollund man is the median thickened ridge (Helbaek 1950, 336). This plant is found on arable land, especially on basic or neutral soils (Clapham, Tutin and Warburg 1964, 81).

Chapter 20

The Food Values of Utilized Plants

The wide range of cultivated and utilized plants described in the preceding chapters may cause one to wonder why so many plants of each category – cereals, pulses, oil-seeds, fruits, nuts and wild plants – were utilized. To some extent the answer must lie in a study of their food values as a part of the diet, and in their suitability for different food preparations. It is difficult to reconstruct the form in which these vegetable products were consumed as food in prehistoric times. Occasionally one is allowed a brief insight: for example in the analysis of the stomach contents of the Tollund and Grauballe men, or in the examination of the small 'buns' found chiefly in the Swiss lake-side settlements; it is tempting to take these isolated examples as being typical where no other evidence exists, of the gruel and bread of our prehistoric ancestors. Most of the fruits and nuts were probably consumed without any form of food preparation, fresh from the tree or dried, though some fruits were clearly also made into some form of wine. There is evidence of making olive oil in the Aegean in prehistoric times. The wide range of wild plants were used in a multitude of different ways – as supplements or substitutes for the cereals in times of dearth, as condiments and flavourings, as pot-herbs, or as root vegetables, quite apart from any medicinal value they may have also had.

THE CEREALS

These were chiefly used, once they had been threshed, winnowed and ground to coarse groats on saddle querns, either for making into porridge or bread. They were important principally as a source of carbohydrate and for providing a daily supply of vitamins B and E (cf. Drummond and Wilbraham 1940, 99 and 103). The cereals have the compositions shown in the table opposite (cf. Winton 1935, 60, 254–5).

Perhaps the best indicators that have survived of the composition of porridge or gruel are the stomach contents of the Tollund, Grauballe and Borremose men, from the Danish iron age. All these men died in unnatural circumstances so that their last meals may not have been typical of the normal diet of the time, but they do give an indication of the utilization of cereals and wild seeds which were available. Briefly, they were

	Protein	*Fat*	*Nitrogen-free extract (carbohydrate)*	*Fibre*	*Ash* %
Bread wheat	13·3	2·3	80·4	2·0	2·0
Emmer wheat	13·28	1·91	69·42	11·31	4·07
Einkorn wheat	14·67	2·19	64·02	13·55	5·57
Naked barley	13·5	2·5	80·3	1·6	2·1
Oats	15·9	7·7	73·2	1·0	2·2
Rye	12·0	1·9	82·1	1·9	2·1

mainly composed of lax-eared six-row barley, oats, flax and a mixture of seeds of wild plants, of which the following were present in appreciable quantities: *Rumex*, *Polygonum*, *Chenopodium*, *Spergula*, *Camelina* and *Viola*. In the case of Grauballe man seeds of some fifty-eight wild species were present, mainly in small quantities (Helbaek 1958A, 84 f. See also table in chapter 2, p. 18).

BREAD

This was possibly first made in the form of unleavened bannocks, consisting of bruised cereals mixed together with water and possibly oil (such as are made in Khuzistan today). The discovery that bread could be leavened may have been quite accidental – as a result of watching the spontaneous fermentation which occurs if the flour-and-water mixture is allowed to stand in a moderately warm place for some hours – due to the absorption of wild yeast from the air. Bread which is unfermented has very little taste (Tibbles 1912, 401). Unleavened bread made from wheat flour mixed with salt and water, kneaded and baked in thin layers until crisp, contains very little moisture, is very compact and of high nutritive value. Because of the gluten content of wheat flour (especially that of bread wheat) wheaten loaves are the lightest and most nutritious form of bread. Tolerable bread can also be made from barley and rye – although their loaves are much more dense in texture (see pl. 46). They are greatly improved and lightened by the addition of a little wheat flour. It is interesting to compare the food values of bread made from flour of different cereals with that for the grains themselves:

The composition of varieties of bread: nutrients %
(after Tibbles 1912, 424–6)

	Water	*Protein*	*Fat*	*Carbohydrate*	*Fibre*	*Salt*	*Ash* %
Wheaten bread	33·02	7·94	1·95	56·76	0·24	0·50	1·05
Barley bread	12·39	5·90	0·90	71·00	5·63	—	—
Rye bread	35·70	9·00	0·60	53·20	0·50	—	1·50

Flat oatcakes can be made from oatmeal – they become more spongy and digestible if it is made with sour leaven (Tibbles 1912, 147). Ancient Irish unleavened oatcakes were prepared by 'mixing the meal with sweet milk or butter milk so as to make a stiff dough, which was fashioned into flat cakes and baked, supported in an upright position before the fire by means of a three-pronged forked stick'. Honey was also sometimes added to the dough (O'Curry 1873, ccclxii). Piers Plowman lived on 'cake of oats' during a year of scarcity (Ashley 1928, 96).

In the *Description of England* by William Harrison prefixed to Holinshed's *Chronicles* (1577) there is an account of different forms of bread in medieval England:

> The bread throughout the land is made of such grain as the soil yieldeth, nevertheless the gentility commonly provide themselves sufficiently of wheat for their own table whilst their household and poor neighbours are forced to content themselves with rye or barley, yea and in time of dearth, many with bread made either of beans, peas or oats, or of all together and some acorns among. ASHLEY 1928, 58

Some seventy-two different types of bread were made by the ancient Greeks with the additions of milk, oil, honey, cheese and wine. *Azumos* was a delicate unleavened biscuit; *Escarites* was made from a light paste seasoned with new sweet wine and honey; *Dolyres* and *Typhes* were coarse compounds of rye and barley; *Placites* were puff cakes and *Tyrontes* bread mixed with cheese, consumed by the robust workmen of Piraeus (Soyer 1853, 32 f.).

The Romans also made a number of different types of bread: *Autopyron* was a coarse, dark mixture of bran with a little flour made for the consumption of slaves and dogs. *Athletae* was bread mixed with a soft curd cheese, but otherwise unleavened; *Buccellatum* was a biscuit, or dried bread given to the troops, and *Artophites* was a light, leavened bread made from the best wheaten flour and baked in a mould. Many carbonized loaves of this type of bread were found at Pompeii in a baker's oven (see pl. 47; Soyer 1853, 36–7).

The Romans used millet for making their leaven; they mixed it with sweet wine and left it to ferment. They also made leaven from wheat bran soaked in white wine and dried in the sun. Often leaven was made from the newly mixed dough before the salt had been added, a piece being taken from the mass and allowed to go sour.

Prehistoric loaves of bread have seldom been found: all that are known at present are rather small, resembling buns or bread rolls rather than the large loaves which we are familiar with today. In the Swiss lake-side settlements a number of finds of these 'buns' have been made: at Robenhausen nearly 3·6 kg of carbonized buns were found; at Wangen Mr. Löhle discovered similar buns measuring 2·5–3·8 cm high and 10–13 cm in diameter, composed of 'grains of corn more or less crushed. In some specimens the halves of grains of barley are clearly discernible. The underside of these cakes is sometimes flat, sometimes concave, and there appears no doubt that the mass of dough was baked by being laid on hot stones and covered with glowing ashes' (Keller, trans. Lee, 1866, 49, 63). Bread made from millet is reported from bronze age contexts at Irgenhausen, Switzerland (ibid. 58), and Marmariani, Thessaly, Greece (Vickery 1936, 49, 55). The buns from iron age Glastonbury, England, were examined by Clement Reid and he found them to be composed of 'whole unbroken wheat grains with a noticeable proportion of glumes and fragments of awn . . . this small cake looks as though it had been kneaded out of a mixture of wheat and something sticky, probably honey. It does not appear to have been much baked, as there is no sign of crust or of burning' (Bulleid and Gray 1916, 629). Helbaek also examined one of the buns from this site and managed to isolate fragments of wheat, barley, wild oats, chess and a seed of *Atriplex patula*. He concludes 'it seems evident that the buns and bread were made from coarsely ground cereals, including impurities of the field' (Helbaek 1952B, 212, and pl. XXIIIa, b).

In times of scarcity bread has also been made from bean and pea flour; this should first be soaked in water, then drained and mixed with salt and yeast, being left to ferment before baking. These loaves are greatly improved by the addition of equal quantities of wheat flour to the paste (Tibbles 1912, 147).

There is evidence that various wild plants were probably utilized in prehistoric times when scarcity of cereals forced them to be mixed with wild seeds for bread making; thus acorns appear to have been mixed with six-row barley in a pithos in level IV at Sitagroi and these two ingredients were found between the upper and nether quern stones at Raskopanitza near Manole, Plovdiv, Bulgaria, in the late bronze age levels (both Renfrew, unpublished). Acorn bread was resorted to by the French peasants during the famines of 1546 according to Bishop Du Bellay of Mans, who complained about their sorry condition to Francis I (Soyer 1853, 24). Among other seeds which were probably also used as substitutes for cereals those of *Polygonum convolvulus* are perhaps the commonest to be found in prehistoric contexts.

The cereals were also used, at least in later prehistoric times, for brewing alcoholic beverages. There is evidence that emmer and spelt wheat were used in preparing beer in ancient times – emmer has been found in residues inside beer jars in Egypt (Täckholm, Täckholm and Drar 1941, 249); and sprouted grains of spelt were found mixed with similar grains of bread wheat, rye and hulled barley in a deposit from Roman Isca (Caerleon, South Wales) (see Helbaek 1964B, 158 f.). Finds of sprouted barley grains apparently used for brewing were identified from Eketorp, Öland, Sweden (idem 1966C, 218), and at Østerbølle in Jutland (idem 1938, 216).

PULSE CROPS

These were grown chiefly as a source of vegetable protein – obtained from the seeds. Their protein content is not so concentrated as in meat, fish, eggs and milk (the 'first-class' proteins); consequently when vegetable proteins are relied on exclusively, a greater bulk of food needs to be eaten to obtain the same quantity of protein (Drummond and Wilbraham 1940, 89). The composition of beans and peas is set out in the following table based on Hunter (1951, 122, 126).

Chemical composition of Vicia faba *L. and* Pisum sativum *L.*

	Protein	*Oil*	*Carbohydrate*	*Fibre*	*Ash*
	%	%	%	%	%
Vicia faba	25·4	1·5	48·5	7·1	3·2
Pisum sativum	22·5	1·6	53·7	5·4	2·8

OIL SEEDS

These were recognized at an early stage of agriculture to have nutritional value, and the first crop of this type to be cultivated for food was flax. Its linseed contains from 30 to 40% oil, varying with the variety and the environment in which the crop is grown. As well as oil, linseed also contains 20–25% protein. The linseed cake or remains of the seed after the extraction of the oil contains on average 29·5% crude protein, 9·5% ether extract, 9·1% crude fibre, 5·2% ash and 35·5% nitrogen-free extract, and is thus very nutritious (Hunter 1951, 139).

NUTS

These provide another rich source of fats, protein, and carbohydrate: 100 gm of shelled walnuts (about thirty nuts) contain as much fat as 1·24 kg of lean beef, and as much protein as 28 gm of beef (Tibbles 1912, 681–2). They are also rich in minerals, especially iron and lime (Howes 1948, 17). The trouble with nuts, however, is that they become rancid after a few months, even when stored with great care, and their palatability is thus greatly diminished (ibid., 683). There is ample evidence that nuts were valued in prehistoric times and were collected in the autumn to supplement the diet.

Food values of nut kernels (cf. Howes 1948, 23)

	Water	*Protein*	*Fat*	*Carbohydrate*	*Fibre*	*Ash*
	%	%	%	%	%	%
Almond	4·8	21·0	54·9	14·3	3·0	2·0
Pistachio	4·2	22·6	54·5	15·6		3·1
Walnut	3·4	18·2	60·7	13·7	2·3	1·7
Acorns	6·3	5·2	43·0	45·0		3·6
Beechnuts	9·1	21·7	42·5	19·2	3·7	3·9
Hazelnuts	41·1	9·0	36·0	6·8	10·3	—

FRUITS

In general, fruits are very deficient in proteins and other nitrogenous substances. They are succulent, sweet or acid, and are fairly rich in carbohydrates, chiefly sugars, which may be easily digested. The sugar content increases as the fruits ripen and the starch, which may be abundant in unripe fruits (such as apples), becomes converted to sugar during ripening. Many of the fruits are also rich in pectin and are thus well adapted for making into preserves – as jam or jelly (Winton 1935, 479). It is interesting to note that the sugar content of dried fruits is high: Figs for example consist of 52% sugar (Tibbles 1912, 656). It is as a source of sugar that they must have been most highly valued in prehistoric times – when honey was the only other source of sweetness.

Only one species of fruit has a high fat content – the olive. Its flesh contains 17% of fat, and it was cultivated as a source of vegetable oil at least from the bronze age in Greece.

Average composition of fresh and dried fruits % (after Tibbles 1912, 596–7)

	Refuse	*Water*	*Protein*	*Fat*	*Carbohydrate*	*Fibre*	*Ash*
	%	%	%	%	%	%	%
Apples – dried	—	26·1	1·6	2·2	62·0	6·1	2·0
Figs – dried	—	18·8	4·3	0·3	68·0	6·2	2·4
Olives	17·9	67·0	2·5	17·1	5·7	3·3	4·4
Pears – dried	—	16·5	2·8	5·4	66·0	6·9	2·4
Grapes — fresh	25·0	77·4	1·3	1·6	14·9	4·3	0·5
,, — raisins	10·0	14·6	2·6	3·3	73·6	2·5	3·4
Cherries – fresh	5·0	80·9	1·0	0·8	16·5	0·2	0·6
Plums – fresh	5·0	78·4	1·0	—	20·1		0·5
Hawthorn haws	20·0	60·7	1·6	0·5	14·85		0·7

Many of the seeds of wild plants recovered in palaeoethnobotanical samples are edible; it is only very rarely that poisonous species are encountered such as *Lolium temulentum* L. From the point of view of their place in the diet they may be divided into those species whose seeds are rich in carbohydrate and may be used as a substitute for the cereals – the wild grasses, and species of *Polygonum*, *Spergula*, *Chenopodium* for example; and seeds of plants which have a high oil content – *Capsella bursa-pastoris*, *Brassica campestris*, *Camelina sativa* for example. The composition of a few of the common weed seeds are shown in the following table:

	Water	*Protein*	*Fat*	*Nitrogen-free extract*	*Fibre*	*Ash*
	%	%	%	%	%	%
Agrostemma githago	11·2	16·1	5·9	57·0	6·3	3·3
Brassica rapa	—	22·2	36·4	26·5	10·8	4·1
Camelina sativa	10·0	25·9	31·8	35·1	11·5	9·2
Chenopodium album	—	16·82	8·12	49·98	21·45	6·98

The palaeoethnobotanist is concerned chiefly with these food plants already mentioned and it will be noted that they do not include the *green vegetables* – so important a source of vitamins – especially Vitamin A. The difficulty is that leaves do not survive as often as seeds in archaeological deposits, and it is only when weed seeds or seeds of wild plants whose leaves are known to be relatively palatable are discovered that we can hazard a guess at a source of green vegetable food, for example *Rumex acetosa* L., *Brassica campestris* and *Galeopsis tetrahit*. It may happen that plants with edible seeds may also have palatable leaves and occasionally also roots which may be eaten too – for example *Brassica campestris*.

Thus a consideration of the food values of the different categories of plants whose seeds and fruits were used for food in prehistoric times reveals that they were important for their various different components – the cereals as a source of carbohydrate, the pulses as a vegetable source of protein – most important when animal products were scarce, the nuts and olives and linseed all as sources of vegetable oil, and also for their protein content (especially the nuts), and fruits as a source of sweetness. Seeds of wild plants may indicate sources of green vegetables: they may also indicate flavourings – poppy seed, wild mustard, fenugreek, anise and coriander, or starvation foods – acorns, tubers of *Agropyron repens* (L.) Beauv. for example; as well as of weeds growing accidentally in the field.

Chapter 21

Conclusion

The origins and development of agriculture in the Near East, its spread to and development within Europe in prehistoric times

This chapter brings together in outline the results of palaeoethnobotanical researches over the last hundred years which have contributed to our present state of knowledge about the origins and developments of agriculture in the Near East and Europe. The picture is by no means complete, but it is useful to draw together the information we now have for the prehistory of the different crops in different areas. The chapter is divided into broad geographical regions and periods of development, and it sets out to give an outline of the evidence from key sites in the different cultures for the evolution of agriculture in Europe and the Near East.

The Near East before 6000 B.C.

There is little doubt that agriculture first evolved in the region broadly known as the Near East, and that it had become well established before 6000 B.C. over a wide area (see Map III). At present it appears that it may have arisen independently at a number of places within the area of distribution of the wild forms of wheat and barley as we know them today. Certainly palaeoethnobotanical finds of cultivated plants are known from south-west Iran, Iraqi Kurdistan, Syria, Palestine, and Central Anatolia before 6000 B.C.

The earliest finds in south-west Iran come from the Bus Mordeh and Ali Kosh phases at the site of Ali Kosh, Deh Luran, where Helbaek has found cultivated emmer wheat, mixed with some einkorn, hulled barley (possibly two-row since all the grains are straight), and naked six-row barley, some wild flax and a single lentil. The wild plants collected comprise chiefly wild legumes (especially in the earliest phase), wild grass seeds, and the fruits of prosopis, caper, and *Pistacia atlantica*. Already by the Bus

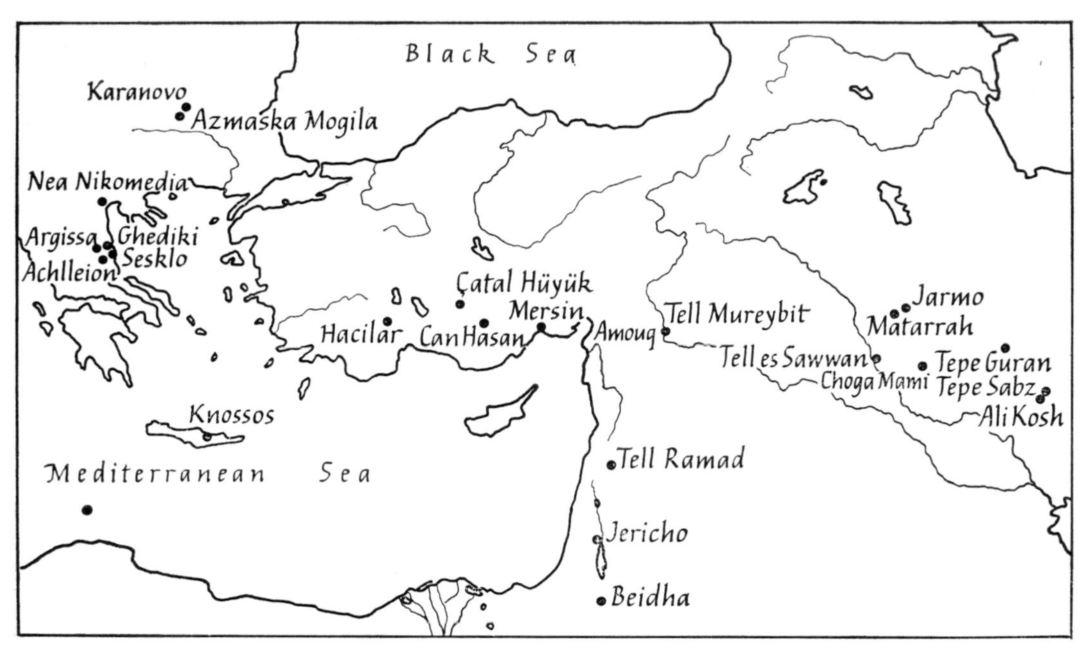

Map 3 Distribution map of sites with palaeoethnobotanical remains dating to before 5000 B.C. in the Near East.

Mordeh phase (7500–6750 B.C. in radiocarbon years) two species of wheat and two of barley appear to have been domesticated in this region (see Helbaek 1969, 383 f.).

Further north, in the Zagros foothills in Iraqi Kurdistan, lies the settlement of Jarmo occupied around 6750 B.C. (on carbon-14 evidence). Here a number of seeds were identified by Helbaek, who found wild and domesticated forms of einkorn and emmer while the bulk of the find was made up of *Hordeum spontaneum*. Interestingly the find of several rachis segments still attached together suggests that the barley had not so brittle a rachis as in the wild species and may indicate that it had already come into cultivation. Also present at this site were the several pulses – field pea, lentil and blue vetchling. Among the wild plants were grains of *Aegilops* sp. and nutlets of *Pistacia atlantica*. Nutlets of *Pistacia atlantica* were also found at the neighbouring site of Tepe Sarab (see Helbaek 1960A, 99 f.). At Tepe Guran wild and cultivated two-row barley have been found (Meldgaard, Mortensen and Thrane 1963, 112).

Further south and west at Tell Mureybit, north Syria, the excavations of van Loon revealed a settlement where the plant remains consisted of entirely wild cereals: van Zeist and Casparie identified grains of the twin-grained *Triticum boeoticum* ssp. *thaoudar*, and a few grains of *Hordeum spontaneum* in levels dated by carbon 14 to between 8050 and 7542 B.C. This is the only permanent settlement site so far found in the Near East which can be shown to have subsisted on wild cereals. The analyses of the animal bones give no sign that domesticated animals had been kept here (van Zeist and Casparie 1968, 44 f.).

In southern Syria the excavations of the pre-pottery neolithic B settlement at Tell Ramad in 1963 yielded remains of cultivated plants dating to around 7000 B.C. These include hulled two-row barley, emmer, einkorn, club wheat, and lentil (van Zeist and Bottema 1966, 179 f.). Again we have the combination of wheats and barley found together. The find of club wheat is most interesting, being the earliest report of a hexaploid wheat species; it may imply quite an involved genetic origin (see chapter 5 above).

At Jericho in Palestine remains of cultivated plants have been found in both pre-pottery neolithic levels: in level A Hopf has identified domesticated emmer wheat and hulled two-row barley. The carbon-14 dates for the pre-pottery A levels at Jericho suggest that this settlement may have begun before 8000 B.C. and on this basis the domesticated emmer wheat and hulled two-row barley would appear to be the earliest finds of domesticated cereal plants yet known. In the pre-pottery B levels einkorn, emmer, hulled two-row barley, peas, lentils and horsebean, *Vicia faba*, were found. This find is of interest, particularly for the last-named species, which has not previously been reported from such early contexts (see Hopf 1969, 355).

At Beidha in south Jordan the excavations of another pre-pottery neolithic B settlement have yielded remains of wild and cultivated emmer wheat, large grains of *Hordeum spontaneum*, and some grains of naked barley (possibly six-row) which must have been cultivated. A number of wild pulse seeds were collected, including species of *Vicia*, *Medicago* and *Onobrychis*. The nutlets of *Pistacia atlantica* were also found here in quantity (see Helbaek 1966B, 62 f.).

The early, aceramic settlement at Haçilar in the Burdar plain of west-central Anatolia yielded finds of wild einkorn, a small form of emmer, *Hordeum* cf *spontaneum*, some naked six-row barley, some lentils and a few weed seeds (idem, 1966B, 352; 1964, 121; 1970, 198) dating to about 7000 B.C.

Thus from south-west Iran to central Anatolia we find evidence that crop husbandry had become established before 6000 B.C. It was based mainly on the cultivation of einkorn and emmer wheat and of hulled two-row, and naked six-row barley. At some of the sites pulses were also present – in particular lentils and peas, and the seeds of wild vetch and medic. Seeds of wild flax were also represented in the Iranian finds.

The development of agriculture in the Near East after 6000 B.C.

In Khuzistan, the area of south-west Iran where the Ali Kosh finds were made, there are a number of palaeoethnobotanical finds in contexts immediately succeeding those already discussed. At Ali Kosh itself the Mohammad Jaffar phase of occupation yielded finds of emmer, hulled barley, wild flax and lentil, a number of wild grass seeds including numerous wild oats and seeds of canary grass, and quantities of plantain seed, mallow and wild legumes. The fruits of the wild caper and particularly of prosopis were also gathered. This phase is dated by carbon-14 to 6000–5500 B.C.

In the same plain of Deh Luran, not far from Ali Kosh, is the site of Tepe Sabz where the following species have been reported for levels dated between 5500 and 5000 B.C.: einkorn, emmer, hexaploid wheat, hulled two- and six-row barley, naked six-row barley, flax, lentil, grass pea. All these appear to have been cultivated here. The finds of exceptionally large linseeds in the Sabz phase at this site have led Helbaek to suggest that we are here faced with evidence of irrigation for the first time (Helbaek 1969, 408). The hexaploid wheat found here is bread wheat, *Triticum aestivum*, making its first appearance in this region. This species appears in Iraq and Anatolia during the first half of the sixth millennium (see below). Also notable are the large seeds of lentil, and the appearance in the plant list of hulled six-row barley (which in its earliest cultivated form appears to have been naked – see chapter 6).

In the same area lies Tepe Musiyan, and from the excavations there in 1963 two samples of seeds were recovered from levels dating to the Mehmeh phase (4500–4000 B.C.). Here the same spectrum of crops including linseed, lentil and bread wheat was recovered, again possibly indicating the practice of irrigation in this hot, dry plain (see Helbaek 1969, 411).

Further south in Khuzistan, not far from Susa lies the site of Djaffarabad where finds of palaeoethnobotanical material have come from the Susa A levels dating to about 3500 B.C. They include hulled six-row barley, einkorn, emmer, grass pea and lentil (Renfrew, unpublished).

In neighbouring Iraq the finds from Tell es-Sawwan on the Tigris compare well with those from Tepe Sabz, with which they are roughly contemporary. Here Helbaek has identified einkorn, emmer, bread wheat, hulled two- and six-row barley, naked six-row barley and flax. The fruits of prosopis and caper were also collected.

In later palaeoethnobotanical samples from Mesopotamia it is clear that emmer and bread wheat occur most widely of all the cereals. Emmer is known from ʿUbaid levels at Ur (c. 3500 B.C.), Tell Chragh (late ʿUbaid/early Uruk), Tell Qurtass (c. 2000 B.C.) together with bread wheat, and they are found together in the Isin Larsan and Horian levels at Tell Bazmosian (2100–1500 B.C.). They also occur in the late Assyrian levels at Nimrud. The chief form of barley cultivated is the hulled two-row species which is found almost as widely distributed as the wheats. One interesting find was the impression of *Panicum miliaceum* on a pottery lid from Jemdet Nasr (c. 3000 B.C.), the only record for the species in this area (Helbaek 1959B, 371).

Finds of linseed are known from Arpachiyah (5000–4500 B.C.), Ur (3500 B.C.), Khafajah B and from the late Assyrian levels at Nimrud (idem 1960B, 186 f.).

The most common of all the pulse crops in Mesopotamia was the lentil. It was found at Tel Chragh (3500 B.C.), at Tell Qurtass (2100–1800 B.C.). in the same valley, and in the Isin Larsan and Horian deposits at Tell Bazmosian (2100–1500 B.C.), and at Nimrud. The other pulses are far less common; grass peas were found in the Samarran plant remains at Choga Mami (5500–5000 B.C.) and occur again only at Nimrud (715–600 B.C.). Bitter vetch is found only at Tell Qurtass and at Nimrud. The chickpea makes its first appearance in this area in the Isin Larsan levels at Tell Bazmosian. It is also found in the later Horian levels here, and again at Nimrud. The late Assyrian finds at Nimrud also contained a number of other species which are not represented elsewhere in this area: Helbaek (1966D, 613 f.) reports finding grape, olive, fig, date, pomegranate, cucumber, prosopis and hazelnut from this site, together with a great range of weed seeds.

We see, in Mesopotamia and in southern Iran in the period immediately following the establishment of agriculture, that flax comes into cultivation, perhaps indicating irrigation in certain areas. Also the hulled form of six-row barley, hexaploid bread wheat and lentils are all in full cultivation in this region. Chickpeas were not added to the crop spectrum until just before 2000 B.C. The establishment of orchard husbandry is not evident until the very end of this series of finds at Nimrud (715–600 B.C.).

In Syria there are two finds from the period after the establishment of agriculture: in ʿAmouq A in the Plain of Antioch samples of silica skeletons and grain impressions yielded evidence for the cultivation of emmer and hulled barley, dating to about 5750 B.C.

From Hama, in sherds dating to the second half of the third millennium B.C., Helbaek has found traces of cultivated grape pips, together with hulled six-row barley, two impressions of emmer grains with one of einkorn wheat, and a single lentil seed (Helbaek 1948, 206). This is probably the earliest evidence in the Near East for the cultivation of the vine.

In Palestine the most important plant finds after the pre-pottery neolithic period come from Tell ed-Duweir (Lachish), where samples belonging to both the early bronze age and iron age deposits have been found. The early bronze age deposits contain large amounts of emmer, hulled barley and olive stones. Also present but less numerous were grape pips, pistachio nutlets, chickpeas, bitter vetch, lentils, grass peas and hawthorn pips (see Helbaek 1958B, 310). This indicates the early development of orchard husbandy in this region, as compared with Mesopotamia. Essentially the same picture is found for the iron age levels here, where olive stones and bread wheat were the chief finds but einkorn and emmer wheat were also cultivated, and the tradition of viticulture still continues. Other finds from these levels include seeds of fenugreek, chickpea, vetch and horsebean.

A sequence of early and middle bronze age finds has also been established for Jericho (Hopf 1969, 356 f.). The early bronze age deposits contained grains of hulled two-row barley, einkorn, emmer, and bread wheat, the pips and whole fruits of grapes, seeds of figs, date stones and also lentils and chickpeas. Interestingly the bulbs of onions (*Allium* cf. *ampeloprasum*) were also recovered from this context. The middle bronze age levels at Jericho contained much the same spectrum of species: hulled two-row barley, naked barley, einkorn, emmer, and bread wheat; the pips and fruits of grapes, seeds of the fig, fragments of the fruit wall and seeds of the pomegranate. Peas, lentils and horse beans appear to have been the main pulses cultivated, and again onion bulbs were found.

It is appropriate to include here the two bronze age plant finds from Cyprus. From the sherds recovered at the site of Kalopsidha an impression of a single grape pip was found dating to about 1950 B.C. (Helbaek 1966E, 119). The rest of the grain impressions from this site belong to the period 1575–1400 B.C. and comprise bread wheat, barley including hulled six-row type, grape pips, lentils (the most numerous species amongst the impressions), and nutlets of *Pistacia atlantica*, besides a number of grass and weed seeds (ibid., 119 f.). Carbonized grains, seeds and fruits have also been found at Apliki dating to 1300 B.C. The plant list here includes bread wheat, hulled six-row barley, horsebean, lentil, grape pips, olive stones, almonds, and seeds of coriander as well as a number of wild plants. Thus the palaeoethnobotanical finds from Cyprus correspond closely to those found at Hama, Lachish and Jericho from roughly contemporary levels. In the iron age funerary pyres from Salamis, Cyprus, grapes, olives, almonds, hazelnuts, bullace, rosehips, acorns, figs, horsebeans, grass peas, chickpeas, lentils and bitter vetch were found with some badly charred wheat and barley grains (Renfrew, in press).

In central Anatolia the sequence of finds after the earliest one so far examined at Hacilar, begins with the rich finds from Çatal Hüyük levels VI–II carbon-14 dated to 5850–5600 B.C. Here Helbaek (1964A, 121 f.) has found einkorn, of the twin-grained variety, emmer and bread wheat. This is the earliest occurrence of bread wheat in Anatolia, and roughly contemporary with the Tell es-Sawwan find in Mesopotamia; the find from Tepe Sabz dates to 5200 B.C. The other hexaploid wheat found in early contexts is club wheat found at Tell Ramad in the pre-pottery neolithic B levels. These finds are all of exceptional interest as they indicate that some degree of plant breeding had already been practised since there is no wild form of hexaploid wheat (see chapter 5). Also present at Çatal Hüyük were deposits of hulled six-row barley, pea, bitter vetch, acorn, almond, hackberry and nutlets of *Pistacia atlantica.* Essentially the same picture is revealed by the late neolithic finds from Hacilar, where quantities of twin-grained einkorn, emmer, bread wheat, naked six-row barley and hulled two- and six-row barley represent the range of cereals cultivated, and the pulses represented are lentils, bitter vetch and purple pea. Hulled six-row barley and field peas were also found at Can Hasan in contexts dated to 5250 B.C. (Renfrew 1968).

The later development of agriculture in Anatolia is shown by the palaeoethnobotanical finds from the bronze age at Troy and Beycesultan. The finds from Troy are reported by Wittmack (1890, 614) and include twin-grained einkorn, bitter vetch, horsebeans and possibly some grains of *Triticum durum*, besides seeds of *Fumaria* sp. Grape pips were found in the Troy II levels (Helbaek 1961, 80).

The late bronze age finds at Beycesultan (ibid., 77 f.) date to the thirteenth century B.C. The cultivated crops here include einkorn, bread wheat and club wheat but with a predominance of emmer wheat. Some hulled barley and a couple of grains of naked barley were also found. Lentils and bitter vetch formed the chief pulse crops. A single grape pip was found. Otherwise the seeds are of wild plants: mustard, cleavers, cowherb, Syrian scabious and knot grass, for example. Once again the first evidence for viticulture in this area dates to the bronze age.

The outlines of crop production in the prehistoric Near East can thus be established. It began in many separate areas with the cultivation of wheat and barley, possibly the cultivation of lentils and peas also, and soon the cultivation of flax also as a source of oilseed. Later there is evidence of the chickpea in the crop spectrum, and the increased use of horsebeans as a pulse crop. The collection of fruits goes back to the mesolithic, but it is noticeable that nutlets of *Pistacia atlantica* were much favoured by the early farmers. The cultivation of the grape and the olive are known in the western part of the Near East from early bronze age times, occurring later in Mesopotamia. Among the other fruits exploited during the bronze age were the fig, the date, the pomegranate and the almond.

Let us now turn our attention to Europe and examine the evidence for the beginnings and development of agriculture here. The account will not be as comprehensive as that for the Near East, since very many more palaeoethnobotanical finds have been made there, but finds will be selected to show the main outlines of agricultural development during prehistoric times in the major regions.

The earliest agriculture in Europe

It appears that the first farmers reached south-east Europe soon after 6000 B.C. To Greece they brought with them a spectrum of crops very similar to that found in the Near East, with emmer wheat being the chief cereal cultivated at Ghediki, Sesklo, Argissa, Achilleion in Thessaly and Nea Nikomedaria in Macedonia. The range of species included single-grained einkorn, hulled and naked two-row barley, hulled six-row barley, possibly broomcorn millet, oats, peas, lentils and vetch, acorns and *Pistacia atlantica* nutlets. A wild olive stone was reported from Thessaly (Renfrew 1966, 26; Hopf 1962, 101 f.). At Knossos in Crete the plant remains also include grains of bread wheat (Evans 1964, 140), suggesting that the agriculture here was introduced from Anatolia.

The cultivation of cereals and pulses began early also in Bulgaria, which may have received its first crops direct from the Near East, or via Greece. The finds here differ from those in the earliest neolithic levels in Greece in that the only cereals yet found are einkorn and emmer wheat, and there is some evidence that lentils were also important, though the small size of seeds may cause one to doubt whether they were fully domesticated. So far only a few samples for this period have been examined from Tell Azmak and Karanovo and it may be unwise to draw too strong a conclusion about the absence of other species at this time. In the succeeding Vesselinovo culture levels at Yassatepe near Plovdiv (dated to about 4500 B.C.) hulled and naked two-row barleys were found, and these gave way in importance to hulled six-row barley in this region in Gumelnitsa culture times (see chapter 3 above).

In Yugoslavia the Starčevo culture levels at Vršnik (dated by carbon 14 to around 5000 B.C.) have yielded large quantities of einkorn together with emmer, a hexaploid wheat, and half a grain of hulled six-row barley (Hopf 1961, 41).

Recent finds of grain impressions from Starčevo, and carbonized seeds from Obre I and Kakanj, have added *Triticum compactum*, *Pisum sativum* var. *arvense*, *Lens esculenta*,

Cornus mas and *Pyrus malus* to the plant list for the Starčevo culture (Renfrew, unpublished).

These finds comprise the earliest evidence for agriculture in the Balkans. From this region it is clear that agriculture spread along two main routes into Central and Northern Europe. The first led along the Danube and into the Rhine valley through Central Europe from Hungary and Czechoslovakia to Poland, Germany and the Low Countries. The bearers of the Danubian I culture spread rapidly over this area to reach the southern shores of the North Sea before 4000 B.C. Many of their village sites have yielded palaeoethnobotanical finds. The material from the following sites gives a good range of the plants which they cultivated and collected: Köln Lindenthal and Rödingen in the Rhineland, Rosdorf near Göttingen in Lower Saxony, and Bylany in Czechoslovakia. The main cereals represented are einkorn, emmer, naked and hulled six-row barley; peas and lentils were also commonly found. Less frequent are finds of bread wheat, spelt wheat, broomcorn millet and rye. Flax and opium poppy are also found and the several fruits were collected: hazelnuts, cornelian cherries, crab apples, raspberries, rosehips, and wild grapes. Seeds of *Bromus secalinus*, *Chenopodium album*, *Polygonum convolvulus*, *Polygonum lapathifolium*, *Polygonum persicaria* and *Lapsana communis* were also recovered. The same range of cereals is reported from Danubian sites in Poland – for example at Inowrocław. Klichowska has identified the following species in impressions in daub: hulled six-row barley, broomcorn millet, emmer wheat and club wheat (Klichowska 1969, 395 f.).

The earliest evidence of agriculture in Denmark comes from impressions in pottery at Store Valby, an early neolithic settlement. Here Helbaek (1954A, 202 f.) found imprints of naked barley, club wheat, einkorn and emmer wheat and a seed of *Galium*.

The second of the routes by which agriculture was introduced into Europe was along the Mediterranean coasts to Spain and thence along the Atlantic coast to the British Isles where it arrived before 3500 B.C. This route is not so well documented with palaeoethnobotanical finds as that through Central Europe. Several silica skeletons of grains have been found in samples of daub from the Italian Impressed Ware neolithic sites: einkorn, emmer and barley are all represented at Torre Canne, Puglia, for example (Renfrew, unpublished), and emmer and barley are frequently found, as at Palese, Puglia; and Villagio Leopardi, Abruzzio. In the middle neolithic levels of the Arene Candide cave, Liguria, silica skeletons of thirty-two grains of barley and five of emmer wheat have been recognized (Renfrew, unpublished).

The earliest find from Spain is that at Coveta de l'Or, Alicante. Here in Impressed Ware neolithic contexts Hopf has found carbonized grains of einkorn, emmer, club wheat and naked six-row barley (Hopf and Schubart 1965, 20 f.).

The earliest finds of cultivated plants in Britain come from the grain impressions in pottery found at Windmill Hill, Wiltshire. Here the bulk of the material belongs to emmer wheat, but there is some einkorn and a total of ten grains of naked and hulled barley. Both apple seeds and linseed have been found here also (Helbaek 1952B, 197).

A single grain impression of club wheat is known from the neolithic at Maiden Castle, Dorset (Jessen and Helbaek 1944, 17).

The development of prehistoric agriculture in Europe

We have already, in chapter 3, examined something of the development of agriculture in Thessaly, noting the apparent purity of the wheat crops in the late neolithic, and the increased interest in collecting fruits from that period leading to the eventual establishment of orchard husbandry, especially of olives and vines, which was well established in the bronze age of the Aegean. In this region too we saw the characteristic transition from a wheat-dominated cereal agriculture to one chiefly interested in hulled six-row barley. The establishment of horsebeans as a crop in Greece goes back to the early bronze age levels at Lerna in the Argolid (Hopf 1964, 4), and was well established by late Mycenaean times at Iolkos (Renfrew 1966, 33). One interesting find from Early Helladic Lerna is a large number of flax seeds, the only find of this species from the Aegean, so far.

In Bulgaria the late neolithic Gumelnitsa culture also showed a marked interest in barley cultivation, as we have seen, and it is interesting to note that six-row barley and emmer wheat continue to be cultivated in equal quantities throughout the bronze age at Ezerovo II near Varna, and Raskopanitsa near Plovdiv (Renfrew, unpublished). There is one find of flax from the Gumelnitsa levels at Kapitan Dimitrievo III (Arnaudov 1948/1949). In general the bronze age finds in Bulgaria show a continuity with those of the Gumelnitsa culture of the late neolithic period. Rye and horsebeans are not found in this part of the Balkans until very much later, in historic times.

There is little evidence for the establishment of orchard husbandry in prehistoric Bulgaria, but in the bronze age deposits from Bosnia we find grape pips reported from Ripač near Bihać on the Save (Stummer 1911, 291), where they were found with broomcorn millet and lentils (Neuweiler 1905, 116). Grape pips from the iron age have also been found in this area of northern Yugoslavia at Donja Dolina (Stummer 1911, 291).

As we move into Central Europe we find new species appearing in the plant lists of the late neolithic settlements. At Lengyel, Hungary, all three forms of millet are reported: *Panicum miliaceum, Setaria italica* and *Echinochloa crus-galli.* Horsebeans, lentils, grass peas, flax, acorns and cornelian cherries were also found here (Neuweiler 1905, 116). In the Michelsberg culture settlement at Ehrenstein, Hopf (1968, 7 f.) reports finding einkorn, emmer, bread and spelt wheat, naked and hulled forms of six-row barley as the cultivated cereals, together with the following seeds and nuts: beech mast, hazelnuts, crab apples, raspberry and dewberry, strawberry, bullace, cornelian cherry, and a number of weed species. In particular this find of grains of *Triticum spelta* is very interesting; it belongs to the small number of finds from the neolithic of Central Europe (the others being from Riedsaschen in south-west Germany, Ojców, Ksianżice and Zberzynek in Poland (see Schultze-Motel and Kruse 1965, 588). Rye was another cereal first cultivated in late neolithic times in Central Europe, although, like spelt wheat, it did not rise to a position of importance before the bronze age. It was particularly popular in the north of Europe in iron age times. The early finds come from Vösendorf, Austria (Werneck 1961, 71), and from five sites in Poland (Tempír 1964, 88). Finds from the bronze age at Tòszeg, Hungary, consisted of emmer wheat, hulled two-row barley and bitter vetch – this is one of the most northerly occurrences of the latter species which has its centre of distribution in the southern Balkans. The find of two-row barley is also unusual for this period in Central Europe, where six-row barley was the most common form. In the late bronze age of Hungary at Sághegy (Hallstatt A–B), Tempír has identified horsebeans, six-row barley, emmer, club wheat and lentils, showing the horsebean as a late bronze age crop in this area.

The lake-side settlements of Switzerland and north Italy give us a wide range of early and later bronze age finds of grains, seeds and fruits. The finds from Robenhausen on Lake Pfaffikon give us a good range of species for the early bronze age in Switzerland: broomcorn and Italian millet, peas, flax, opium poppy, apples, pears, strawberries, raspberries, blackberries, cherries, bullace, sloes, bilberries, water chestnuts, acorns, beech mast, and a great range of wild plants (see Neuweiler 1905, 111). A similar range is reported from the southern side of the Alps at Vallegio am Mincio (Villaret-von Rochow 1958, 96 f.). Here we find einkorn and emmer cultivated, together with hulled six-row barley and the two forms of millet and flax. Among the large number of wild plants the most notable are the pips of wild grapes and the stones of cornelian cherries, both typical finds for this southern part of the Alpine region. Figs were also collected at this site.

For the late bronze age of this region there are a series of finds from the site at Mörigen, where broomcorn millet and Italian millet, peas, lentils, horsebeans, opium poppy and flax were cultivated, and beech mast, hazelnuts, acorns, apples, strawberries, raspberries, blackberries, dewberries, sloes and cherries were collected. In this area too horsebeans appear to have come late into cultivation.

The earliest finds of cultivated oats, *Avena sativa*, appear to be the impressions found in corded ware vessels from Schraplau and Calbe in the Saale valley (Matthias and Schultze-Motel 1967, 146). Other species found in sherds belonging to this late neolithic horizon in Germany include six-row barley, emmer, einkorn, club wheat, cornelian cherry, apple and the seeds of *Bromus secalinus* L. From the site of Göttingen–Walkemühle (dating to around 1350 B.C.) Willerding (1969b, 392 f.) has identified club wheat and hulled six-row barley, showing that these two species continued to be cultivated well on into the bronze age. In the iron age of this region at Göttingen–Schillerwiese (ibid.) finds of einkorn, emmer, hulled barley, peas and vetch have been made.

In Scandinavia, after the first introduction of agriculture in the T.R.B. neolithic in Denmark, the cultivation of emmer, einkorn, bread/club wheat and barley continued, for example at Barkaer and Bundsö. At the latter site *Chenopodium album* seeds, raspberries and crab apples were also gathered (Helbaek 1954B, 253). At the late neolithic site of Nørre Sandegaard, Bornholm, *Panicum miliaceum* is found for the first time in this area (ibid.). Impressions of *Triticum spelta* and *Panicum miliaceum* grains have been found in south Sweden in sherds dating to the late bronze age (Hjelmqvist 1969, 260 f.). By the early iron age in Denmark oats and rye appear as cultivated crops. Among the 15 litres of carbonized grains and seeds from Fjand barley, oats and rye were the main cereals and seeds of *Chenopodium* were collected in very large quantities (1½ litres in one pure deposit). The burnt house at the contemporary site of Alrum, Jutland, contained large amounts of barley, oats and bread/club wheat, and about a litre of seeds of *Polygonum lapathifolium*. Seeds of flax and of *Camelina linicola* and *Spergula arvensis* were also found here (Helbaek 1954B, 254–5). At Dalshöj on Bornholm, in a house dating to the early Roman iron age, 30 litres of carbonized grains were recovered consisting mainly of emmer, einkorn and barley. There were also seeds of Italian millet and flax as well as a series of weed seeds (idem., 1954B, 255). At Vallhagar, in the late iron age in Sweden, einkorn, emmer, spelt, bread and club wheat were found together with rye and barley (probably the hulled six-row variety), and a large number of seeds of wild plants (idem 1955B, 653 f.).

Turning from Scandinavia to Britain, we find a similar development of prehistoric agriculture, with the increasing importance of barley during the bronze age and the arrival of *Triticum spelta*, with cultivated oats and rye at the end of the prehistoric period. Millets do not appear ever to have been cultivated in Britain. Flax which appeared in the Windmill Hill culture was still cultivated during the succeeding early bronze age – for example there is a beaker from Handley Down, Dorset, which has fifteen impressions of linseeds on its surface (idem 1952B, 205). During the bronze age hulled six-row barley appears to have been the most important cereal, and emmer is the chief cultivated form of wheat, though some einkorn and bread wheat are also known (ibid., 207). In the early iron age a number of new species make their debut in the plant record: spelt wheat was found at Hembury, Devon, and Itford Hill, Sussex, and in the vast deposit of carbonized grain from Fifield Bavant, Wiltshire. At the latter site one-third consisted of spelt wheat, with only a few grains of club wheat and no emmer. The bulk of this find was of hulled six-row barley, with a single grain of rye. Rye is also reported from the neighbouring site of Winkelbury (ibid., 210). The richest palaeoethnobotanical finds for this period come from the marsh villages of Glastonbury and Meare, Somerset, where emmer and spelt wheat were the cultivated cereals and horsebeans formed the chief pulse crop. Amongst the fruits collected here were sloe, blackberry, rosehips, dewberries and haws (Reid 1916, 627).

This survey of the development of plant husbandry in the Near East and Europe is completed by the later finds from the west Mediterranean region. In peninsular Italy the finds from the iron age graves in the Forum Romanum show the development of agriculture here. Einkorn, emmer and spelt wheat were cultivated, as were barley and broomcorn millet. Horsebeans and grass peas were also grown and viticulture had become established.

This survey does not pretend to be comprehensive, but it presents the main outlines of the development of crop husbandry in each region in prehistoric times. For the more detailed survey of the evidence for the development of individual crops the reader is referred to the detailed account of each species in the preceding chapters.

As more work is done these simple outlines may well have to be altered. In the Near East it is likely that plant remains will be found setting the early beginnings of agriculture back well before 10,000 B.C. In Europe the development of agriculture and especially the early exploitation of crops such as oats and rye should become much clearer.

There are a number of outstanding problems to be considered in the future. One of the most important concerns the distinction between wild and domesticated species: at the moment the processes of domestication and their reflection in the morphology of the seeds are not clearly understood and so they give rise in the literature to phrases such as 'cultivated wild barley'. How long does a plant have to be cultivated before it becomes domesticated and how do we draw this distinction are questions which are currently being asked by and of palaeoethnobotanists.

Another problem which has long been neglected concerns the pulse crops. Very little is known at present of their progenitors or of their processes of domestication. Clearly they are of equal antiquity to the cereal crops, but so far they do appear to have been relatively overlooked.

In the process of identification of seeds our techniques may well develop in the near future with a much greater emphasis being placed on cell structure and morphology rather than on the broader surface features which we use today.

Once the identifications are completed it is also the palaeoethnobotanist's responsibility to interpret his findings and relate them to their archaeological contexts. This is a demanding task, but if it is not undertaken by the palaeoethnobotanist who understands the requirements of the plants, it is much less likely to bear a close relation with reality. It is in this sphere that archaeologist and palaeoethnobotanist must co-operate closely together. The finds can only satisfactorily be interpreted by having a good understanding of how and where they were found. The interpretation will not only show the plants cultivated and gathered by the community being studied, it may indicate methods of cultivation, types of storage, processes of food preparation, and the utilization of a spectrum of plant seeds as food by a given community.

Palaeoethnobotany provides a challenge to archaeologists, botanists and agricultural historians; and as the subject develops it can be expected to throw new light on the economic developments of prehistoric agricultural communities.

Bibliography

ALLAN, W. (1965) *The African Husbandman*. London.

ALLBAUGH, L. G. (1953) *Crete: a Case Study of an Undeveloped Area*. New Jersey.

ALLCHIN, F. R. (1969) Early cultivated plants in India, in Dimbleby, G. W. & Ucko, P.: *The Domestication and Exploitation of Plants and Animals*. London.

ANDERSON, E. (1942) Prehistoric maize from Canyon del Muerto, *American Journal of Botany* 29, 832–5.

ANDRÉ, J. (1961) *L'Alimentation et la Cuisine à Rome*. Paris.

ARNAUDOV, N. (1936) Über prähistorische und subrezente pflanzenreste aus Bulgarien, *ТРУНОВЕ НА БЪЛГАРСКОТО ПРИРОДО—МЗПИТАТЕЛНО ДРУЖЕСТВО КНИГ АХVII*. Sofia.

ARNAUDOV, N. (1937/1938) Untersuchung über pflanzenreste aus den Ausgrabungen bei Sadowetz in Bulgarien, *Annuaire de l'Université de Sofia, Faculté Physico-mathématique* XXIV, Livre 3, *Sciences Naturelles*. Sofia.

ARNAUDOV, N. (1940/1941) Über die Neuendeckten Prähistorischen pflanzenreste aus Südbulgarien, *Jahrbuch der Universität sveti Climent Ochridski in Sofia, Physico-Mathematischen* Sofia *Facultät* XXXVII, Bd. 3, Naturwissenschaft. Sofia.

ARNAUDOV, N. (1948/1949) *ПРЕМИСТОРИУСКИ РАСТЕЛИ МАТЕРИАЛИ*, *Annuaire de l'Université de Sofia, Faculté de Sciences* XLV, Livre 3, *Sciences Naturelles*. Sofia.

ARTAMONOV, M. I. (1965) Frozen tombs of the Scythians, *Scientific American* 212 (5).

ASHLEY, W. (1928) *The Bread of our Forefathers*. Oxford.

BAKER, H. G. (1965) *Plants and Civilization*. Belmont. California.

BATTAGLIA, R. (1943) *La palafitta del Lago di Ledro nel Trentino*. Trento.

BEHRE, K-E. (1969) Untersuchungen des botanischen Materials, in *Ausgrabungen in Haithabu* 2. Schleswig.

BELL, G. D. H. (1965) The comparative phylogeny of the temperate cereals, in Hutchinson, Sir Joseph: *Essays on Crop Plant Evolution*, Ch. IV. Cambridge.

BERTSCH, KARL and BERTSCH, FRANZ (1949) *Geschichte unserer Kulturpflanzen*. Stuttgart.

BIALOR, P. A. and JAMESON, M. H. (1962) Palaeolithic in the Argolid, *American Journal of Archaeology* 66.

BIFFEN, R. H. (1934) Report on grain from the Fayum, in Caton-Thompson, G. & Gardner, E. W.: *The Desert Fayum*. Gloucester.

BOULGER, G. S. (undated) *Some Familiar Trees*. London.

BOWDEN, W. M. (1959) The taxonomy and nomenclature of the wheats, barleys and ryes and their wild relatives, *Canadian Journal of Botany* 37, 657 ff.

BRETAUDEAU, J., BARTON, J. G. and LE FAOU, A. (1966) *Trees, a Guide to the Trees of Great Britain and Europe*. London.

BROTHWELL, D. R. and BROTHWELL, P. (1969) *Food in Antiquity*. London.

BROUWER, W. and STAHLIN, A. (1955) *Handbuche der Samenkunde*. Frankfurt.

BULLE, H. (1909) *Orchomenos*, Abhandlungen der Kaiserlichen bayerischen Akademie (philosophisch-philologische Klasse). Munich.

BULLEID, A. and GRAY, H. ST G. (1916) *Glastonbury Lake Village* 2. Glastonbury.

BUSCHAN, G. (1895) *Vorgeschichtliche Botanik der Kultur- und Nutzpflanzen der alten Welt*. Breslau.

CALLEN, E. O. (1963) Diet as revealed by coprolites, in Brothwell, D. & Higgs, E. S.: *Science in Archaeology*. London (2nd ed. 1969).

DE CANDOLLE, A. (1884) *Origin of Cultivated Plants*. London.

CARSON, G. P. and HORNE, F. R. (1962) The identification of barley varieties, in Cook, A. H.: *Barley and Malt*. London and New York.

CARTER, G. F. (1945) *Plant Geography and Culture History in the American Southwest*. New York.

CATON-THOMPSON, G. and GARDNER, E. W. (1934) *The Desert Fayum*. Gloucester.

CHILDE, V. G. (1929) *The Danube in Prehistory*. London.

CHILDE, V. G. (1956) *The Dawn of European Civilization*. London.

CHOWDHURY, K. A. (1965) Plant remains from pre- and proto-historic sites and their scientific significance, *Science and Culture* 31, 177–8.

CLAPHAM, A. R., TUTIN, T. G. and WARBURG, E. F. (1964) *Excursion Flora of the British Isles*. Cambridge.

CLARK, H. H. (1967) The origin and early history of the cultivated barleys – a botanical and archaeological synthesis, *Agricultural History Review* XV, pt. 1, 1–18.

CLARK, J. G. D. (1952) *Prehistoric Europe. The Economic Basis*. London.

COBLEY, S. (1963) *An Introduction to the Botany of Tropical Crops*. London.

COFFMAN, F. A. (1946) The origins of cultivated oats, *Journal of American Society of Agronomists* 38.

CONDIT, I. J. (1947) *The Fig*. Waltham, Mass.

COSTANTIN, J. and BOIS, D. (1910) Sur les graines et tubercules des Tombeaux péruviens de la Period Incasique, *Revue Générale de Botanique* 22, 242–6. Paris.

DARYLL FORDE, C. (1934) 1966 *Habitat, Economy and Society*. London.

DETEV, P. (1050) Le Tell Baniata près de Kapitan Dimitrievo, *Godishnik Narodne Musej Plodiv* II, 1–23. Plodiv.

DETEV, P. (1954) Tell près du village de Bikovo, *Godishnik M.O. Plodiv* I, 511 f. Plodiv.

DETEV, P. (1959) Matériaux de la préhistoire de Plovdiv, *Godishnik Narodne Muzej Plodiv* III. Plodiv.

DIMBLEBY, G. W. (1967) *Plants and Archaeology*. London.

DIOSCORIDES (1959 ed.) *Greek Herbal* Book IV, trans. Goodyear 1655, edited Gunther 1959.

DIXON, D. M. (1969) Archaeological, botanical, textual and artistic evidence for the cultivation of cereals in Ancient Egypt, in Dimbleby, G. W. & Ucko, P.: *The Domestication and Exploitation of Plants and Animals*. London.

DOROFEEV, V. F. (1968) The variability and breeding value of Armenian wheats, *Euphytica* 17, 451–61.

DRUMMOND, J. C. and WILBRAHAM, A. (1940) *The Englishman's Food.* London.

DZAMBAZOV, N. (1964) Recherches sur la culture paléolithique et mésolithique en Bulgare, *Archaeologia* VI, part 3. Sofia.

ECONOMIC AND SOCIAL ATLAS OF GREECE (1964) Athens.

ERNLE, LORD (1961) *English Farming Past and Present,* 6th ed. London.

EVANS, SIR A. (1921 f.) *The Palace of Minos.* London.

EVANS, J. D. (1964) Excavations in the neolithic settlement of Knossos 1957–60, part I. *British School Annual Athens.* 59.

FONNER, R. L. (1957) *Soc. Amer. Archaeol. Mem.* 14, 303–4.

FREEMAN, M. B. (1943) *Herbs for the Medieval Household for Cooking, Healing and Divers Uses.* New York.

GAUL, J. H. (1948) *The Neolithic Period in Bulgaria.* Cambridge, Mass.

GEORGIEV, G. I. (1961) Kulturgruppen der Jungstein- und der Kupferzeit in der Ebene von Thrazien (Südbulgarien), in *Europe à la fin de l'âge de la pierre,* ed. Böhm, J. & De Laet, S. Prague.

GEORGIEV, G. I. (1964) Recherches sur la culture des Tells, *Archaeologia* VI, part 3. Sofia.

GEORGIEV, G. I. The Azmak mound in southern Bulgaria, *Antiquity* XXXIX. Cambridge.

GEORGIEV, G. I. and MERPERT, N. J. (1966) The Ezero mount in south east Bulgaria, *Antiquity* XL. Cambridge.

GILL, N. T. and VEAR, K. C. (1958) *Agricultural Botany.* London.

GILMORE, M. R. (1931) Vegetal remains of the Ozark Bluff-Dweller culture, *Pap. Mich. Acad. Sci. Arts and Let.* 14, 83–102. Michigan.

GIMBUTAS, M. (1956) *The Prehistory of Eastern Europe,* part I: Mesolithic, Neolithic and Copper Age Cultures in Russia and the Baltic Area. Cambridge, Mass.

GODWIN, H. (1956) *History of the British Flora.* Cambridge.

GODWIN, H. (1965) The beginnings of agriculture in north west Europe, in Hutchinson, Sir J.: *Essays on Crop Plant Evolution.* Cambridge.

GOOR, A. (1965) The history of the fig in the Holy Land from ancient times to the present day, *Economic Botany* 19. New York.

GRAHAM, J. W. (1962) *The Palaces of Crete.* Princeton.

HANDBOOK OF BULGARIA (1920) H.M.S.O. London.

HARLAN, J. R. (1967) A wild wheat harvest in Turkey, *Archaeology* 20, No. 3.

HARLAN, J. R. and ZOHARY, D. (1966) The distribution of wild wheats and barleys, *Science* 153, 1074–80. Washington.

HARMS, H. VON (1922) Übersicht der bisher in altperuanishen Graben gefundenen Pflanzenreste, in *Festschrift Eduard Seler,* 157–86. Stuttgart.

HAWKES, J. (1968) *The Dawn of the Gods.* London.

HECTOR, J. M. (1936) *Introduction to the Botany of Field Crops,* Vol I *Cereals*: Vol. II *Non-cereals.* Johannesburg.

HEER, O. (1866) Treatise on the plants of the Lake Dwellings, in Keller, F., trans. Lee, J. E.: *The Lake Dwellings of Switzerland and other parts of Europe.* London.

HEHN, V. (1911) *Kulturpflanzen und Haustiere in ihrem Übergang aus Asien nach Griechenland und Italien sowie in das übrige Europa.*

HELBAEK, H. (1938) Planteavl, *Aarboger for Nordisk Oldkyndighed og Historie.* Copenhagen.

HELBAEK, H. (1948) Les empreintes de céréales, in Riis, P. J.: *Hama.* Copenhagen.

HELBAEK, H. (1950) Tollund mandens sidste maaltid, *Aarbøger for Nordisk Oldkyndighed go Historie*, 311–41. Copenhagen.

HELBAEK, H. (1951) Ukrudtsfrø som Naeringsmiddel; førromersk Jernalder, *Kuml.* 1951, 65–74. Athens.

HELBAEK, H. (1952A) Spelt (*Triticum spelta* L.) in bronze age Denmark, *Acta Archaeologica* 23, 97–107. Copengagen.

HELBAEK, H. (1952B) Early crops in southern England, *Proceedings of the Prehistoric Society* XVIII, 194 f. London.

HELBAEK, H. (1952C) Preserved apples and panicum in the prehistoric site at Nøore Sandegaard in Bornholm, *Acta Archaeologia* 23, 108 f. Copenhagen.

HELBAEK, H. (1953A) Queen Icetis' wheat, *Dan. Biol. Medd.* 21, No. 8. Copenhagen.

HELBAEK, H. (1953B) Appendix I, in Gjerstad, E.: *Early Rome* I. Lund.

HELBAEK, H. (1953C) Archaeology and agricultural botany, *Annual Report of Institute of Archaeology* 9, 44 f. London.

HELBAEK, H. (1954A) Store Valby Komavl i Danmarks Første Neolitiske Fase, *Aarbøger for Nordisk Oldkyndighed og Historie*, 202–4. Århus.

HELBAEK, H. (1954B) Prehistoric food plants and weeds in Denmark. A survey of archaeobotanical research 1923–54, *Danmarks Geoligiske Unders* II, No. 80, 250–61. Copenhagen.

HELBAEK, H. (1955B) The botany of the Vallhagar Iron Age field, in Stenberger, M.: *Vallhagar, a Migration Period site on Gotland, Sweden.* Stockholm.

HELBAEK, H. (1956) Vegetables in the funeral meals of pre-urban Rome, Appendix I in Gjerstad, E.: *Early Rome* II. *Acta Inst. Roman Suerciae* scr. 4. 27: 2, 287–94. Lund.

HELBAEK, H. (1958A) Grauballemandens Sidste Måltid, *Kuml.* 1958, 83–116. Århus.

HELBAEK, H. (1958B) Plant economy in ancient Lachish, in Tufnell, O.: *Lachish* IV. London

HELBAEK, H. (1959A) How farming began in the Old World, *Archaeology* 20, No. 3. Washington.

HELBAEK, H. (1959B) Domestication of food plants in the Old World, *Science* 130, 365 f.

HELBAEK, H. (1959C) Notes on the evolution and history of *Linum*, *Kuml.* 1959. Århus.

HELBAEK, H. (1960A) The palaeoethnobotany of the Near East and Europe, in Braidwood, R. J. & Howe, B.: *Prehistoric Investigations in Iraqi Kurdistan.* Chicago, 99–118. Chicago.

HELBAEK, H. (1960B) Ecological effects of irrigation in ancient Mesopotamia, *Iraq* XXII. London.

HELBAEK, H. (1060C) Cereals and weed grasses in Phase A, in Braidwood, R. J. & Braidwood, L. J.: *Excavations in the Plain of Antioch* I. Chicago.

HELBAEK, H. (1960D) Comment on *Chenopodium album* as a food plant in prehistory. *Bericht des Geobotanischen Institut der Eidg. Techn. Hochschule, Stiftung Rübel* 31. Heft 1959. Zurich, 16–19.

HELBAEK, H. (1960E) Ancient crops in the Shahrzoor valley in Iraqi Kurdistan, *Sumer* XVI, 79–81. Baghdad.

HELBAEK, H. (1961) Late Bronze Age and Byzantine crops at Beycesultan in Anatolia, *Anatolian Studies* XI, 77–97.

HELBAEK, H. (1962A) Late Cypriote vegetable diet at Apliki, *Opuscula Atheniensia* IV, 171–83. Lund.

HELBAEK, H. (1962B) Les grains carbonisés de la 48ème couche des fouilles de Tell soukas, *Les Annales Archaeologiques de Syrie* XI–XII. Damascus.

HELBAEK, H. (1964A) First impressions of the Çatal Hüyük plant husbandry, *Anatolian Studies* XIV, 121 f.

HELBAEK, H. (1964B) The Isca Grain. A Roman plant introduction in Britain, *The New Phytologist* 63.

HELBAEK, H. (1965A) Early Hassunan vegetable food at Tell Es-Sawwan near Samarra, *Sumer* XX 45 f. Baghdad.

HELBAEK, H. (1965B) Isin-Larsan and Horian food remains at Tell Bazmosian in the Dokan Valley, *Sumer* XIX, 27–35. Baghdad.

HELBAEK, H. (1966A) Pre-pottery neolithic farming at Beidha, *Palestine Exploration Quarterly* 98 (I), 61 f.

HELBAEK, H. (1966B) 1966. Commentary on the phylogenesis of *Triticum* and *Hordeum, Economic Botany* 20, 350 f. New York.

HELBAEK, H. (1966C) Vendeltime farming products at Eketorp on Öland, Sweden, *Acta Archaeologica* 37, 216–21. Copenhagen.

HELBAEK, H. (1966D) The plant remains from Nimrud, in Mallowan, M. E.: *Nimrud and its Remains*. London, 615 f.

HELBAEK, H. (1966E) What farming produced at Cypriote Kalopsidha, in Astrom, P.: *Excavations at Kalopsidha and Ayios Iakovos in Cyprus*. Studies in Mediterranean Archaeology II. Lund.

HELBAEK, H. (1969) Plant collecting, dry-farming and irrigation in prehistoric Deh Luran in Hole, Frank *et al.*: *Prehistory and Human Ecology of the Deh Luran Plain: An Early Village Sequence from Khuzistan, Iran*. Mem. Mus. Anthrop. Univ. Michigan No. 1 (Ann Arbor), 244–383. Michigan.

HELBAEK, H. The plant husbandry of Hacilar, in Mellaart, J.: *Excavations in Hacilar*, I. 189–244. Edinburgh.

HESIOD (1936 ed.) *Works and Days*, transl. by White, H. G. Evelyn, in *Hesoid, the Homeric Hymns and Homerica*. London.

HEURTLEY, W. A. (1939) *Prehistoric Macedonia*. Cambridge.

HILL, J. (1941) *Wild Foods of Britain*. London.

HJELMQVIST, H. (1969) Dinkel un Hirse aus der Bronzezeit südschwedens nebst einigen Bermerkungen über ihre spätere Geschichte in Schweden, *Botaniska Notiser* 122. Lund.

HOLE, F. and FLANNERY, K. (1967) The prehistory of southwestern Iran: a preliminary report, *Proceedings of the Prehistoric Society*, XXXIII, 147 f. London.

HOMER (1950 ed.) *The Iliad*, trans. Rieu, E. V. London.

HOPF, M. (1955) Formveranderungen von Getreide – Körnern beim Verkohlen, *Berichte der Deutschen Botanischen Gesellschaft* 68.

HOPF, M. (1957) Botanik und Vorgeschichtc, *Jahrbuch Röm. German. Zentralmuseums Mainz* 4.

HOPF, M. (1961) Untersuchugsbericht über Cornfunde aus Vršnik *ЗБОРНИК НА ШТИПСКИОТ НАРОДЕН МУЗЕI КН II за 1960-61 ПОСЕБЕН ОТИСОК*. Stip.

HOPF, M. (1962) Bericht über die Untersuchung von Samen und Holzkoheresten von der Argissa-Maghula aus der präkeramischen bis mittelbronzezeitlichen Schichten, in Milojčić, V., Boessneck, J., & Hopf, M.: *Die Deutschen Ausgrabungen auf der Argissa-Maghula in Thessalien*. Bonn.

HOPF, M. (1964) Nutzpflanzen vom Lernäischen Golf, *Jahrbuch Röm. German. Zentralmuseums Mainz* 11. Mainz.

HOPF, M. (1965) Untersuchungen aus dem botanischen Labor am R.G.Z.M. II. Getreidekorn-Abdrücke als Schmuckelements in neolithischer Keramik? *Jahrbuch Röm. German. Zentralmuseums Mainz* 12. Mainz.

HOPF, M. (1968) Früchte und Samen, in Zürn, H.: *Das Jungsteinzeitliche Dorf Ehrenstein (Kries Ulm)*. Staatlichen Amtes für Denknalpflege ser. A, Heft 10/11. Stuttgart.

HOPF, M. (1969) Plant remains and early farming in Jericho, in Dimbleby, G. W. & Ucko, P.: *The Domestication and Exploitation of Plants and Animals*. London.

HOPE, M. and CATALÁN, M. P. (1970) Neolithische Getreide du funde in der höhle von Nerja (Prov Málaga). *Madrider Mitteilungen* II, 18–34. Heidelberg.

HOPF, M. and SCHUBART, H. (1965) Getreidefunde aus der Coveta de l'Or, Alicante, *Madrider Mitteilungen* 6. Heidelberg.

HOWES, F. N. *Nuts. Their Production and Everyday Uses*. London.

HUBBARD, C. E. (1968) *Grasses*. London.

HUNTER, H. (1951) *Crop Varieties*. London.

HUNTER, H. (1951) *The Barley Crop*. London.

HUTCHINSON, R. W. (1962) *Prehistoric Crete*. London.

JESSEN, K. and HELBAEK, H. (1944) *Cereals in Great Britain and Ireland in Prehistoric and Early Historic Times*. Kgl. Dan. Vidensk. Selsk. Biol. Skrifter. Copenhagen.

JONES, V. H. (1936A) A summary of data on aboriginal cotton of the Southwest, (*Symposium on Prehistoric Agriculture*) *University of New Mexico* Bulletin 296. Anthropological Series I (5), 51–64. Albuquerque.

JONES, V. H. (1936B) The vegetal remains of Newt Kash Hollow Shelter, in Webb, W. S. & Funkhouser, W. D.: *Rock Shelters in Menifee County, Kentucky. The University of Kentucky Reports in Archaeology and Anthropology* 3, No. 4. Lexington.

JORDANOV, D. (1963) *Flora Reipublicae Popularis Bulgaricas* I. Sofia.

KAZAROW, G. J. (1914) Sveti-Kyrillovo, S. Bulgaria, *Praehistoricher Zeitschrift* VI, Heft 1/2.

KELLER, F. trans. Lee, J. E. (1866) *The Lake Dwellings of Switzerland and other parts of Europe*. London.

KIHARA, H. (1919) Über cytologische Studien bei einigen Getreidearten. I. Spezies-Bastard des Weizens und weizenroggen-Bastard, *Bot. Mag. Tokyo* 35, 19–44.

KIHARA, H. (1924) Cytologische und genetische Studien bei wichtigen Getreidearten mit besonderer Rucksicht auf das Verhalten der Chromosomen und die Sterilität in den Bastarden, *Mem. Coll. Sei. Kyoto* ser. B, I, 1–200. Kyoto.

KIHARA, H. (1944) Die Entdeckung des DD-Analysators beim weizen, Vorlänfige Mitteilung. *Agr. & Hort.* 19, 889–90.

KLICHOWSKA, M. (1969) Odciski roslinne na neolitycznej ceramice ze stanowiska nr 3 w rybitwach pow, Inowrocław, *Spawozdania Archeologiczne* XXI. Poznan.

KNÖRZER, K-H. (1966) Über Funde römischer Import-früchte in Novaesium (Neuss/Rh.), *Bonner Jahrbuch* 166. Bonn.

KNÖRZER, K-H. (1971) Prähistorische Mohnsamen im Rheinland, *Bonner Jahrbuch* 171, 34–9.

KNÖRZER, K-H. (1971B) Urgeschichtliche unkräuter im Rheinland. *Vegatatio* 23, 1–2.

KOHL, G. and QUITTA, H. (1966) Berlin radiocarbon measurements II, *Radiocarbon* 8.

KOKKOROS, P. and KANELLIS, A. (1960) Découverte d'un crâne d'homme paléolithique dans la péninsule chalcidique, *L'Anthropologie* 64.

KOROSEĆ, J. (1958) *Neolitska Naseobina u Danilu Bitinju*. Zagreb.

KUNTH, C. (1826) Examen botanique, in Passalacqua, J.: *Catalogue raisonné et historique de antiquités découvertes en Egypte*. Paris.

LAMB, W. (1936) *Excavations at Thermi in Lesbos*. London.

LEONARD, W. H. and MARTIN, J. H. (1963) *Cereal Crops*. New York/London.

LEVI, D. (1956) *Bollettino d'Arte* 1956, 225.

LISSITSINA, G. N. (1970) Plantes satives du Proche-Orient et du Sud de l'Asie Centrale aux VIIIe–Ve millénaires avant notre ère, *СОВЕТСКАЯ АРХЕОЛОГИЯ*. 3. Moscow.

LOGOTHETIS, B. C. (1962) *ΑΙ ΑΥΤΟΦΥΕΙΣ ΑΜΠΕΛΟΙ ΩΣ ΠΡΟΤΟΓΕΝΕΣ ΑΜΠΕΛΟΥΡΓΙΚΟΝ ΥΛΙΚΟΝ ΕΝ ΕΛΛΑΔΙ. ΕΠΕΤΗΡΙΣ ΤΗΣ ΓΕΟΠΟΝΙΚΗΣ. ΚΑΙ ΔΑΣΟΛΟΓΙΚΗΣ ΣΧΟΛΗΣ.* Thessalonika.

MACKEY, J. (1954) The taxonomy of hexaploid wheat, *Svensk Botanisk Tidskrift* 48.

MANGELSDORF, P. C. and SMITH, C. E. JNR (1949) New archaeological evidence on evolution in maize, *Botanical Museum Leaflet 13 Harvard University*, 213–47.

MANGELSDORF, P. C., MACNEISH, R. S. and GALINAT, W. C. (1956) Archaeological evidence on the diffusion and evolution of maize in N. E. Mexico, *Botanical Museum Leaflet 17 Harvard University*, 125–50.

MARTIN, J. H. and LEONARD, W. H. (1967) *Principles of Field Crop Production*. New York/London.

MASEFIELD, G. B., WALLIS, M., HARRISON, S. G. and NICHOLSON, B. E. (1969) *The Oxford Book of Food Plants*. Oxford.

MATTHIAS, W. and SCHULTZE-MOTEL, J. (1967) Kulturpflanzenabdrücke an Schnurkeramischen Gefäsen aus mitteldeutschland, *Jahresschrift für mitteldeutsche Vergeschichte* 51. Halle.

MCFADDEN, E. S. and SEARS, E. R. (1946) The origin of *Triticum spelta* and its free-threshing hexaploid relatives, *Journal of Heredity* 37 (3), 81–90, (4), 107–16.

MELDGAARD, J., MORTENSEN, P. and THRANE, H. (1963) Excavations at Tepe Guran, Luristan, *Acta Archaeologia* XXXIV.

MILOJČIĆ, V. (1958) Die neuen mittel- und altpaläolithischen Funde von der Balkanhanbinsel, *Germania* 36.

MILOJČIĆ, V. (1959) Ergebnisse der deutschen Ausgabungen in Thessalien 1953–1958, *Jahrbuch Röm. German. Zentralmuseums Mainz* 6.

MILOJČIĆ, V., BOESSNECK, J. and HOPF, M. (1962) *Die Deutschen Ausgrabungen auf der Argissa-Magula in Thessalien. I. Das präkeramische Neolithicum Sowie die Tier- und Pflanzenreste* (*Beiträge zur ur- und frühgeschichtlichen Archäologie des Mittelmeer-Kulturaumes*, Band 2). Bonn.

MIREAUX, E. (1959) *Daily Life in the Time of Homer* (trans. Sells, I.). London.

MORTIMER, J. (1707) *The Whole Art of Husbandry, or the Way of Managing and Improving of Land . . .*

MUENSCHER, W. C. (1955) *Weeds*. New York.

MURRAY, J. (1970) *The First European Agriculture*. Edinburgh.

MUSÉE ARCHEOLOGIQUE DE PLOVDIV (1964) Editions Bulgarski Houdojnik Sofie.

MUSIL, A. (1963) *Identification of Crop and Weed Seeds*. Agriculture Handbook No. 219. Washington D.C.

MYLONAS, G. E. (1929) *Excavations at Olynthas Part I: The Neolithic Settlement*. Baltimore.

MYLONAS, G. E. (1957) *Ancient Mycenae*. London.

NATHO, I. and ROTHMAIER, W. Bandkeramische Kulturpflanzenreste aus Thüringen und Sachsen, *Beitrage z. F.d.L. III* 73–98. Berlin.

NETOLITZKY, F. (1912) Hirse und Cyperus aus dem prähistorischen Ägypten, *Beihefte z. Bot. Zentralbl.*

NETOLITZKY, F. (1931) Ein Cruciferensamen aus dem vorgeschichtlichen Greichenland, *Bul. fac. stünte Cernauti 1930.*

NEUWEILER, E. (1905) Die Prähistorischen Pflanzenreste Mitteleuropas, *Vierteljahresschrift der naturforschenden Gesellschaft in Zürich.*

NEUWEILER, E. (1924) Die Pflanzenwelt in der jüngeren Stein- und Bronzezeit der Schweiz, *Mitteilungen der Antiquarischen Gesellschaft in Zürich.*

O'CURRY, E. (1873) *Manners and Customs of the Ancient Irish* I. London.

OLDHAM, C. (1948) *Brassica Crops and allied cruciferous crops.* London.

PARKINSON, S. T. and SMITH, G. (undated) *Impurities of Agricultural Seed.* London.

PEACE HANDBOOKS (1920) Vol. IV: *The Balkan States.* H.M.S.O. London.

PERCIVAL, J. (1900) *Agricultural Botany Theoretical and Practical.* London.

PERCIVAL, J. (1921) *The Wheat Plant.* London.

PERCIVAL, J. (1936) Cereals of Ancient Egypt and Mesopotamia, *Nature* 138.

PERCIVAL, J. (1943) *Wheat in Great Britain.* London.

PETERSON, R. F. (1965) *Wheat. Botany, Cultivation and Utilization.* New York.

PIERPOINT JOHNSON, C. (1862) *The Useful Plants of Great Britain. A Treatise upon the Principal Native Vegetable capable of Application as Food, Medicine or in the Arts and Manufactures.* London.

PIGGOTT, S. (1965) *Ancient Europe from the Beginnings of Agriculture to Classical Antiquity.* Edinburgh.

PLATON, N. (1964) *Ergon tis archaiologikis etaireias,* 141–2. Athens.

PLINY (1961 ed.) *Natural History,* XVII–XIX, trans. H. Rackham. London.

POLUNIN, O. and HUXLEY, A. (1965) *Flowers of the Mediterranean.* London.

PORTER, J. (1925) The bean crops, in Paterson, W.: *Farm Crops* I. London.

PURSEGLOVE, J. W. (1968) *Tropical Crops. Dicotyledons I.* London.

REID, C. (1916) Plants, wild and cultivated in Bulleid, A. and St. George Gray, H., *Glastonbury Lake Village* 2, 625 f.

REISS, W. and STÜBEL, A. (1880) *The Necropolis of Ancón in Peru* (3 vols.). Berlin/New York/London.

RENFREW, JANE M. (1965) Appendix IV: Grain impressions from the Iron Age sites of Wandleburg and Barley, in Cra'ster, M. D.: Aldwick, Barley: Recent work at the Iron Age site, *Proceedings of Cambridge Antiquarian Society* LVIII.

RENFREW, JANE M. (1966) A report on recent finds of carbonized cereal grains and seeds from prehistoric Thessaly, *Thessalika* 5, 21 f. Volos.

RENFREW, JANE M. (1968A) A note on the neolithic grain from Can Hasan, *Anatolian Studies* XVIII.

RENFREW, JANE M. (1968B) Appendix X: The cereal remains, in Renfrew, C. & Evans, J. D.: *Excavations at Saliagos near Antiparos.* Oxford.

RENFREW, JANE M. (1969) The archaeological evidence for the domestication of plants: methods and problems, in Dimbleby, G. W. & Ucko, P.: *The Domestication and Exploitation of Plants and Animals.*

RENFREW, JANE M. (1971) Carbonized seeds and fruits from the funeral pyres of Salamis 6th–5th centuries B.C., in Kargeorghis, V.: *The Necropolis of Salamis* Vol. II.

RILEY, R. (1965) Cytogenetics and the evolution of wheat, in Hutchinson, Sir Joseph: *Essays on Crop Plant Evolution*. Cambridge.

ROCHEBRUNE, A. T. DE (1897) Recherches d'ethnographie botanique sur la flore des sépultures Péruviennes d'Ancón, *Actes Société Linnnaeus Bordeaux* 3,

RODDEN, R. J. and RODDEN, J. M. (1964) A European link with Çatal Hüyük: Uncovering a seventh millennium settlement in Macedonia. Part I, *Illustrated London News*, 11 April.

RUDENKO, S. I. (1970) *Frozen Tombs of Siberia*. London.

SAFFORD, W. E. (1917) Food plants and textiles of Ancient America, *Proceedings of the Nineteenth Internation Congress of Americanists*, 12–30. Washington.

SAFFRAY, DR (1876) Les antiquités péruviennes à l'exposition de Philadelphia, *La Nature*, 4, 401–7. Paris.

SAKAMURA, T. (1918) Kurze Mitteilung über die Chromosomenzahlen und der Verwandtschaftsverhaltnisse der *Triticum*-arten, *Bot. Mag. Tokyo* 32, 151–4.

SALISBURY, SIR EDWARD (1961) *Weeds and Aliens*. London.

SAMPSON, D. R. (1954) On the origin of oats, *Harvard University Botanical Museum Leaflets* 16.

SANDERS, I. T. (1949) *Balkan Village*. Kentucky.

SAVULESAI, T. (1958) *Flora of the People's Republic of Roumania*. Bucharest.

SCHIEMANN, E. (1946) *Weizen, Roggen, Gerste Entstehung, Geschichte und Verwertung*. Jena.

SCHIEMANN, E. (1951) New results on the history of cultivated cereals, *Heredity* 5, Part 3, 305–18.

SCHIEMANN, E. (1953) *Vitis* in Neolithicum der Mark Brandenburg, *Der Zuchter* 23. Heft 10/11.

SCHULTZE-MOTEL, J. (1968) Literatur über archäologische Kulturpflanzenreste (1965–67), *Die Kulturpflanze* XVI. Berlin.

SCHULTZE-MOTEL, J. and KRUSE, J. (1965) Spelz (*Triticum spelta* L.) andere Kulturpflanzen und Unkräuter in der frühen Eizenzeit Mitteldeutschland, *Die Kulturpflanze* XIII. Berlin.

SCHWANITZ, C. (1966) *The Origin of Cultivated Plants*. Cambridge, Mass.

SEARS, E. R. (1956) The systematics, cytology and genetics of wheat, *Handbuch der Pflanzenzuchtung*. Berlin and Hamburg.

SOYER, A. (1853) *The Pantropheon or History of Food and its Preparation from the Earliest Ages of the World*. London.

STAMPFUSS, R. (1942) Die ersten altsteinzeitlichen Höhlenfunde in Greichenland, *Mannus* 34,

STANTON, T. R. (1955) *Oat Identification and Classification*. U.S. Dept. Agriculture Technical Bulletin No. 1100. Washington, D.C.

STUMMER, A. (1911) Zur Urgeschichte der Rebe und des Weinbaues, *Mitteilungen der Anthropologischen Gesellschaft in Wien* 41.

TÄCKHOLM, V., TÄCKHOLM, G. and DRAR. M. (1941) *Flora of Egypt I*. Cairo.

TAKAHASHI, R. (1955) The origin and evolution of cultivated barley, *Advanced Genetics* VII, 227 f.

TEMPÍR, Z. (1964) Beitrage zur ältesten geschichte des pflanzenbaus in Ungarn, *Acta Archaeologica Hungaricae* XVI, 1–2.

TEMPÍR, Z. (1966) Výsledky paleoetnobotanického Studia pestováni zemedelskych rostlin na úzeémi C.S.S.R., *Vedecké Prace Ceskoslovenskeho Zemedelskeho Muzea* 6.

THEOCHARES, D. (1958) *ΕΚ ΤΗΣ ΠΡΟΚΕΡΑΜΕΙΚΗΣ ΘΕΣΣΑΛΙΑΣ ΠΡΟΣΩΡΙΝΗ ΑΝΑΣΚΑΦΙΚΗ ΕΚΘΕΣΙΣ*, *Thessalika* A'. Volos.

THEOCHARES, D. (1962) *ΑΠΟ ΤΗ ΝΕΟΛΙΘΗ ΘΕΣΣΑΛΙΑ: 1. Thessalika Δ'*. Volos.

THEOCHARES, D. (1967) *Η ΑΥΓΗ ΤΗΣ ΘΕΣΣΑΛΙΚΗΣ ΠΡΟ-Ι-ΣΤΟΡΙΑΣ ΑΡΧΗ ΚΑΙ ΠΡΩΙΜΗΕΞΕΛΙΞΗ ΤΗΣ ΝΕΟΛΙΘΙΚΗΣ*. Volos.

THEOPHRASTUS (1916 ed.) *Enquiry into Plants* (2 vols.) trans. Hort, Sir Arthur. London.

TIBBLES, WILLIAM (1912) *Foods: Their Origin, Composition and Manufacture*. London.

TOWLE, M. A. (1961) *The Ethnobotany of Pre-Columbian Peru*. New York.

TROELS-SMITH, J. (1959) The Muldbjerg Dwelling Place: an Early Neolithic archaeological site in the Aamosen Bog, West-Zealand, Denmark. *Annual Report Smithsonian Institute*.

TSOUNTAS, C. (1908) *ΑΙ ΠΡΟΪΣΤΟΡΙΚΑΙ ΑΚΡΟΠΟΛΕΙΣ ΔΙΜΙΝΙΟΝ ΚΑΙ ΣΕΣΚΛΟΝ*. Athens.

TSOUNTAS, C. and MANATT, J. I. (1897) *The Mycenean Age*. London.

TURNER, C. (1968) A note on the occurrence of *Vitis* and other new plant records from the Pleistocene deposits at Hoxne, Suffolk. *New Phytologist* 67, 333–4.

TURRIL, W. B. (1952) Wild and cultivated Olives, *Kew Bulletin*, 437 f.

UPHOF, J. C. T. (1959) *Dictionary of Economic Plants*. Weinheim.

VERMEULE, E. (1964) *Greece in the Bronze Age*. Chicago.

VICKERY, K. F. (1936) *Food in Early Greece*. Illinois.

VILLARET-VON ROCHOW, M. (1958) Die pflanzenreste der Bronzezeitlichen pfahlbauten von Valeggio am Mincio, *Bericht über die Geobotan*. Forschungsinst. Rübel in Zürich für das Jahr 1957. Zurich.

VILLARET-VON ROCHOW, M. (1969) Fruit size variability of Swiss prehistoric *Malus sylvestris*, in Ucko, P. & Dimbleby, G. W.: *The Domestication and Exploitation of Plants and Animals*. London.

VILLARET-VON ROCHOW, M. (1971) *Avena ludoviciana* Dw. in Schweizer Spätneolithikum, ein Beitrage zur Abstammung des Saathafers (*Avena Sativa* L.), *Bericht. Deutsch Bot. Ges.* Bol. 84, 45 (1971).

WACE, A. J. B. and THOMPSON, M. S. (1912) *Prehistoric Thessaly*. Cambridge.

WARD, H. M. (1908) *Trees* Vol. IV: *Fruits*. Cambridge.

WARREN, P. (1969) Minoan village on Crete, *Illustrated London News* 8 February.

WARREN, S. HAZZLEDINE (1911) On a prehistoric interment near Walton-on-the-Naze. *Essex Naturalist* 16, 198–208.

WATERBOLK, H. T. and VAN ZEIST, W. (1967) Preliminary report on the neolithic bog settlement of Niederwil, *Palaeohistoria* 12, 559–80.

WATSON, W. (1969) Early cereal cultivation in China, in Dimbleby, G. W. & Ucko, P.: *The Domestication and Exploitation of Plants and Animals*. London.

WERNECK, H. L. (1961) Ur- und frühgeschichtliche Sowie mittelalterliche Kulturpflanzen und Hölzer aus den Ostalpen und dem Südlichen Böhmerwald, *Archaeologia Austriaca* 30.

WHITAKER, T. W. and DAVIS, G. N. (1962) *Cucurbits*. London and New York.

WHITE, K. D. (1970) *Roman Farming*. London.

WILLERDING, U. (1969A) Ursprung und Entwicklung der Kulturpflanzen in vor- und frühgeschichtlicher Zeit, *Deutsche Agrargeschichte* I, 188 f.

WILLERDING, U. (1969B) Pflanzenreste aus frühgeschichtlichen Siedlungen des Gottinger Gebietes, *Neue Ausgrabungen und Forschungen in Niedersachsen* 4. Hildersheim.

WILLERDING, U. (1970) Vor- und frühgeschichtliche Kulturpflanzenfunde in Mitteleuropa, *Neue Ausgrabungen und Forschungen in Niedersachsen* 5. Hildersheim.

WILSON, H. K. (1955) *Grain Crops*. New York.

WINTON, A. L. and K. B. (1932) *The Structure and Composition of Foods* Vol. I. *Cereals, Starch, Oil Seeds, Nuts, Oils, Forage Plants*. New York/London.

WINTON, A. L. and K. B. (1935) *The Structure and Composition of Foods* Vol. II. *Vegetables, Legumes, Fruits*. New York/London.

WITTMACK, L. (1880) Antike Samen aus Troja und Peru, *Monatsschr. Ver. Beförd. Gartenbau i. Preussen.*

WITTMACK, L. (1888) Die Nutzpflanzen der alten Peruanes, in *Congrès International de Americanistes* 7, 325–49. Berlin.

WULLF, T. (1910) Report, in Frödin, O.: *En svenzk pålbygnad från Stenålderu. Foruvännen.*

YACOVLEFF, E. and HERRERA, F. L. (1934/35) El mundo vegetal de los antiguos peruanos, *Revista del Museo Nacional* 3, 241–322; 4, 29–102. Lima.

YARNELL, R. A. (1964) Aboriginal relationships between culture and plant life in the Upper Great Lakes region, *Anthropological Papers* 23. Museum of Anthropology, University of Michigan.

YARNELL, R. A. (1969) Palaeo-ethnobotany in America, in Brothwell, D. & Higgs, E.: *Science in Archaeology*. London.

VAN ZEIST, W., and BOTTEIMA, S. (1966) Palaeobotanical investigations at Ramad, *Annales Archéologiques Arabes Syriennes* XVI, 179–80.

VAN ZEIST, W., and CASPARIE, W. A. (1968) Wild Einkorn wheat and barley from Tell Mureybit in Northern Syria, *Acta. Bot. Neerl.* 17 (1), 44–53.

VAN ZEIST, W., (1970) Prehistoric and Early Historic food plants in the Netherlands, *Palaeohistoria* XIV, 42–173.

VAN ZEIST, W. and BOTTEMA, S. (1971) Plant Husbandry in early neolithic Nea Nikomedeia, Greece. *Acta Bot. Neerl.* 20 (5).

ZOHARY, D. (1963) Spontaneous brittle six-row barleys, their nature and origin, *Barley Genetics I. Proceedings of the 1st International Barley Genetics Symposium*. Wageningen.

ZOHARY, D. (1969) The progenitors of wheat and barley in relation to domestication and agricultural dispersal in the Old World, in Dimbleby, G. W. & Ucko, P.: *The Domestication and Exploitation of Plants and Animals*. London.

ZOHARY, M. (1940) *Forests and forest remnants of* Pistacia atlantica *Desf. in Palestine and Syria.*

ZOHARY, M. (1951) *The Arboreal Flora of Israel and Transjordan and its Ecological and Phytogeographical Significance*. Institute Paper 26, Imperial Forestry Institute, Oxford.

ŽUKOVSKIJ, P. M. (1950) Cultivated plants and their wild relatives (trans. Hudson, P. S. 1962). Commonwealth Agricultural Bureau, Farnham, Bucks.

Glossary

ABSCISSION	A layer of separation which causes the ripened floret to be dispersed by an explosive mechanism as in *Avena fatua.*
ACHENE	A small, dry, one-seeded fruit not splitting when ripe.
ALEURONE	The outer layer of the endosperm, below the testa in cereal caryopses; its cells contain protein grains.
ANATROPOUS	Inverted ovule so that the funicle and micropyle are adjacent.
AWN	A fine bristle-like projection, usually on dorsal side or at the apex of the lemma in *Gramineae.*
AXIS	Main stem of growth.
BASAL HAIRS	Hairs on the base of the lemma and the callus in *Gramineae.*
BASAL INTERNODE	Lowest segment of the rachis: in barley these have a distinctive form.
BEAK	Extended keel of a glume, especially in wheat: or a narrow prolongation of a fruit.
BERRY	A fleshy fruit without a hard layer of pericarp around the seeds, usually with several seeds.
CALLUS	Hard basal projection at the base of the floret or spikelet, in *Gramineae.*
CALYX	Outer whorl of perianth of a flower composed of sepals.
CAPSULE	A dry, dehiscent fruit composed of two or more carpels, splitting or opening by slits or pores when ripe.
CARPEL	One of the units composing the pistil and containing one or more ovules.
CARYOPSIS	A dry one-seeded fruit with the pericarp and testa fused together.
CHALAZA	The basal or prominent part of an ovule where it is united with the funicle.
CHROMOSOMES	Small, deeply staining bodies, found in all nuclei, which determine most of the heritable characters of organisms. Two similar sets of these are normally present in all vegetative cells (diploid number = $2n$), the number usually being constant for a given species. The sexual reproductive cells normally contain half this number or one set of chromosomes only (haploid number = n). Closely related species commonly have the same number of chromosomes, or a multiple of the common basic number, often referred to as x to distinguish it from n. Thus in wheats the diploid group have $2n$; the tetraploids $4n$ and the hexaploids $6n$.

COTYLEDON — Seed leaf forming part of the embryo.

CULM — Stem of grasses and sedges.

CULTIVARS — Varieties and strains of cultivated crops, not forming botanical 'varieties'.

DEHISCENT — Opening in order to shed seeds.

DIPLOID — Having two sets of chromosomes ($2n$).

DRUPE — A more or less fleshy fruit with one or more seeds each surrounded by a dense woody layer of pericarp, forming a stone.

EMBRYO — Rudiment of a plant within a seed consisting of radicle, cotyledons and plumule.

ENDOSPERM — The nutritive tissue deposited outside the embryo in the seed of a flowering plant after fertilization.

EQUILATERAL PANICLE — Panicle in which the branches spread out evenly around the axis.

FALSUM BASE — Bevelled grain base in barley.

FLORET — In *Gramineae*, consists of lemma, palea and enclosed flower or seed.

FOLLICLE — A dry, several-seeded fruit formed from one carpel and dehiscing along the inner side.

FRUIT — A ripened pistil containing seeds. Some so-called fruits include additional parts such as the succulent receptacle in the strawberry.

FUNICLE — The stalk of the ovule.

GENICULATE — Bent like a knee.

GENOME — A basic set of chromosomes; one such set is usually found in a gamete and two in a body cell.

Fig. 126 DRY INDEHISCENT FRUITS. 1 achene (whole and l.s.), 2 cypsela (achene with pappus), 3 caryopsis (*a*) whole, (*b*) l.s., 4 samara (*Fraxinus*), 5 double samara (*Acer*).

Fig. 127 DRY DEHISCENT FRUITS (capsules). 1 follicle, 2 legume, 3 leguminous lomentum, 4 t.s. of same, 5 siliqua, 6 cruciferous lomentum, 7 t.s. of same, 8 silicule, 9 pyxis, 10 poricidal capsule, 11 capsule with teeth, 12 septicidal capsule, 13 loculicidal capsule, 14 septifragal capsule (loculicidal form), 15 septifragal capsule (marginicidal form).

Fig. 128 SCHIZOCARPIC FRUITS. 1 Schizocarp of an umbellifer (*a*) whole, (*b*) t.s. of same, 2 schizocarp (nutlets) of a labiate, 3 schizocarp of *Geranium*, 4 schizocarp of *Malva* (*a*) single carpel, (*b*) lateral view of same, (*c*) plan view of same.

Fig. 129 FLESHY FRUITS. 1 berry, 2 l.s. of berry, 3 drupe, 4 l.s. of drupe, 5 etairio of druplets, 6 l.s. of same.

Fig. 130 FALSE FRUITS. 1 Mulberry (*Morus*,) 2 the same: enlargement to show fleshy calyx, 3 Fig (*Ficus*), 4 l.s. of fig, 5 portion of fig, enlarged, 6 Rose hip (*Rosa*), 7 l.s. of rose hip, 8 Strawberry (*Fragaria*), 9 l.s. of strawberry, 10 t.s. of apple (*Pyrus*), 11 l.s. of apple.

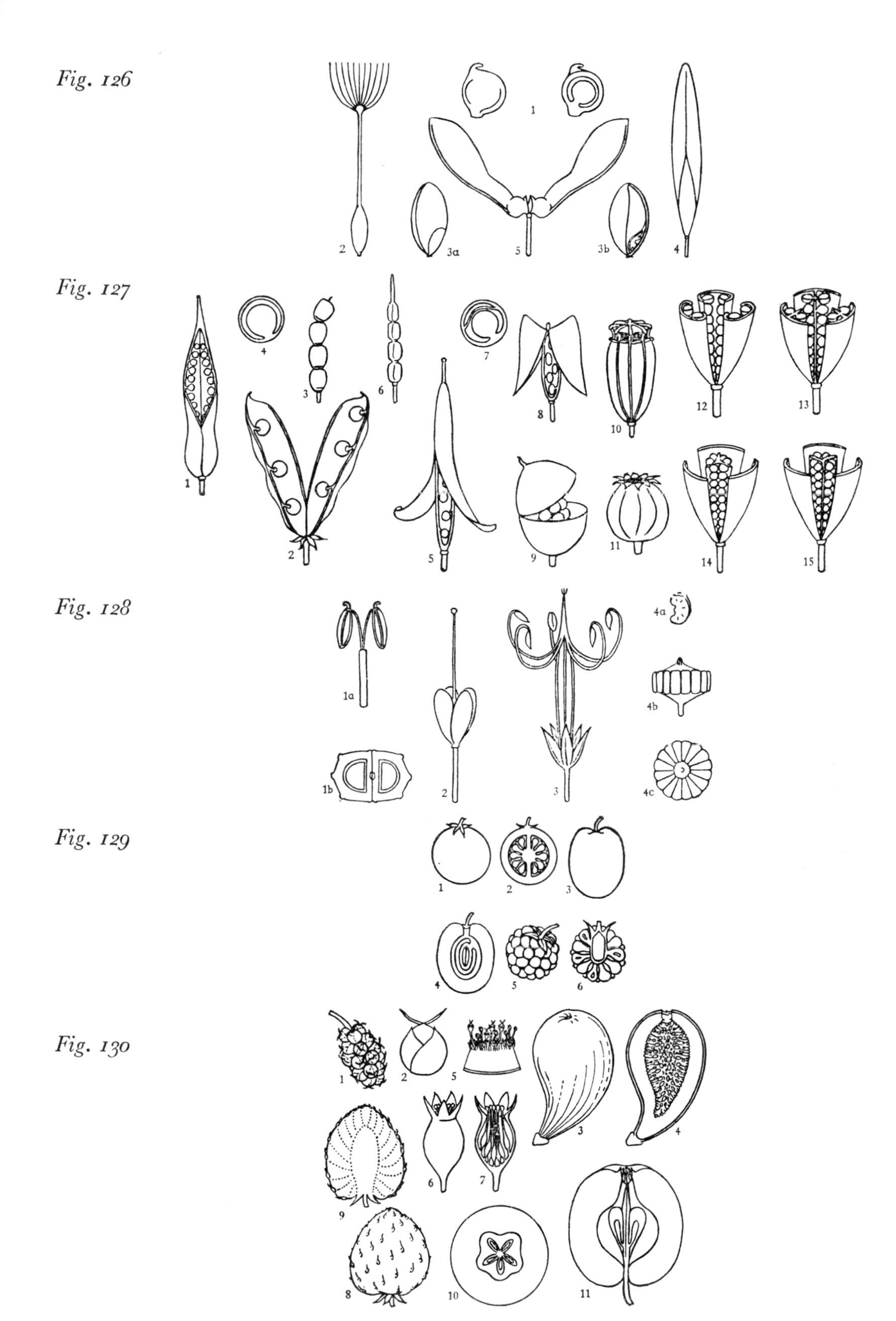
Fig. 126
1
2
3a
5
3b
4
Fig. 127
4
7
1
3
6
8
10
12
13
2
5
9
11
14
15
Fig. 128
1a
4a
4b
1b
2
3
4c
Fig. 129
1
2
3
4
5
6
Fig. 130
1
2
5
3
4
6
7
9
8
10
11

GLABROUS	Without hairs.
GLUME	One, or usually two empty bracts at the base of spikelets in *Gramineae.*
GLUME SHOULDER	In wheat glumes, the upper margin of the broad fold below the beak.
HEXAPLOID	Having six sets of chromosomes.
HILUM	Scar left on the seed by the stalk of the ovule (funicle).
INDEHISCENT	Not opening to release seeds or spores.
INTERNODE	Part of stem between two successive nodes.
KEEL	Sharp edge, resembling the keel of a boat.
LEGUME	Fruit or pod of the *Leguminosae,* consisting of a single carpel opening round the margin along both sutures.
LEMMA	Lower of the two bracts enclosing the caryopsis in *Gramineae.*
LEMMA NERVES	Prominent veins visible on lemmas of some cereals, for example barley.
LOCULUS	Compartment or cell especially of an ovary.
LODICULES	Pair of minute papery scales overlying the embryo, beneath the lemma base in *Gramineae.*
LOMENTUM	A dry fruit, usually elongated, which breaks transversely into single-seeded portions.
MICROPYLE	Minute hole in the seed coat, the opening of the ovule through which the pollen tube enters: often visible on the seed.
NODE	Part of the stem, or rachis, where the branch, leaf or spikelet may arise.
NUT	A fruit containing a single seed and having a woody pericarp usually derived from a syncarpous ovary.
ORIFICE, OSTIOLE	Opening to the flask-like receptacle forming the fleshy outside of the fruit of figs.
PALEA	The upper of the two bracts enclosing the caryopsis in *Gramineae.*
PANICLE	Racemose inflorescence with lateral branches, often applied to other branched inflorescences.
PEDICEL	Stalk of individual flower or spikelet in *Gramineae.*
PENTAPLOID	Having five sets of chromosomes.
PERICARP	Wall of ripe ovary enclosing seeds; may be fused or divided into epicarp, mesocarp and endocarp.
PERIGYNOUS TUBE	Flowers in which the perianth and stamens are borne on an outgrowth of the receptacle.
PLUMULE	The bud in the embryo from which develop the stem and leaves.
PUBESCENT	Bearing short hairs.
PYRIFORM	Pear-shaped.
RACEME	An unbranched racemose inflorescence in which the flowers are borne on pedicels.
RACEMOSE	Inflorescence in which the youngest flowers are towards the apex.
RACHILLA	The axis of the spikelet in grasses.
RACHIS	The axis of a spike.
RACHIS SEGMENT	Division of rachis consisting of one node and the attached section of internode.

RADICLE	Rootlet in the embryo.
RAPHE	Ridge along one side of an anatropus ovule, formed by the fusion of the funicle with the outside of the ovule.
RENIFORM	Kidney-shaped.
RETICULATE	Marked with a network of surface ridges or markings.
SCHIZOCARP	A dry syncarpous fruit breaking into one-seeded when ripe.
SCUTELLUM	The cotyledon of the embryo of a grass; in wheat, the shield-shaped structure at the back of the embryo, in contact with the endosperm.
SESSILE	Without a stalk.
SILICULA	A short siliqua often wider than its length, cf. *Capsella*.
SILIQUA	Capsule of two carpels divided into two loculi by a thin partition (the replum) opening from below by two valves. The capsule is usually at least three times longer than its width, cf. *Cruciferae*.
SPIKE	A simple raceme with sessile flowers.
SPIKELET	Unit of the ear consisting of two glumes enclosing florets comprising lemma, palea and caryopsis; in *Gramineae*.
SPIKELET FORK	In palaeoethnobotanical material, the remains of the bases of the glumes of hulled wheats still adhering to the rachis segment.
SYNCONIUM	Fruit of a fig comprising large, pear shaped receptacle with numerous hard pip-like carpels borne on the inside.
TERMINAL SPIKELET	Spikelet at the apex of the ear.
TESTA	Outer seed coat.
TETRAPLOID	Having four sets of chromosomes.
TUBERCULE	A small, spherical swelling on the fruiting perianth of *Rumex*.
UNILATERAL PANICLE	Panicle with branches standing fairly erect and on one side of rachis only.
VENTRAL GROOVE	Deep longitudinal groove down the ventral side of caryopsis in *Gramineae*.
VERUM BASE	Grain base with a deep transverse groove in barley.

Index

Numbers in italic in the index refer to the main treatment of the topic.

R

Plates

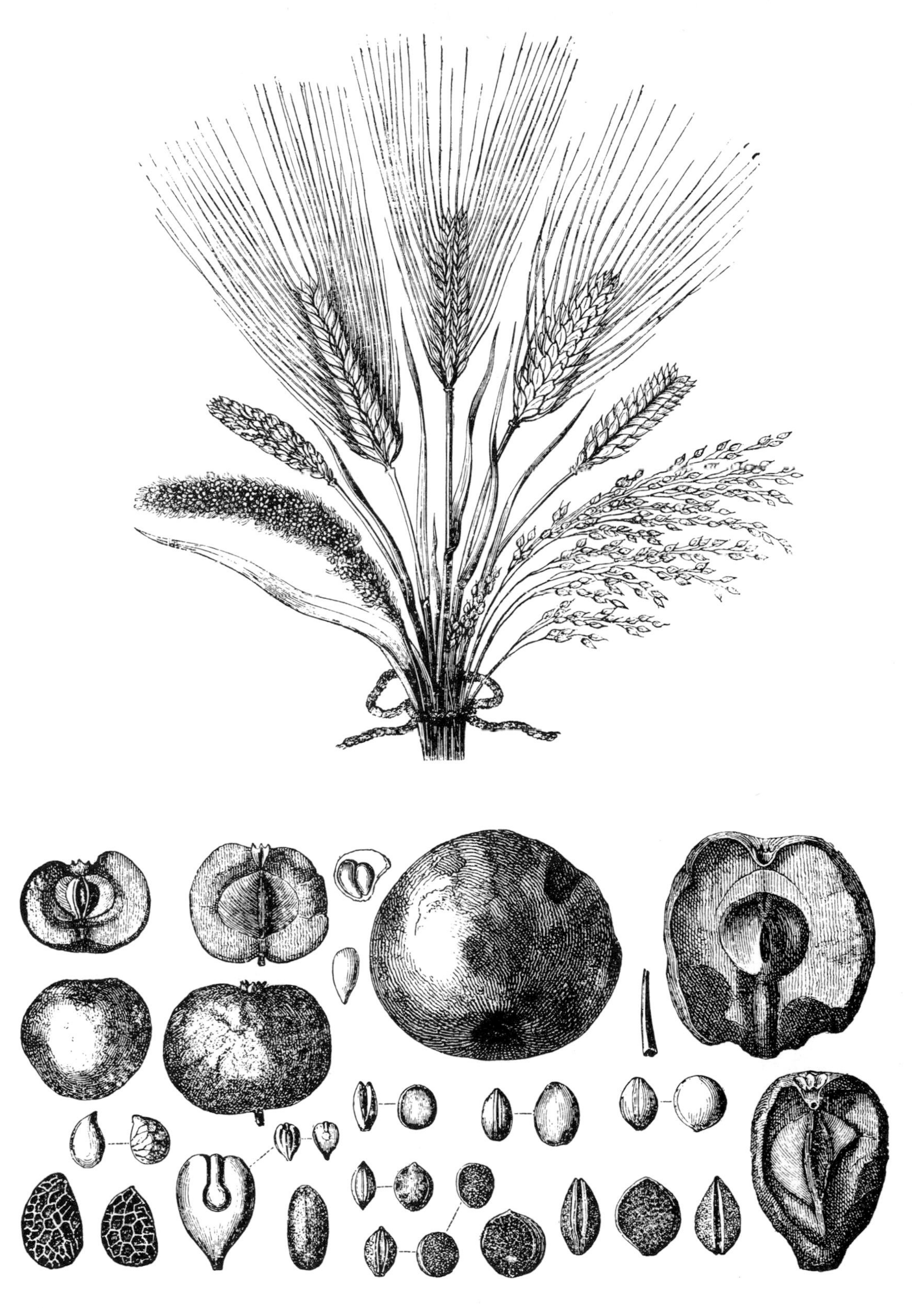

Plate 1
Cereals and fruits found in the Swiss lakeside dwellings, illustrated by O. Heer in 1866.

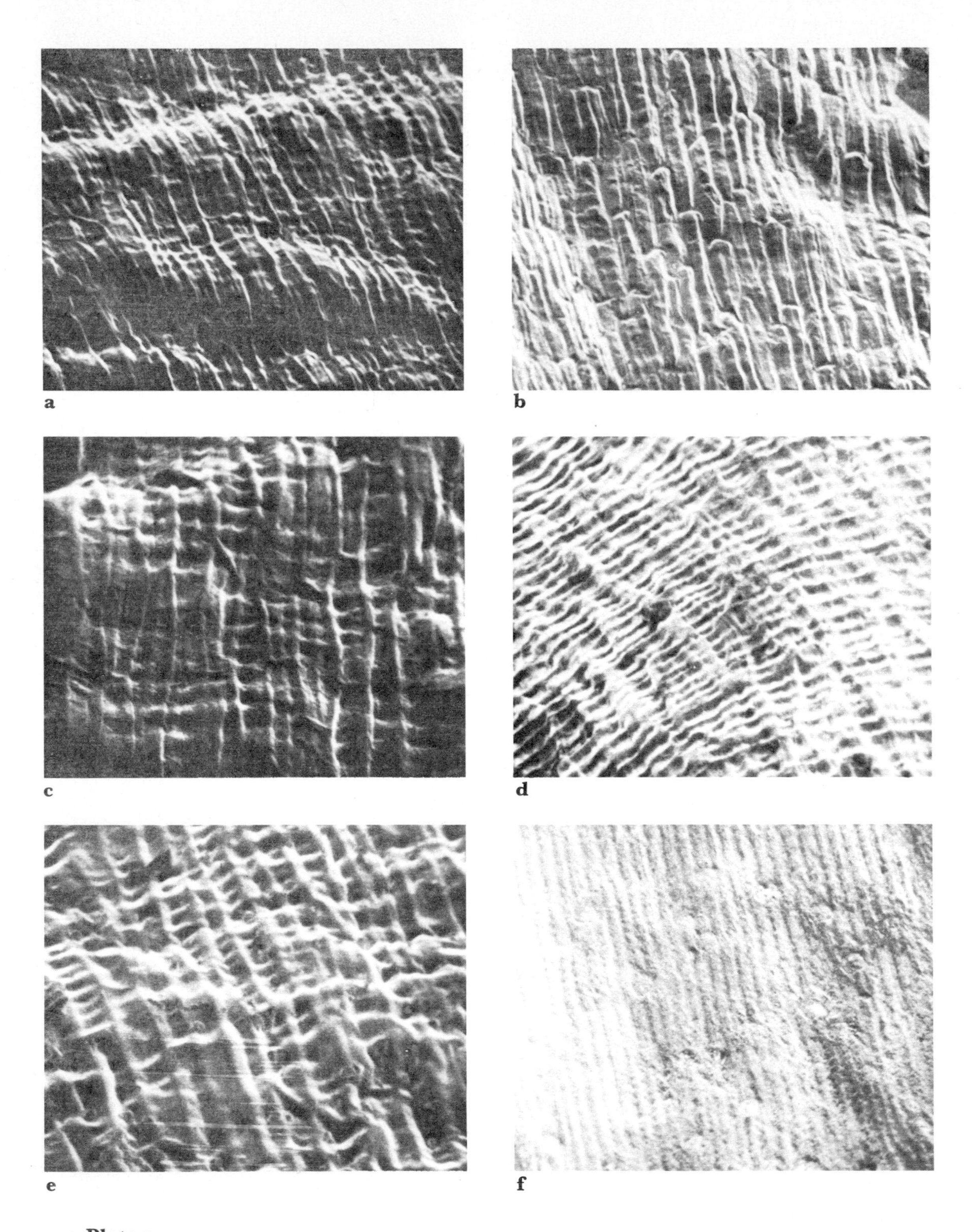

Plate 2
Electron scanning micrographs of the epidermis of modern grains of **a.** *Triticum boeoticum*, **b.** *T. monococcum*, **c.** *T. dicoccoides*, **d.** *T. dicoccum*, **e.** *Secale cereale*, **f.** *Hordeum distichon* (lemma); all × 310.

Plate 3
Comparison of fresh grains of oats, wheat and hulled barley with carbonized grains of the same species from the Roman fort at Ambleside, Westmorland; × 12.

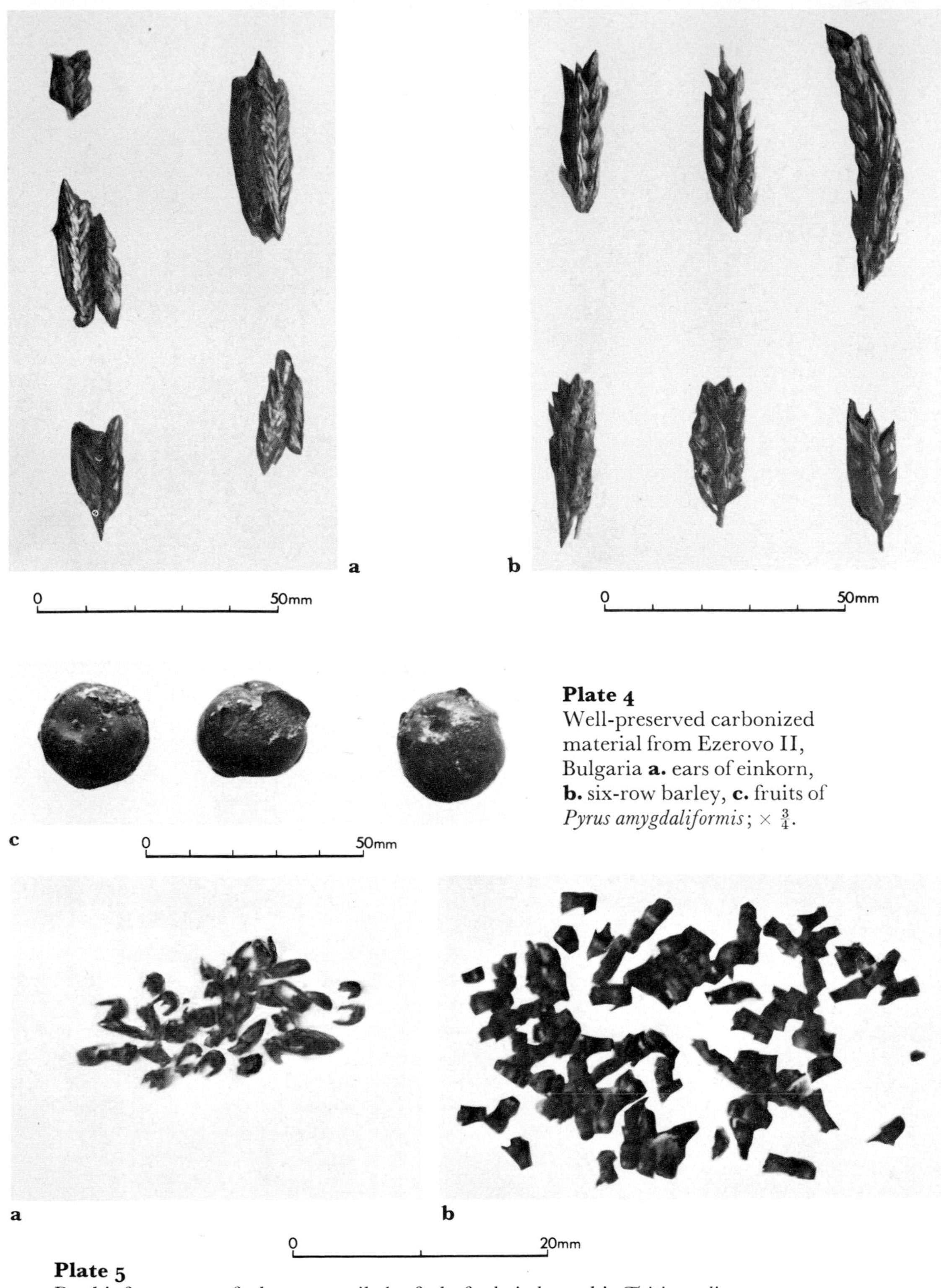

Plate 4
Well-preserved carbonized material from Ezerovo II, Bulgaria **a.** ears of einkorn, **b.** six-row barley, **c.** fruits of *Pyrus amygdaliformis*; $\times \frac{3}{4}$.

Plate 5
Rachis fragments of wheat: **a.** spikelet forks for brittle-rachis *Triticum dicoccum*, **b.** fragments of touch rachis of *T. aestivo-compactum* recovered by flotation from Stratum IV at Sitagroi, N. Greece; $\times$ 2.

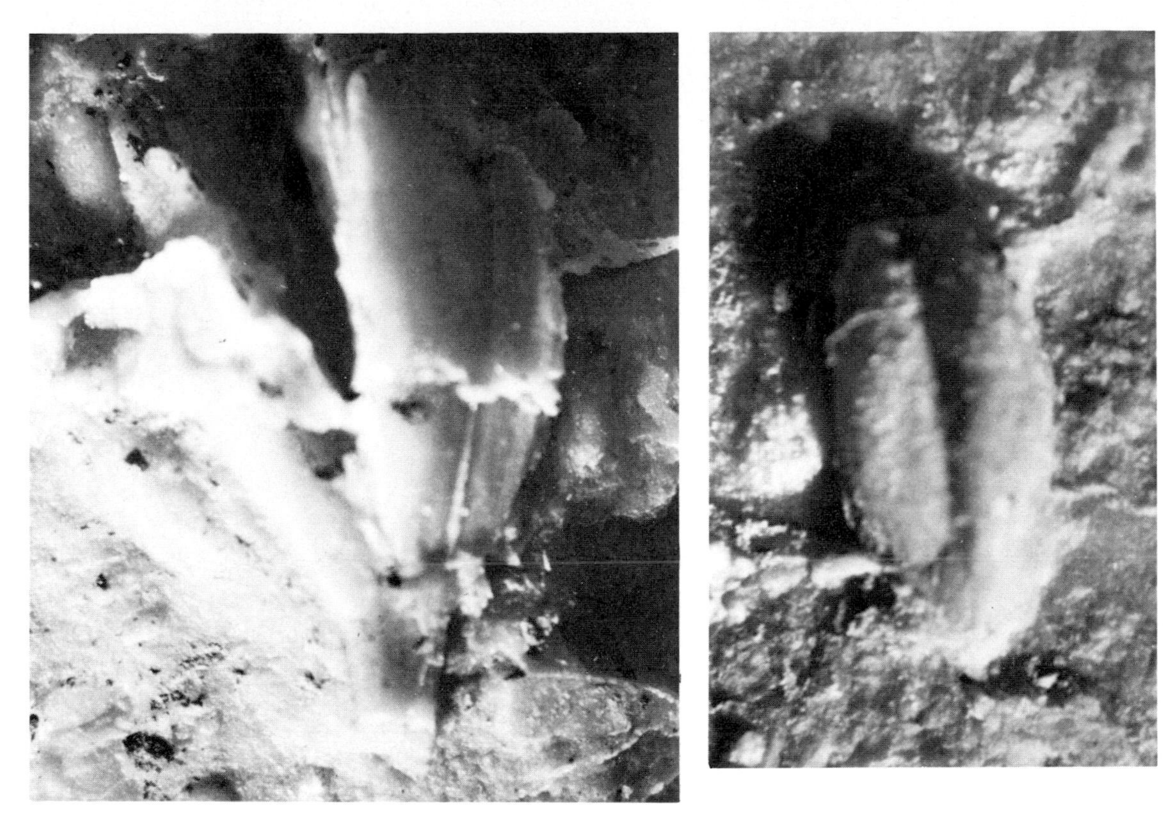

Plate 6
Grain impressions in neolithic pottery, Mildenhall, Suffolk.
Left: part of a spikelet of emmer. Right: ventral side of emmer grain; both × 10.

Plate 7
Left: modern mud bricks being sun-dried near Nova Zagora, Bulgaria.
Right: silica skeletons of grains in pisé mud at Saliagos, Cyclades; × 2.

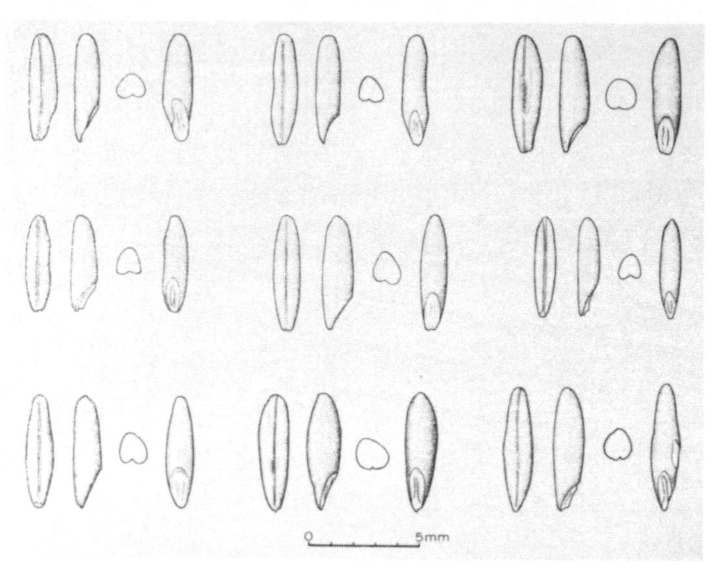

Plate 8
Grains of wild *Triticum boeoticum* from Tell Mureybit (Van Zeist and Casparie 1968) × 3.

Plate 9
Modern spikelet of *T. boeoticum*; × 6.

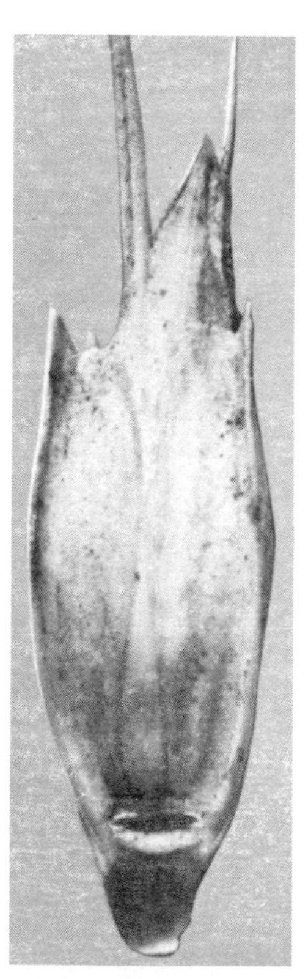

a

b

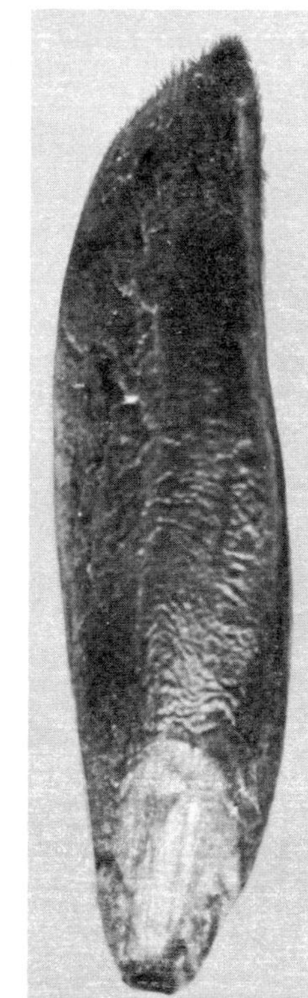

c

Plate 10
Modern *T. monococcum*
a. spikelet; × 6.
b. caryopsis, lateral view; × 10.
c. caryopsis, dorsal view; × 10.

Plate 11
Modern *T. dicoccoides* spikelet; × 3.

Plate 12
Left: spikelet of *T. dicoccum*; × 8.
Right: lateral view of caryopsis of *T. dicoccum*; × 9.

Plate 13
Spikelet of *T. spelta* showing the rachis segment attached to the face of the spikelet; × 5.

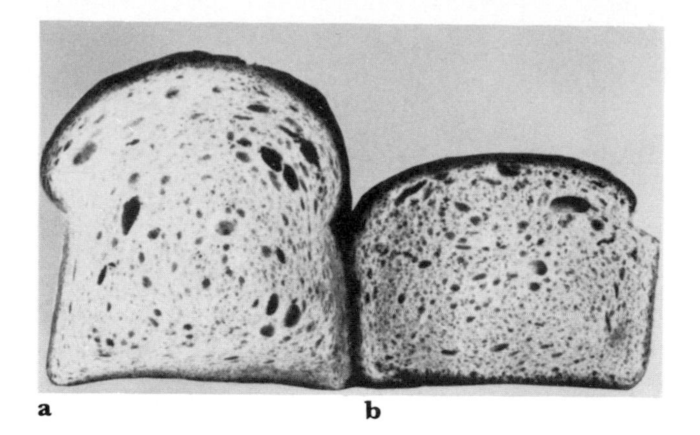

Plate 14
Loaves of bread made with equal quantities of **a.** strong and **b.** weak wheat flour.

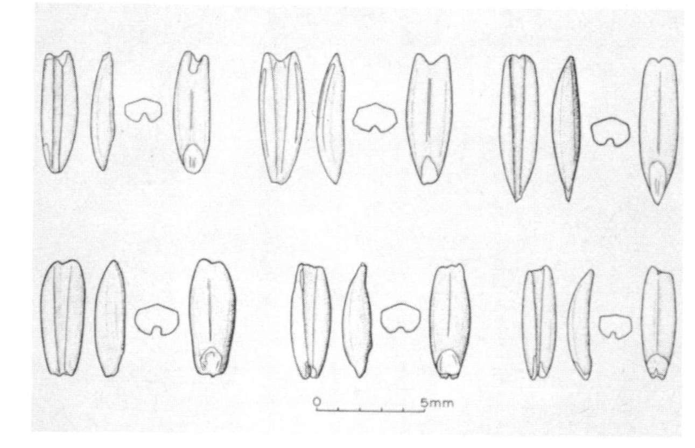

Plate 15
Hordeum spontaneum,
Tell Mureybit; $\times 2\frac{1}{2}$.

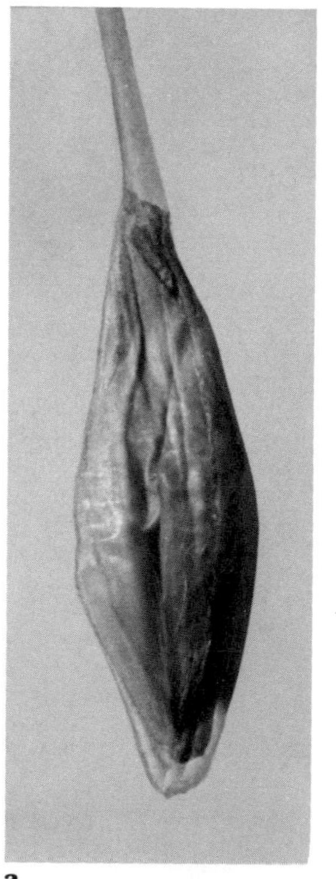

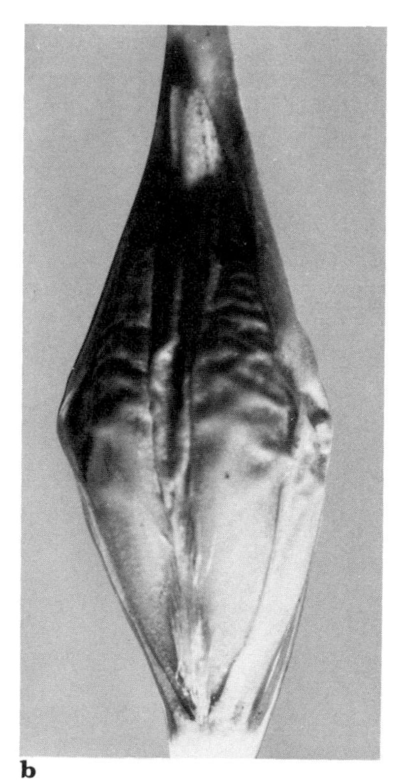

Plate 16
Hordeum vulgare, ventral views of:
a. asymmetrical lateral grain; $\times$ 7.
b. symmetrical median grain; $\times$ 8.

Plate 17
Ears of lax and dense six-row barley viewed from the side and the end of the ear. The lax form used to be known as *Hordeum tetrastichum* but is now included in *Hordeum vulgare*; the dense ear belongs to *H. vulgare* var *hexastichum*. Reproduced by kind permission of N.I.A.B. × 1.

Plate 18
Ears of two- and six-row barley compared. Reproduced by permission of N.I.A.B., × ½.

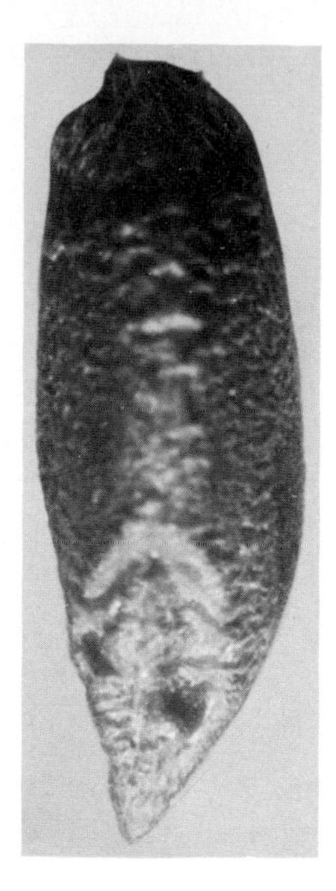

Plate 19 Caryopsis of *Secale cereale*; × 9.

20 The earliest oat grain in Europe from the aceramic neolithic levels at Achilleion, Thessaly, Greece; × 8½.

21 Floret of *Avena strigosa*; × 5.

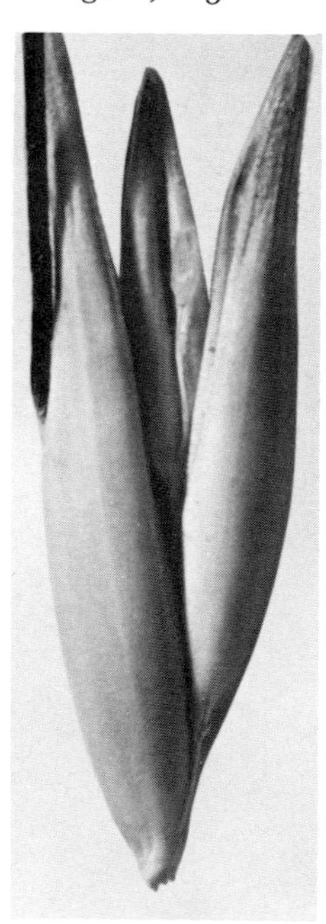

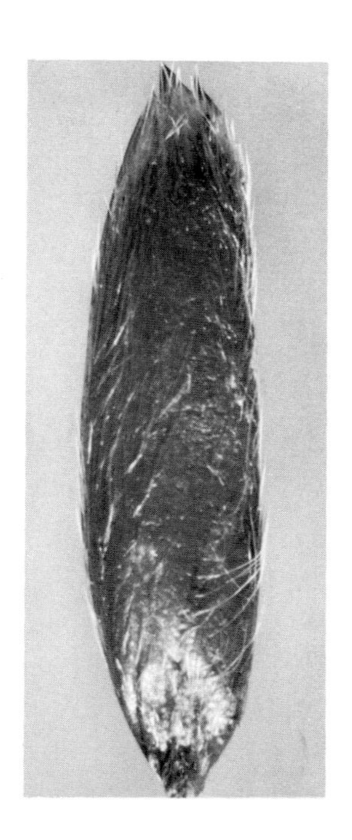

a **b**

Plate 22 Primary and secondary florets of *Avena fatua*; × 5.

23 **a.** floret of *Avena sativa*.
b. single secondary floret *Avena sativa* in ventral view; both × 7.

24 Caryopsis of *Avena sativa*; × 7.

Plate 25
Part of a panicle of *Panicum miliaceum*; × ½.

Plate 26
Spike-like panicles of *Setaria italica*; × ¼.

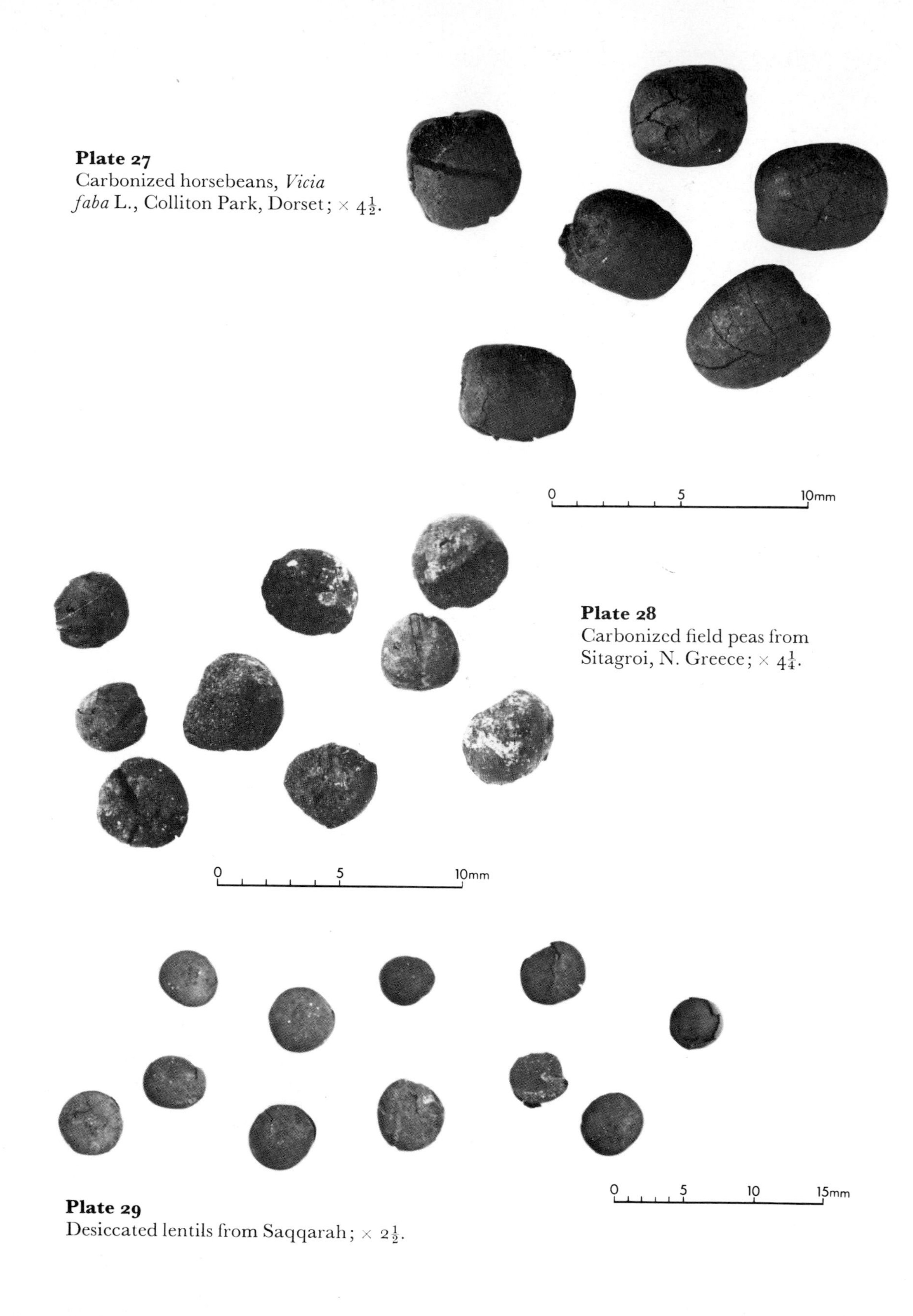

Plate 27
Carbonized horsebeans, *Vicia faba* L., Colliton Park, Dorset; $\times 4\frac{1}{2}$.

Plate 28
Carbonized field peas from Sitagroi, N. Greece; $\times 4\frac{1}{4}$.

Plate 29
Desiccated lentils from Saqqarah; $\times 2\frac{1}{2}$.

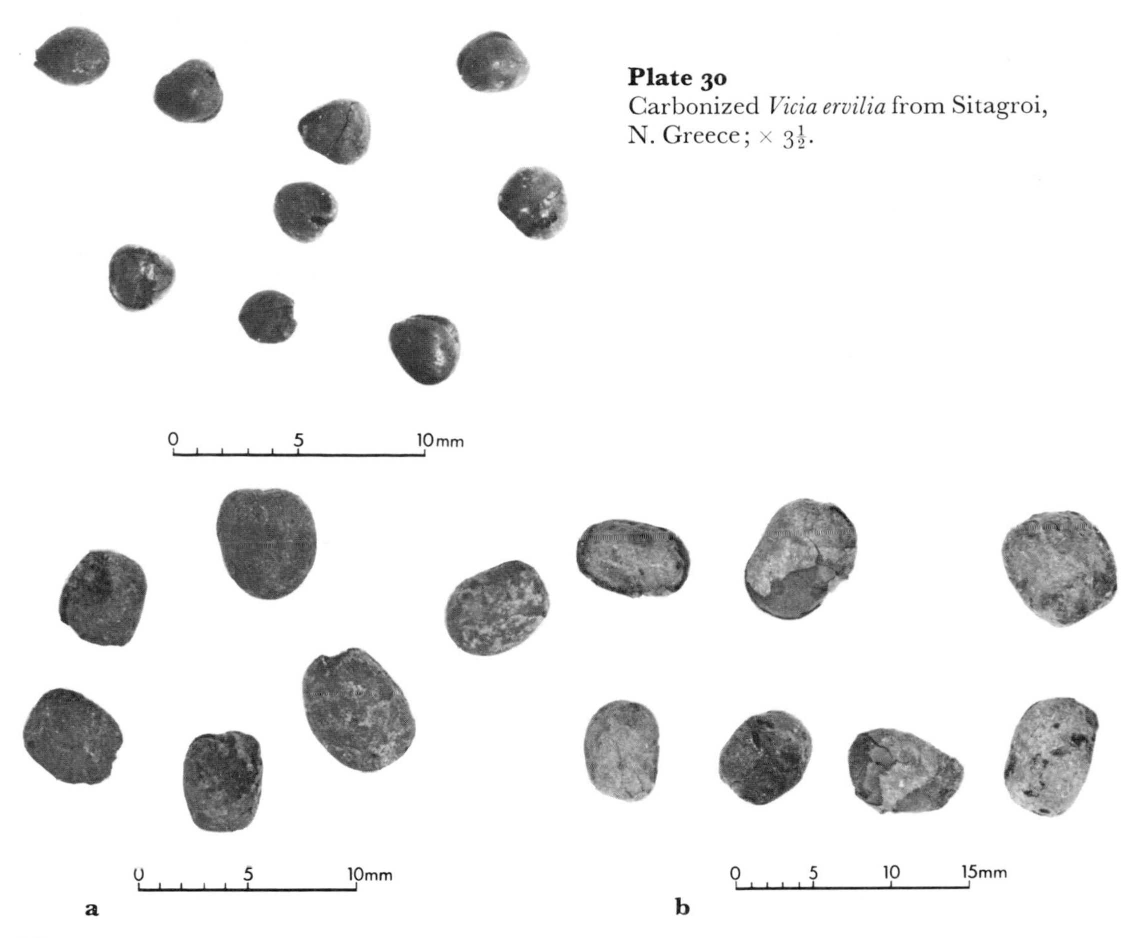

Plate 30
Carbonized *Vicia ervilia* from Sitagroi, N. Greece; × $3\frac{1}{2}$.

Plate 31
Carbonized *Lathyrus sativus* from **a.** Salamis, Cyprus; × 3, and **b.** Kephala, Kea; × 2.

Plate 32
Carbonized *Cicer arietinum* from Salamis, Cyprus; × 3.

Plate 33
Modern seed of *Linum usitatissimum* L.; × 8.

Plate 34
a. carbonized grape pips from Salamis, Cyprus; × 3.
b. empty grape skins, Early Minoan Myrtos, Crete; × 3.

Plate 35
Carbonized olive stones from Salamis, Cyprus; × 3.

Plate 36
Carbonized figs from House Q at Rachmani: above × $\frac{2}{3}$; below × $2\frac{1}{2}$.

Plate 37
Carbonized stones of *Cornus mas* L. from Sitagroi, N. Greece; × $1\frac{1}{2}$.

Plate 38
Stones of *P. insititia* from Salamis, Cyprus; × $3\frac{1}{2}$.

Plate 39
Acorn cotyledons from the aceramic levels at Achilleion, Thessaly, Greece; × 2.

Plate 40
Almond from Salamis, Cyprus; × $3\frac{3}{4}$.

Plate 41
Nutlets of *Pistacia atlantica* Desf. from the aceramic neolithic levels at Sesklo, Thessaly, Greece; × $2\frac{1}{2}$.

Plate 42
Nut of *Pistacia vera* L. from the late neolithic levels at Sesklo, Thessaly, Greece; × 5.

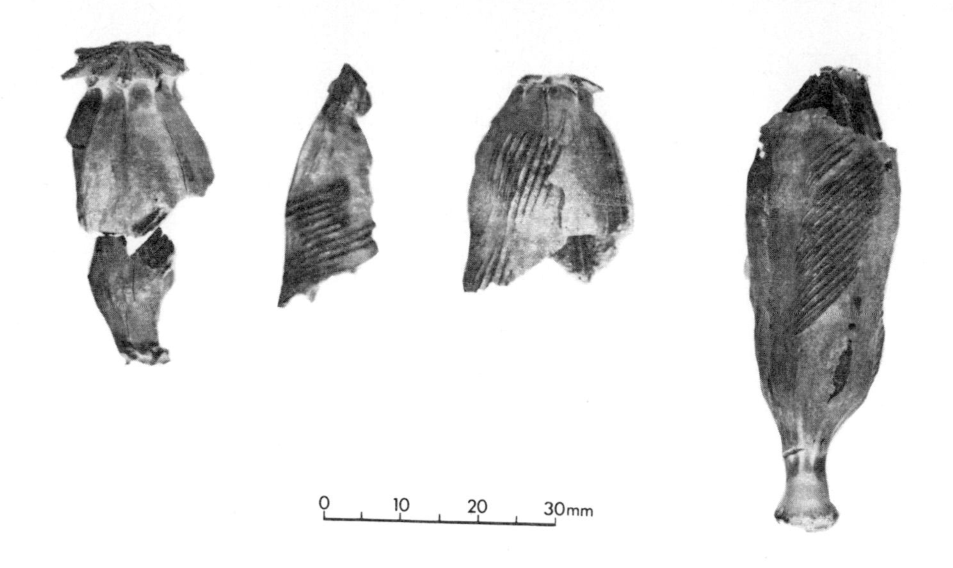

Plate 43
Above: dried opium poppy capsules bought in the bazaar Ahwaz, Spring 1969; × 1.

Below: terracotta statuette of Goddess with 3 opium poppy capsules on her head. Gazi, Greece, 1400–1200 B.C. (Hawkes 1968).

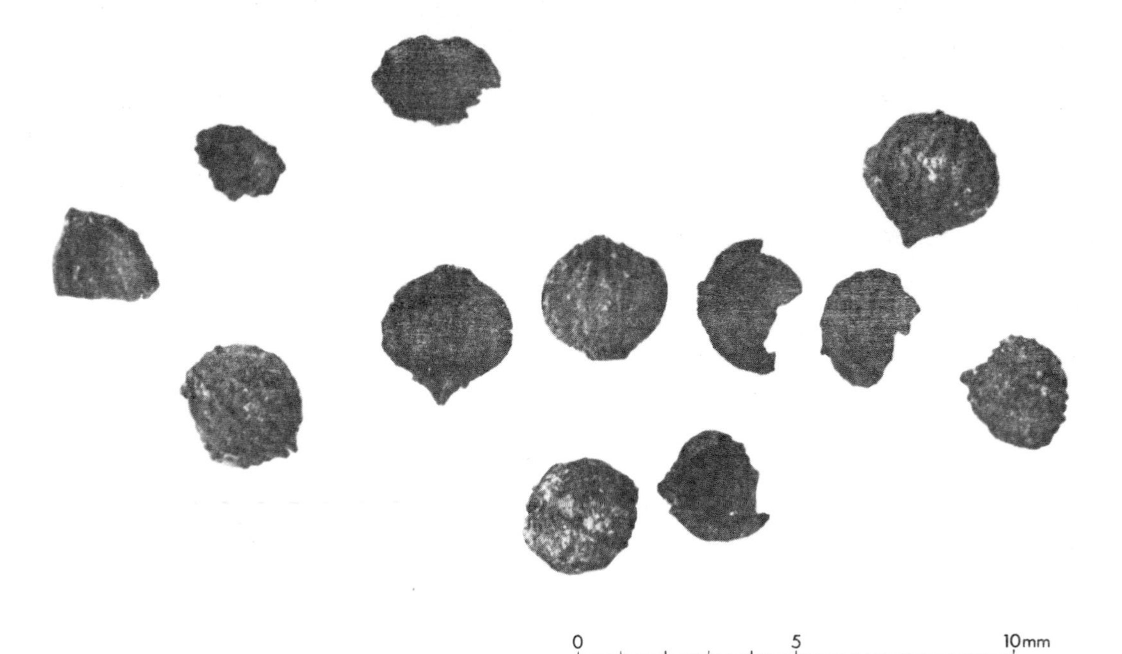

Plate 44
Coriander 'seeds' from the early bronze age level at Sitagroi, N. Greece; × 6.

Plate 45
Seeds of *Polygonum aviculare* L. from the neolithic levels at Sitagroi, N. Greece; × 6.

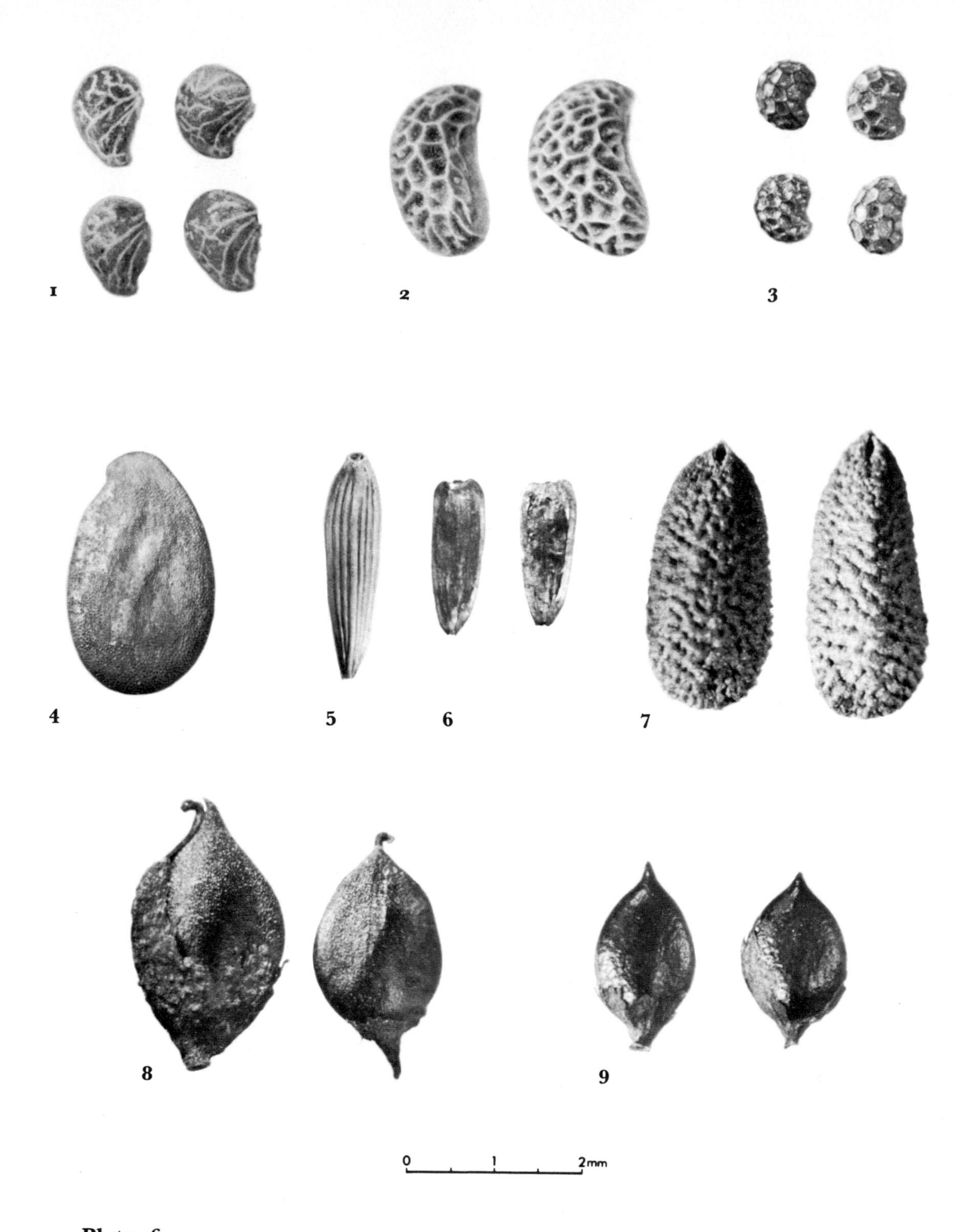

Plate 46

Seeds of 1 *Fragaria vesca*, 2 *Rubus idaeus*, 3 *Papaver somniferum*, 4 *Linum usitatissimum*, 5 *Lapsana communis*, 6 *Sambucus nigra*, 7 *Achillea millefolium*, 8 *Polygonum hydropiper*, and 9 *Polygonum persicaria* from Niederwil, Switzerland (Waterbolk and Van Zeist 1967) × 15.

Plate 47
Seeds of 1 *Fagus silvatica* (× 3), 2–3 *Polygonum convolvulus* (× $13\frac{1}{2}$), 4 *Agrostemma githago* (× $13\frac{1}{2}$), 5–6 *Rubus fruticosus* (× $13\frac{1}{2}$), 7 *Chenopodium album* (× 27), 8–9 *Juglaas regia* (× $1\frac{1}{2}$), 10–11 *Scleranthus annuus* (× 18), 12 *Spergula arvensis* (× 27), 13–14 *Vicia tetrasperma* (× $13\frac{1}{2}$), 15–16 *Prunus spinosa* (× 3), from Viking Haithabu (Behre 1969).

a

b

Plate 48

a. loaves of bread made from equal quantities of wheat, barley and rye flour showing their comparative baking qualities (Peterson 1965).

b. carbonized loaf of bread from Pompeii found in the bakers shop belonging to Modestus (Brion and Smith 1960).